Model theory is a branch of mathematical logic that has found applications in several areas of algebra and geometry. It provides a unifying framework for the understanding of old results and more recently has led to significant new results, such as a proof of the Mordell–Lang conjecture for function fields in positive characteristic. Perhaps surprisingly, it is sometimes the most abstract aspects of model theory that are relevant to these applications.

This book gives the necessary background for understanding both the model theory and the mathematics behind the applications. Aimed at graduate students and researchers, it is unique in that it contains surveys by leading experts covering the whole spectrum of contemporary model theory (stability, simplicity, o-minimality and variations), and introducing and discussing the diverse areas of geometry (algebraic, diophantine, real analytic, p-adic and rigid) to which the model theory is applied.

The book begins with an introduction to model theory by David Marker. It then broadens into three components: pure model theory (Bradd Hart, Dugald Macpherson), geometry (Barry Mazur, Ed Bierstone and Pierre Milman, Jan Denef), and the model theory of fields (Marker, Lou van den Dries, Zoé Chatzidakis).

Mathematical Sciences Research Institute
Publications

39

Model Theory, Algebra, and Geometry

Mathematical Sciences Research Institute
Publications

Volumes 1–4 and 6–27 are available from Springer-Verlag

Model Theory, Algebra, and Geometry

Edited by

Deirdre Haskell

Holy Cross College

Anand Pillay

*University of Illinois
at Urbana–Champaign*

Charles Steinhorn

Vassar College

CAMBRIDGE
UNIVERSITY PRESS

Deirdre Haskell
Holy Cross College
Department of Mathematics
Worcester, MA 01610
United States

Anand Pillay
University of Illinois
Department of Mathematics
1409 W. Green St.
Urbana, IL 61801
United States

Charles Steinhorn
Vassar College
Department of Mathematics
Poughkeepsie, NY 12604
United States

Mathematical Sciences
Research Institute
1000 Centennial Drive
Berkeley, CA 94720
United States

Series Editor
Silvio Levy

MSRI Editorial Committee
Joe P. Buhler (chair)
Alexandre Chorin
Silvio Levy
Jill Mesirov
Robert Osserman
Peter Sarnak

The Mathematical Sciences Research Institute wishes to acknowledge
support by the National Science Foundation.

CAMBRIDGE UNIVERSITY PRESS
Cambridge, New York, Melbourne, Madrid, Cape Town, Singapore,
São Paulo, Delhi, Dubai, Tokyo

Cambridge University Press
The Edinburgh Building, Cambridge CB2 8RU, UK

Published in the United States of America by Cambridge University Press, New York

www.cambridge.org
Information on this title: www.cambridge.org/9780521143493

First published 2000
This digitally printed version 2010

A catalogue record for this publication is available from the British Library

Library of Congress Cataloguing in Publication data

Model theory, algebra, and geometry / [edited by] Deirdre Haskell, Anand
Pillay, Charles Steinhorn.   vii, 227 p. ; 24 cm. (Mathematical Sciences
Research Institute publications ; 39)
ISBN 0-521-78068-3 (hbk.)
1. Model theory.   2. Geometry.   3. Fields (Algebra)   I. Haskell, Deirdre,
1963–   II. Pillay, Anand.   III. Steinhorn, Charles.   IV. Series.
QA9.7.M595 2000
511'.8–dc21   00–020294

ISBN 978-0-521-78068-1 Hardback
ISBN 978-0-521-14349-3 Paperback

Model Theory, Algebra, and Geometry
MSRI Publications
Volume 39, 2000

Contents

Model Theory, Algebra, and Geometry
MSRI Publications
Volume **39**, 2000

Overview

DEIRDRE HASKELL, ANAND PILLAY, AND CHARLES STEINHORN

Note: All bibliographic references are to articles in this volume.
See the individual articles for more references.

Even the most reserved among model theorists would no doubt agree that the
subject has grown dramatically over the last thirty years. This period has produced a substantial and beautiful abstract theory as well as a range of remarkable
applications that extend into several areas of mathematics and incorporate the
most sophisticated theoretical developments in the field. During the last decade
in particular, results obtained by model theorists have attracted the attention of
researchers outside logic and have opened up broad avenues for interaction. The
Model Theory of Fields program held at the Mathematical Sciences Research
Institute from January to June 1998 sought to capitalize on the intense activity
that has taken place in the discipline by bringing together for an extended period
of time model theorists and mathematicians working in the areas of some of the
most exciting applications.

Model theory's stock-in-trade is the analysis of the so-called *definable subsets*
of a mathematical structure. The definable subsets of classical mathematical
structures have long occupied a central position in algebraic and geometric inves-
tigations; the constructible sets in algebraic geometry and the semialgebraic sets
in real geometry provide two notable examples. Although the model-theoretic
viewpoint may have supplied these areas with some basic results, it did not offer
enough until recently for practitioners to notice that the objects that they study
with their own sophisticated methods could be illuminated by model theory.

For an area of mathematics to contribute significantly to another, the new
point of view must add to the understanding of the objects in the area of application, and not, for example, merely provide convenient new terminology. In the
case of the model theory of fields, the subject around which the articles in this
volume are organized, the deep advances that have been made in abstract model

Haskell was partially supported by NSF grant DMS-9401328. Pillay was partially supported
by NSF grant DMS-0969628. Steinhorn was partially supported by NSF grant DMS-9704869.

theory over the last thirty or forty years have generated concepts and methods that have found meaning and application in classical mathematical contexts, ultimately leading to important new results in these areas.

This volume collects articles arising from lectures presented at the Introductory Workshop that inaugurated the MSRI model theory of fields program (January 1998). Section 1 of this overview introduces each article, providing a sense of our conception of the workshop minicourses. Sections 2 and 3 are intended to give a reader unfamiliar with Model Theory a gentle conceptual introduction to its main themes. We hope that these two modest aims shall whet the reader's appetite to explore the excellent contributions that follow, for which we herewith express our gratitude to the authors.

We also thank Silvio Levy for compiling the contributions into a book, and the Director and Deputy Director of MSRI for their encouragement of this project.

1. The Organization of This Volume

The goal of this volume, like the workshop on which it is based, is to serve as a guide to current developments in model theory and its many geometric applications in the model theory of fields. It attempts to provide the reader with a unified introductory account of contemporary pure model theory, the model theory of fields, and the different aspects of geometry in which the model theory has found its most significant recent applications.

The articles in this volume are organized around three themes which roughly speaking comprise the minicourses on which the workshop was based: the model theory of fields, dimension theory, and geometry. The expression "model theory of fields" connotes the analysis of various classes of fields, including identifying elementary classes, finding axioms for these classes, and determining the definable sets in fields belonging to each class (relative quantifier elimination). This tradition goes back to Alfred Tarski and Abraham Robinson. The more recent developments are in many instances informed by concepts from pure model theory such as forking and orthogonality. "Dimension theory" deals with the conceptual apparatus of pure model theory and the associated body of results. The term "dimension theory" is employed because this apparatus typically involves the assignment of dimensions to definable sets in suitable structures. "Geometry" refers (somewhat imprecisely) to those areas of mathematics in which most of the new applications of model-theoretic methods and analyses have been made.

The diversity of material within each of these topics suggested the somewhat unconventional approach that each minicourse be divided among several speakers. The organization of this volume reflects that of the workshop in that the articles forming each of the three minicourses are grouped together. Within each group the order of the papers is roughly parallel. Although by no means perfect, this parallelism is intended to emphasize how the several components of each minicourse correlate. It also should help a reader interested in a particular aspect of the subject to locate relevant material more easily. Bearing these organizational themes in mind, we now briefly review the contents of the articles.

David Marker's first article, *Introduction to model theory*, serves as a preamble to the three minicourses. It provides an excellent survey of elementary model theory that prepares a non-logician well for what is to come. It also covers some of the classical model theory of the first classes of fields to be studied and understood from the model-theoretic point of view: the algebraically closed and real closed fields.

The next three articles survey the model theory of fields. They examine fields with additional structure that have been treated successfully by model theory, reflecting the range of model-theoretic phenomena encountered in the abstract theory and also including several of the algebraic objects that arise in the applications. Lou van den Dries' paper, *Classical model theory of fields*, deals with the fields of real and p-adic numbers, and expansions of these fields obtained by adding analytic structure. Fields enriched by adjoining a derivation are discussed in Marker's *Model theory of differential fields*. Some of the most important recent applications of model theory have been to Diophantine questions, and Marker briefly indicates how differential fields enter into Hrushovski's model-theoretic proof of the Mordell–Lang conjecture for function fields of characteristic zero (see Section 5 of his article). In *A Survey of the model theory of difference fields*, Zoé Chatzidakis focuses on fields to which a distinguished automorphism has been adjoined. Hrushovski also has applied model-theoretic results concerning difference fields to Diophantine questions, and in Section 4 of her article Chatzidakis outlines his proof of the Manin–Mumford conjecture, including the explicit bounds his argument yields.

The next two articles deal with "dimension theory," that is, pure model theory. Dugald Macpherson's article, *Notes on o-minimality and variations*, concentrates on the body of pure model-theoretic results dating from the early 1980's that place the theory of semialgebraic and subanalytic sets from real geometry into an abstract context. The article by Bradd Hart, *Stability and its variants*, provides an introduction to the vast body of work on stable and simple theories. This aspect of pure model theory has its origins in Morley's proof of the Łoś conjecture in the 1960's and took flight as a full-fledged theory with the deep work of Shelah in the 1970's around the classification of first order theories according to whether or not their class of models has a "structure theory". Even though o-minimality and stability/simplicity have developed separately, the two theoretical frameworks share strong conceptual similarities that emerge upon comparing the two articles. We shall comment further upon this in Section 3 below. Moreover, several of the technical notions that play a significant role in the abstract theory have clear meaning and significance in several of model theory's most striking applications in other areas of mathematics. This theme emerges in several articles in the volume.

The final three contributions to the volume, drawn from the introductory workshop's geometry minicourse and written by experts from outside model theory, accomplish several goals: they supply the mathematical background for many of the most notable applications of the model theory of fields, offer an introductory discussion of several of these developments, and raise provocative questions and/or suggest directions in which model theory might play a future role.

The article by Edward Bierstone and Pierre Milman, *Subanalytic geometry*, provides an introductory account of semialgebraic and subanalytic subsets of $\mathbb{R}^n$ (or $\mathbb{C}^n$) that indicates how this theory relates to o-minimality (see also van den Dries' article). It further suggests classes between the semialgebraic and subanalytic sets, based on local behavior of analytic mappings, that are "tame" from an algebraic or analytic point of view in much the same spirit as Grothendieck's vision of "tame topology". Jan Denef's article, *Arithmetic and geometric applications of quantifier elimination for valued fields*, surveys how the rich interaction between model theory and p-adic and rigid analytic geometry has yielded important number-theoretic and geometric applications. He also introduces some of the most recent exciting developments along the intersection between p-adic and rigid analytic geometry on the one hand, and model theory on the other. In Section 4 of his article, Denef discusses new invariants for algebraic varieties in which motivic measure and integration takes the place of p-adic integration. Barry Mazur's article, *Abelian varieties and the Mordell–Lang Conjecture*, the concluding paper in the volume, focuses on the mathematics surrounding the Mordell–Lang conjecture (see also Marker's article, *Introduction to the model theory of differential fields*). The article situates the Mordell–Lang conjecture in historical perspective as a counterpart in higher-dimensions to the Mordell conjecture proved by Faltings. Section 6 includes a proof of the Mordell–Lang conjecture over number fields in the rank one case, using Chabauty's method of embedding the situation in the p-adics and the theory of p-adic Lie groups. In Section 7 Mazur discusses the reduction of the Mordell–Lang conjecture to the number field case, and in Section 8 he mentions several questions regarding effectivity issues in the number of solutions of Diophantine equations that are very much in the spirit of model theory and are intended to suggest further interplay between the two fields.

2. Structures and Definable Sets

Our thematic overview of the volume begins with a brief primer of the objects studied in model theory. (Marker's *Introduction to model theory* offers a thorough treatment.) For a model theorist, a mathematical structure $\mathcal{M}$ is a set M equipped with a set of operations on M, a set of relations on M, and a set of distinguished elements of M. One example is provided by the natural numbers $\mathbb{N}$ with the usual operations of addition and multiplication and the constant element 0: the structure $(\mathbb{N}, +, \cdot, 0)$. The ordered field of real numbers with addition, multiplication, the distinguished elements 0 and 1, and the binary relation $<$, that is, the structure $(\mathbb{R}, +, \cdot, 0, 1, <)$, provides another. At first glance, these examples might seem to be nothing more than mild variants on the kinds of structures dealt with by algebraists.

The model-theoretic point of view is distinguished by the unified perspective it provides for mathematical structures viewed with respect to the degree of generality described above. Thus on some level, the structures in the last paragraph are, to a model theorist, to be treated no differently than structures as diverse as the ordered field of real numbers augmented by all (partial) functions $f : [0, 1]^n \to \mathbb{R}$ where f is the restriction to $[0, 1]^n$ of a function analytic in a neighborhood of $[0, 1]^n$, or the difference field $(\mathbb{C}(t), +, \cdot, \sigma)$ where σ is the "shift operator" defined by

$$\sigma \upharpoonright \mathbb{C} = \text{identity} \quad \text{and} \quad \sigma(t) = t + 1.$$

Many further examples of fields enriched by additional structure figure in the articles in this volume and several assume central roles.

What is it then that provides this unified perspective? It is not just the scope and generality of structures in the above sense that makes the model-theoretic point of view distinctive. Rather it is the objects that model theory attaches to these structures, and the tools and methods that model theory employs to analyze and understand these objects that sets the subject apart. Just as a complex analyst, for example, concentrates on holomorphic functions on the field of complex numbers, model theory focuses on a particular class of relations and functions, the so-called definable ones.

For a set to be definable means simply that it can be "defined" by a formula in first-order logic in the language of the structure. But what is meant by this? We shall be informal, referring the reader to Section 1 of Marker's *Introduction to model theory* for the complete details. To each structure $\mathcal{M}$, there corresponds a formal language $\mathcal{L}$ which includes an n-place function symbol $\widehat{f}$ for each n-place function f in $\mathcal{M}$, an n-place relation symbol $\widehat{R}$ for each n-place relation R in $\mathcal{M}$, and a constant symbol $\widehat{c}$ for each distinguished element in $\mathcal{M}$. Each symbol in $\mathcal{L}$ is interpreted in $\mathcal{M}$ by the corresponding function, relation, or distinguished element. The $\widehat{}$ is usually deleted when writing the symbols in $\mathcal{L}$ where no confusion can arise. Formulas in the language $\mathcal{L}$ are the meaningful finite strings of symbols built from the symbols of $\mathcal{L}$, $=$, variables, the logical connectives $\neg$, $\wedge$, $\vee$, and the quantifiers $\exists$ and $\forall$. Here, "meaningful" refers to nothing more than a carefully executed version of the usual kind of symbolic expression that mathematicians write daily.

For those encountering formulas for the first time, two key restrictions should kept in mind: formulas are *finite* in length and quantification is limited to individual elements of the structure.[1] For example, the disjunction $\bigvee_{n \in \mathbb{N}} x^n = e$, indexed by the natural numbers, which captures the torsion elements of a group $(G, \cdot, e)$, is not a first-order formula because the disjunction is infinite, and hence the expression is infinite in length. As another example, quantification over all ideals of a ring $(R, +, \cdot, 0, 1)$ is not permitted since ideals are not *elements* of the ring.

The variables in an $\mathcal{L}$-formula φ are either bound to a quantifier in φ or not, in which case they are called free. Then, roughly speaking, an $\mathcal{L}$-formula φ whose free variables are $x_1, \ldots, x_n$ describes the definable set $X \subset M^n$ in a structure $\mathcal{M} = (M, \ldots)$ consisting of all $(a_1, \ldots, a_n) \in M^n$ which, when substituted for $x_1, \ldots, x_n$, make the formula true when the symbols in $\mathcal{L}$ are interpreted by the corresponding objects in $\mathcal{M}$ and the set-theoretic operations corresponding to the logical symbols $\neg$ (complement), $\wedge$ (intersection), $\vee$ (union), and $\exists$ (projection) are performed as dictated by the formula.

Definable sets also can be given a simple and succinct set-theoretic characterization. We will be somewhat glib here; the precise characterization is given in Proposition 1.3 of Marker's article (page 18). Let $\mathcal{M} = (M, \ldots)$ be a structure. For each $n \geq 1$ let D^n be the smallest collection of subsets of M^n that contains M^n, all n-ary relations in $\mathcal{M}$, and the graphs of all functions $f: M^{n-1} \to M$ in $\mathcal{M}$, and that is closed under taking generalized diagonals, complements, unions, intersections, and projections (from sets in D^m for $m > n$). A set $X \subseteq M^n$, where $n \geq 1$, is definable in $\mathcal{M}$ if $X \in D^n$. Although correct and relatively easy to state, this set-theoretic version of definability reveals neither why definable sets are so natural and useful, nor, for that matter, why this class of sets is so named.

The power the point of view of definability confers comes from the fact that many of the sets that arise in mathematics can be described by formulas in exactly the way that mathematicians ordinarily do so. To illustrate, suppose that $\mathcal{R} = (\mathbb{R}, +, \cdot, -, <, 0, 1, f)$ where $f: \mathbb{R} \to \mathbb{R}$. Then the set of points at which f is continuous or differentiable is definable in $\mathcal{R}$, since the the usual definitions of continuity and differentiability can be formalized in first-order logic in the language $\mathcal{L} = \{+, \cdot, -, <, 0, 1, f\}$. Likewise, most properties of elementary real analysis and topology of definable sets and functions are readily seen to be definable. Examples from a wide variety of contexts both of definable sets and, equally importantly, non-definable sets (such as the torsion elements in an arbitrary group $(G, \cdot, e)$) are provided in Marker's *Introduction to model theory* as well as throughout the volume.

Although first-order definability might appear to impose rather severe limitations on the objects that model theory studies, it does in fact supply rich and interesting classes of sets and functions. In fact, as mentioned in both Marker's

---

[1]Logicians have developed and studied logics that relax either or both of these restrictions with mixed success, but first-order logic appears to provide the best balance between expressibility and manageability.

introductory article and van den Dries' contribution, Gödel's Incompleteness Theorem implies that the definable sets in $(\mathbb{N}, +, \cdot, 0, 1)$ are complicated to the point of "wildness," and thus exhibit such poor model-theoretic behavior as to escape analysis. As will be evident from the diverse range of structures and applications discussed in this volume, manageable, or "tame", behavior does occur regularly enough. And even in contexts in which model-theoretic analysis might appear at first to be of limited value, it sometimes is possible (and desirable) to carry out the analysis in a setting with a richer collection of definable sets *and* with good model-theoretic properties. This point emerges in the striking applications of model theory to Diophantine problems as described in Marker's article on differential fields and Chatzidakis' article.

3. Analysis of Definable Sets and Applications

The preceding discussion of definable sets and structures is a necessary prelude to the most important aspect of the model-theoretic enterprise: the theoretical methods that have been developed for understanding definable sets and the applications that ensue.

As in any area of mathematics, model theorists analyze definable sets by devising measures of simplicity or tractability. Two main threads appear here. The first deals with the complexity of a definable set based on the "structural complexity" of a formula required to define the set. Just as the set-theoretic operation of projection can transport us outside of the class of Borel subsets of Euclidean space, projection, in the guise of existential quantification, typically adds complexity to definable sets. Indeed, mathematicians have often remarked that a proposition with more than three alternations of quantifiers strains the understanding. Thus, the "structural complexity" of a formula might be measured by whether or not it contains quantifiers or by counting the number of alternations of blocks of universal and existential quantifiers appearing in what is called prenex normal form of the formula. This kind of analysis typically shows that the definable sets of a structure satisfy a general hypothesis that subsequently permits the application of a general theoretical framework for the analysis of the definable sets, as we now describe.

The second approach to the analysis of definable sets that model theorists have developed classifies these sets by various abstract yet natural measures that assign to the sets a combinatorial, algebraic, or geometrically motivated notion of dimension. The two principal avenues of "dimension theory", as we have called this second approach to the analysis of definable sets, are stability and simplicity, the subject of Hart's article, and o-minimality and some of its variants as discussed in Macpherson's survey. In each case, central to the development of a dimension theory is a notion of independence, or freeness: for subsets A, B and C of a structure $\mathfrak{M}$, the expression "B is independent from C over A" should mean "C provides no more information about B than A does". For example, for algebraically or real closed fields, the dimension-theoretic definition of independence coincides with algebraic independence. The general theory has proven to be remarkably rich and has demonstrated its mettle in many applications.

This second mode of analysis can in principle ignore the syntactic shape of a defining formula for a set, and thus *prima facie* have little to do with the first approach. Yet some of the most important applications are often found where the two threads cross. As suggested earlier, if the definable sets of the structures in some class yield to the first kind of analysis, model theorists then may be able to show that the class of structures is amenable to the powerful model-theoretic tools and methods afforded by the second kind of analysis. We now take up in more detail these two approaches to understanding definable sets.

Quantifier elimination and generalizations. We begin with a simple illustration of the analysis of definable sets via the "structural complexity" of their defining formulas. Marker's introduction to model theory, his article on differential fields, and the contributions by van den Dries, Denef, and Bierstone and Milman include many more.

A set $X \subseteq M^n$ of a structure $\mathcal{M} = (M, \dots)$ is *quantifier-free definable* if there is a formula in the language of $\mathcal{M}$ not containing quantifiers that defines X. For a field $\mathcal{F} = (F, +, \cdot, 0, 1)$ the quantifier-free definable sets are exactly the constructible sets; boolean combinations of the zero sets of polynomials over F. Generally speaking, the class of definable sets in a field includes many more sets than the constructible sets (e.g., in the field of rational numbers; see Marker's introductory article). If $\mathcal{F}$ is algebraically closed, Chevalley's theorem that the projection of a constructible set is constructible implies — since existential quantification corresponds to projection — that the definable sets are exactly those that are quantifier-free definable. In this case the structure is said to have quantifier elimination. More is true. For a formula $\varphi(v_1, \dots, v_n)$ there is a quantifier-free formula $\psi(v_1, \dots, v_n)$ that defines the same set as φ in *all* algebraically closed fields. This kind of uniformity plays a crucial role in both pure model theory and applications. Among the most well-known examples of theories — that is, consistent sets of first-order sentences — with quantifier elimination are the theory of real closed fields, the theory of differentially closed fields, and, for each p, the theory of p-adically closed fields. Others appear throughout the volume — in particular see van den Dries' and Denef's articles.

Quantifier elimination shows how the power of definability and the simplicity of definable sets can play off each other. The full strength of definability permits the definition of *a priori* complicated sets in the structure; that a set has a quantifier-free description implies on the other hand that it is in some sense simple. Several applications appear in Marker's introductory article, and Denef presents in Section 1 of his article his beautiful use of quantifier elimination for the p-adic numbers $\mathbb{Q}_p$, in the language with $+, \cdot, -, 0, 1$ augmented by predicates for d-th powers for all $d = 2, 3, 4, \dots$, to prove the rationality of several

Poincaré series. (See Section 2 of van den Dries' article for a detailed exposition of quantifier elimination for valued fields).

It should be noted that quantifier elimination is highly sensitive to the language of the class of structures under consideration. An artificial model-theoretic trick shows that if the language of a structure is enriched sufficiently then the structure can be made to have elimination of quantifiers (see Section 4 of Marker's introductory article). This artifice is of little use, though, since the quantifier elimination so obtained reveals nothing about the definable sets. Finding an appropriate language in which a class of structures has quantifier elimination is a difficult and subtle issue. For further discussion of this we refer again to Denef's and van den Dries' articles.

Sometimes quantifier elimination in a particular language fails but yet the definable sets in a structure still have a manageable and useful form. An instructive example of this is *model completeness*. One of the several equivalent definitions is that a theory is model complete if for every formula φ there is an *existential formula* ψ, i.e., a formula consisting of a block of existential quantifiers followed by a quantifier-free formula, such that φ and ψ are equivalent, that is, they define the same sets in all structures satisfying the theory. Model completeness thus serves as the next best substitute for quantifier elimination.

An example may help bring this into sharper focus. The ordered real field $(\mathbb{R}, +, \cdot, 0, 1, <)$ has quantifier elimination, but an adaptation of an old argument of Osgood demonstrates that quantifier elimination fails in the real field augmented by any collection of total analytic functions. In particular, this is true in the real exponential field $(\mathbb{R}, +, \cdot, 0, 1, <, e^x)$, and so model completeness is the best that could be expected. A important theorem of Wilkie proved in 1991 shows that this structure is in fact model complete, and thus every definable set is the projection of a quantifier-free definable set. Results of Khovanskii from the 1970's provide a good understanding of the quantifier-free definable sets, and thus Wilkie's theorem yields a clear picture of all definable sets. The analysis afforded by model completeness in turn suffices to conclude that the real exponential field is o-minimal, and hence its definable sets enjoy the many good geometric properties that the theory of o-minimality provides (see Section 4.3 of van den Dries' articles). We shall say more about o-minimality in the "dimension theory" subsection below.

Yet another form of partial or relative quantifier elimination emerges in the first-order theory of modules. The language of modules over a ring R includes a symbol $+$ for addition, 0, and, for each $r \in R$, a 1-place function symbol for multiplication by r. For a complete first-order theory T of modules (see page 19 in Marker's article for the definition of completeness), every formula is equivalent to a boolean combination of "positive-primitive" formulas, that is, formulas in which a block of existential formulas precedes a conjunction of atomic formulas. This analysis implies that for a module $(M, +, \dots)$ every definable subset of M^k, where $k \in \mathbb{N}$, is a boolean combination of definable subgroups of M^k. In a more general context, the so-called 1-based groups (elaborated further in Section 3.3 of Hart's article and Section 4 of Chatzidakis' article), it also can be shown that all definable sets are boolean combinations of definable subgroups. This fact plays an crucial role in Hrushovski's applications of model theory to the Mordell–Lang and Manin–Mumford Conjectures (see Section 5 of both Marker's and Chatzidakis' articles).

3.1. Dimension theory. We turn now to the second major approach to the analysis of definable sets via what we call "dimension theory". This has constituted perhaps the central theme in the development of pure model theory for almost 40 years. There have been two main strands present here. The first, beginning with the seminal work of Morley in the 1960's and developed profoundly by Shelah in the following decade provides a combinatorial/algebraic account of dimension theory. Stable, and more generally simple theories are the subject of this analysis, which is elaborated in Hart's article. The second strand yields a more topological/algebraic version of dimension theory that comprises the main focus of Macpherson's article. Although these two dimension theories apply to disjoint classes of structures they share several common conceptual features. We here offer some introductory remarks that should highlight these themes.

The beginnings of the dimension-theoretic analysis of definable sets in structures that are stable or simple runs as follows. Let $\mathcal{M} = (M, \dots)$ be a structure. Since equality is included in every language, for every $n \in \mathbb{N}$ and $a_1, \dots, a_n \in M$, the sets $\{a_1, \dots, a_n\}$ and $M \setminus \{a_1, \dots, a_n\}$ must be definable subsets of M in $\mathcal{M}$: they are defined by the formula

$$v = a_1 \vee \cdots \vee v = a_n$$

and its negation, respectively. Thus, the finite and cofinite subsets of the universe of a structure must always be definable. With a slight twist, the structures $\mathcal{M} = (M, \dots)$ that are the least complicated from the viewpoint of definability are those for which the definable subsets of M are precisely the sets that must be definable, that is, the finite and cofinite sets. The twist is that reference must be made not to individual structures but rather to all structures for a particular language satisfying a set of axioms, that is, a theory: a theory T is *strongly minimal* if for every structure $\mathcal{M} = (M, \dots)$ satisfying T the definable subsets of M are exactly the finite and cofinite sets. Elimination of quantifiers for the theory of algebraically closed fields shows that every definable subset of the universe of an algebraically closed field is defined by a formula which is a boolean combination of polynomial equalities in one variable. As the set given by such a formula is finite or cofinite, the theory of algebraically closed fields is strongly minimal.

As mentioned earlier, by adjoining new symbols to the language, a structure is endowed with a richer collection of definable sets. The balance to strike here is that the richer expressive power of the expanded structure should not yield an intractible definability theory. Thus, if a predicate for the natural numbers is adjoined to the complex numbers, the structure becomes as "wild" as $(\mathbb{N}, +, \cdot, 0, 1)$ itself and so subject to the Gödel phenomenon mentioned earlier. This is not always the case. The theory of differentially closed fields, discussed in depth in Marker's second article, provides an important example. A differential field is a field of characteristic zero equipped with a derivation. The language for this structure is the language of fields with a function symbol for the derivation adjoined. The theory of differentially closed fields is axiomatized by appropriate closure axioms asserting the existence of zeroes for differential polynomials. This theory too has quantifier elimination. It follows that the definable sets satisfy a hypothesis, ω-stability, that ensures a highly manageable model-theoretic dimension theory in which there are definable sets having "transfinite dimension".

For the study of definable sets, strong minimality is relativized to definable sets in a structure. For a structure $\mathcal{M} = (M, \dots)$ and a definable set $X \subseteq M^n$, we say that X is strongly minimal if every definable subset of X in $\mathcal{M}$ is finite or cofinite. Strongly minimal sets can be thought of as "irreducible" sets of dimension one. They form the first layer of a dimensional analysis of definable sets based on what is called *Morley rank* (see Definition 1.16 in Hart's article). In the context of algebraically closed fields, the Morley rank of a definable (= constructible) set agrees with its algebro-geometric dimension. The ω-stable theories can be shown to be exactly those theories for which every definable set has ordinal-valued Morley rank. The most general dimensional analysis, Shelah's theory of forking, applies to the class of stable, or more generally simple theories. The sweep of this theory is remarkably broad given the consequences that flow from it (see Sections 2 and 3 of Hart's article). Among the theories of fields discussed in this volume — see the articles by Marker and Chatzidakis — it embraces algebraically, differentially, and separably closed fields, pseudo-finite fields, and ACFA, the model companion of the theory of difference fields. The crucial point when it comes to Hrushovski's applications to diophantine geometry (see Marker's article on differential fields as well as those by Chatzidakis and by Mazur) is that objects of arithmetic type, such as the torsion points of an abelian variety or the points of an abelian variety over a function field can be embedded in definable groups in enriched structures to which the general theory applies.

Strongly minimal sets and more general versions of "1-dimensional" or "minimal" sets (see Hart's and Chatzidakis' articles) assume an important position in the model-theoretic analysis employed in these applications. Generally speaking, the structure of such sets also determines the structure of finite-dimensional sets, and so understanding the pregeometry given by model-theoretic dependence in "minimal" sets is imperative. For many years the only known examples of such a pregeometry were *trivial*, "module-like" (*locally modular*), or "field-like" (i.e., permit the interpretation in a precise model-theoretic sense of an infinite field) — see Section 4 of Marker's article and Example 1.26 in Hart's article. In the late 1970's, early on in the development of the theory, Zil'ber boldly conjectured that these are the only possible cases. This conjecture exercised a powerful and positive influence on model theory in the 1980's (see the statement of the Zil'ber principle right after Example 1.26 in Hart's article). Zil'ber's conjecture ultimately proved false: in the late 1980's Hrushovski found counterexamples. With the introduction of the notion of a *Zariski geometry*, however, Hrushovski and Zil'ber isolated an important class of strongly minimal sets for which the conjecture holds (see Section 4 of Marker's article on differential fields, as well as Hart's article). Furthermore, the "minimal" sets in the enriched structures that figure in Hrushovski's proof of the Mordell–Lang Conjecture can be shown to be Zariski geometries, and Hrushovski avails himself of this in his proof.

The field of real numbers presents a different situation. The ordering on $\mathbb{R}$ actually is definable in the field language:

$$x < y \quad \Longleftrightarrow \quad \exists u\, (y = x + u^2 \wedge u \neq 0).$$

Hence, the order relation can be adjoined to the real field structure without altering the class of definable sets. It follows that the real field is not stable, or even simple — see Hart's article — and so cannot be analyzed by the machinery described above. Yet, as the ordered field of real numbers, actually the theory of real closed fields, has quantifier elimination, the definable sets are exactly those which are defined by boolean combinations of polynomial equalities and inequalities, that is, the semialgebraic sets. These have been studied with considerable success by real algebraic geometers (see the articles by Bierstone and Milman and van den Dries). Observe in particular that the definable subsets of $\mathbb{R}$ in the real field consist of finitely many open intervals and points. As these sets are those that must be definable in any linearly ordered structure, the definable subsets of $\mathbb{R}$ in the real field are uncomplicated if one adopts the right point of view.

A linearly ordered structure $\mathcal{M} = (M, <, \dots)$ is order-minimal, or *o-minimal* if every definable subset of M is the union of finitely many points and open intervals, that is those that must be definable in the presence of a linear ordering. O-minimal structures — see Macpherson's article for a survey of this subject — permit a dimension-theoretic analysis of definable sets that accords with the geometry of the definable sets. In particular, many of the geometric and analytic properties of semialgebraic sets extend to o-minimal structures, particularly those whose underlying set is $\mathbb{R}$ (see van den Dries' and Macpherson's article). Furthermore, many analogues of theorems from the stable context can be proved under the hypothesis of o-minimality. For example, a version of Zil'ber's conjecture has been proved by Peterzil and Starchenko in the o-minimal setting. Notions of minimality relative to other basic predicates also are mentioned in Section 4 of Macpherson's contribution.

While *a priori* more limited in scope than stability and simplicity, what has made o-minimality successful is that it has been proved that o-minimality is preserved under adjoining many analytically important functions to the real field. Wilkie's theorem that the structure $(\mathbb{R}, +, \cdot, <, 0, 1, e^x)$ is (model complete and) o-minimal was the first dramatic result in this direction and Section 4 of van den Dries' article discusses many others. These theorems have in turn been applied to problems in real analytic and algebraic geometry, and recently have been invoked in work in the representation theory of Lie groups. To illustrate, the well-known semialgebraic fact that there are finitely many homeomorphism types in $\mathbb{R}^m$ of the zero sets of polynomials $p(x_1, \ldots, x_m) \in \mathbb{R}[x_1, \ldots, x_m]$ of some fixed degree d can be extended via o-minimality to establish that the same holds true if "of some fixed degree d" is replaced by "with no more than d monomials" (of arbitrary degree). The proof takes advantage of a form of uniformity in parameters that the model theory provides. Further afield, the o-minimality of expansions of the real field have seen applications in as seemingly distant subjects as neural nets and control theory.

DEIRDRE HASKELL
DEPARTMENT OF MATHEMATICS
COLLEGE OF THE HOLY CROSS
WORCESTER, MA 01610
UNITED STATES
haskell@mathcs.holycross.edu

ANAND PILLAY
UNIVERSITY OF ILLINOIS
DEPARTMENT OF MATHEMATICS
1409 W GREEN ST.
URBANA, IL 61801-2917
UNITED STATES
pillay@math.uiuc.edu

CHARLES STEINHORN
DEPARTMENT OF MATHEMATICS
VASSAR COLLEGE
POUGHKEEPSIE, NY 12604-0257
UNITED STATES
steinhorn@vassar.edu

Model Theory, Algebra, and Geometry
MSRI Publications
Volume **39**, 2000

Introduction to Model Theory

DAVID MARKER

ABSTRACT. This article introduces some of the basic concepts and results from model theory, starting from scratch. The topics covered are be tailored to the model theory of fields and later articles. I will be using algebraically closed fields to illustrate most of these ideas. The tools described are quite basic; most of the material is due either to Alfred Tarski or Abraham Robinson. At the end I give some general references.

1. Languages and Structures

What is a mathematical structure? Some examples of mathematical structures we have in mind are the ordered additive group of integers, the complex field, and the ordered real field with exponentiation.

To specify a structure we must specify the underlying set, some distinguished operations, some distinguished relations and some distinguished elements. For example, the ordered additive group of integers has underlying set $\mathbb{Z}$ and we distinguish the binary function $+$, the binary relation $<$ and the identity element 0. For the ordered field of real numbers with exponentiation we have underlying set $\mathbb{R}$ and might distinguish the binary functions $+$ and $\times$, the unary function $\exp$, the binary relation $<$ and the elements 0 and 1.

Here is the formal definition.

DEFINITION 1.1. A *structure* $\mathcal{M}$ is given by the following data.

(i) A set M called the *universe* or *underlying set* of $\mathcal{M}$.
(ii) A collection of functions $\{f_i : i \in I_0\}$ where $f_i : M^{n_i} \to M$ for some $n_i \geq 1$.
(iii) A collection of relations $\{R_i : i \in I_1\}$ where $R_i \subseteq M^{m_i}$ for some $m_i \geq 1$.
(iv) A collection of distinguished elements $\{c_i : i \in I_2\} \subseteq M$.

Any (or all) of the sets I_0, I_1 and I_2 may be empty. We refer to n_i and m_j as the *arity* of f_i and R_j.

Here are some examples:

(1) The ordered field of real numbers has domain $\mathbb{R}$, binary functions $+, -, \times$, relation $<$, and distinguished elements 0 and 1.

(2) The valued field of p-adic numbers has domain $\mathbb{Q}_p$, binary functions $+, -, \times$, distinguished elements 0 and 1, and a unary relation $\mathbb{Z}_p$, for the ring of integers.

In mathematical logic we study structures by examining the sentences of first order logic true in those structures. To any structure we attach a language $\mathcal{L}$ where we have an n_i-ary function symbol $\hat{f}_i$ for each f_i, an m_i-ary relation symbol $\hat{R}_i$ for each R_i and constant symbols $\hat{c}_i$ for each c_i.

An $\mathcal{L}$-*structure* is a structure $\mathcal{M}$ where we can interpret all of the symbols of $\mathcal{L}$. For example, let $\mathcal{L}$ be the language where we have a binary function symbol $\hat{\times}$ and a constant symbol $\hat{1}$. The following are examples of $\mathcal{L}$-structures:

(1) $\mathcal{M}_1$ has underlying set $\mathbb{Q}$. We interpret $\hat{\times}$ as $\times$ and $\hat{1}$ as 1.

(2) $\mathcal{M}_2$ has underlying set $\mathbb{Z}$. We interpret $\hat{\times}$ as $+$ and $\hat{1}$ as 0.

Of course we also could take the natural interpretation of $\mathcal{L}$ in $\mathbb{Z}$.

(3) $\mathcal{M}_3$ has underlying set $\mathbb{Z}$. We interpret $\hat{\times}$ as $\times$ and $\hat{1}$ as 1.

DEFINITION. If $\mathcal{M}$ and $\mathcal{N}$ are $\mathcal{L}$-structures with underlying sets M and N, respectively, an $\mathcal{L}$-*embedding* $\sigma : M \to N$ is an injective map that preserves the interpretation of all function symbols, relation symbols and constant symbols of $\mathcal{L}$. An $\mathcal{L}$-*isomorphism* is a bijective $\mathcal{L}$-embedding.

We say that $\mathcal{M}$ is a substructure of $\mathcal{N}$ (and write $\mathcal{M} \subset \mathcal{N}$) if $M \subset N$ and the inclusion map is an $\mathcal{L}$-*embedding*.

Formulas in our language are finite strings made from the symbols of $\mathcal{L}$, the equality relation $=$, variables $x_0, x_1, \ldots$, the logical connectives $\exists, \wedge, \vee$, quantifiers $\exists$ and $\forall$ and parentheses. (See the appendix on page 34 for precise definitions.) We interpret $\exists, \wedge, \vee$ as "not", "and" and "or" and $\exists$ and $\forall$ as "there exists" and "for all". I will use $x, y, z \ldots$ and their subscripted varieties as variables and not use the symbol $\hat{}$ when no confusion arises.

Let $\mathcal{L}_r$ be the language of rings, where we have binary function symbols $+, -$ and $\times$ and constant symbols 0 and 1. The language of ordered rings, $\mathcal{L}_{or}$ is $\mathcal{L}_r$ with an additional binary relation symbol $<$. (As usual we will write $x+y$ instead of $+(x, y)$ and $x < y$ for $<(x, y)$.) Here are some examples of $\mathcal{L}_{or}$-formulas:

$$x_1 = 0 \vee x_1 > 0$$
$$\exists x_2 \; x_2 \times x_2 = x_1$$
$$\forall x_1 \; (x_1 = 0 \vee \exists x_2 \; x_2 \times x_1 = 1)$$

Intuitively, the first formula asserts that $x_1 \geq 0$, the second asserts that x_1 is a square and the third asserts that every nonzero element has a multiplicative inverse. We would like to define what it means for a formula to be true in a structure, but these examples already show one difficulty. While in any $\mathcal{L}_{or}$-structure the third formula will either be true or false, the first two formulas express a property which may or may not be true of elements of the structure.

We say that a variable *occurs freely* in a formula ϕ if it is not inside the scope of a quantifier; otherwise we say it is *bound*. For example, x_1 is free in the first two formulas and bound in the third, while x_2 is bound in both formulas.

We call a formula a *sentence* if it has no free variables. For any $\mathcal{L}$-structure each sentence of $\mathcal{L}$ is either true or false. Let ϕ be a sentence. We say that $\mathcal{M}$ is a model of ϕ, and write $\mathcal{M} \models \phi$, if and only if ϕ is true in $\mathcal{M}$.

We often write $\phi(x_1, \ldots, x_n)$ to show that the variables $x_1, \ldots, x_n$ are free in the formula ϕ. We think of $\phi(x_1, \ldots, x_n)$ as describing a property of n-tuples from M. For example, the $\mathcal{L}_{or}$-formula $\exists x_2 \; x_2 \times x_2 = x_1$ has the single free variable x_1 and describes the property "x_1 is a square". If $a_1, \ldots, a_n$ are elements of M we say $\mathcal{M} \models \phi(a_1, \ldots, a_n)$ if the property expressed by ϕ is true of the tuple $(a_1, \ldots, a_n)$.

We say that two $\mathcal{L}$-structures $\mathcal{M}$ and $\mathcal{N}$ are *elementarily equivalent* if for all $\mathcal{L}$-sentences $\mathcal{M} \models \phi \Longleftrightarrow \mathcal{N} \models \phi$.

PROPOSITION 1.2. *If $\mathcal{M}$ and $\mathcal{N}$ are isomorphic, then they are elementarily equivalent.*

PROOF. Show by induction on formulas that if $\phi(x_1, \ldots, x_n)$ is a formula, $\sigma : M \to N$ is an isomorphism and $a_1, \ldots, a_n \in M$, then

$$\mathcal{M} \models \phi(a_1, \ldots, a_n) \iff \mathcal{N} \models \phi(\sigma(a_1), \ldots, \sigma(a_n)). \qquad \square$$

We say that an $\mathcal{L}$-embedding $f : \mathcal{M} \to \mathcal{N}$ is *elementary* if for any $a_1, \ldots, a_n \in M$ and any formula $\phi(x_1, \ldots, x_n)$

$$\mathcal{M} \models \phi(a_1, \ldots, a_n) \iff \mathcal{N} \models \phi(f(a_1), \ldots, f(a_n)).$$

If $\mathcal{M} \subset \mathcal{N}$ we say that $\mathcal{M}$ is an *elementary substructure* if the inclusion map is elementary.

DEFINITION. We say that a set $X \subset M^n$ is *definable* in the $\mathcal{L}$-structure $\mathcal{M}$ if there is a formula $\phi(x_1, \ldots, x_{n+m})$ and elements $b_1, \ldots, b_m \in M$ such that

$$X = \{(a_1, \ldots, a_n) : \mathcal{M} \models \phi(a_1, \ldots, a_n, b_1, \ldots, b_m)\}.$$

We say that X is *A-definable* or *definable over A*, where $A \subseteq M$, if we can choose that $b_1, \ldots, b_m \in A$. For example, if $m = 0$ we say X is $\varnothing$-definable.

For example, $\{x : x > \pi\}$ is definable over $\mathbb{R}$ but not $\varnothing$-definable, while $\{x : x > \sqrt{2}\}$ is $\varnothing$-definable by the formula $x \times x > 1 + 1 \wedge x > 0$. In the field $(\mathbb{Q}_p, +, -, \times, 0, 1)$ if $p \neq 2$ we can define the valuation ring $\mathbb{Z}_p$ by the formula $\exists y \; y^2 = px^2 + 1$.

We can give a very simple characterization of the definable sets.

PROPOSITION 1.3. *Suppose that D_n is a collection of subsets of M^n for all $n \geq 1$ such that $\mathcal{D} = (D_n : n \geq 1)$ is the smallest collection satisfying the following conditions:*

(i) $M^n \in D_n$.

(ii) *For all n-ary functions f of $\mathcal{M}$, the graph of f is in D_{n+1}.*

(iii) *For all n-ary relations R of $\mathcal{M}$, $R \in D_n$.*

(iv) *For all $i, j \leq n$, $\{(x_1, \ldots, x_n) \in M^n : x_i = x_j\} \in D_n$.*

(v) *Each D_n is closed under complement, union and intersection.*

(vi) *If $X \in D_m$ and $\pi : M^n \to M^m$ is the projection map $(x_1, \ldots, x_n) \mapsto (x_{i_1}, \ldots, x_{i_m})$, then $\pi^{-1}(X) \in D_n$.*

(vii) *If $X \in D_n$ and π is as above, then $\pi(X) \in D_m$.*

(viii) *If $X \in D_{n+m}$ and $b \in M^m$, then $\{a \in M^n : (a, b) \in X\} \in D_n$.*

Then $X \subseteq M^n$ is definable if and only if $X \in D_n$.

2. Theories

An *$\mathcal{L}$-theory* is a set of $\mathcal{L}$-sentences. Theories arise naturally as we attempt to axiomatize the properties of mathematical structures. For example, if $\mathcal{L}_r$ is the language of rings we can write down the field axioms as $\mathcal{L}_r$ sentences. We can give the theory of algebraically closed fields (ACF) by taking the field axioms plus, for each $n \geq 1$, the axiom

$$\forall x_0 \, \forall x_1 \, \ldots \, \forall x_{n-1} \, \exists y \, y^n + x_{n-1} y^{n-1} + \cdots x_1 y + x_0 = 0.$$

If T is an $\mathcal{L}$-theory, we say $\mathcal{M} \models T$ if $\mathcal{M} \models \phi$ for all $\phi \in T$. We say that an $\mathcal{L}$-sentence ϕ is a *logical consequence* of an $\mathcal{L}$-theory T (and write $T \models \phi$) if and only if $\mathcal{M} \models \phi$ for all $\mathcal{M} \models T$. For example, ACF $\models \forall x \, \forall y \, \exists z \, x^2 + y^2 = z^2$.

THEOREM 2.1 (GÖDEL'S COMPLETENESS THEOREM, FIRST VERSION). *$T \models \phi$ if and only if there is a formal proof of ϕ using assumptions from T.*

This has a very useful reformulation with an important corollary. We say that an $\mathcal{L}$-theory T is *satisfiable* if and only if there is an $\mathcal{L}$-structure $\mathcal{M}$ with $\mathcal{M} \models T$ and we say that T is consistent if and only if we cannot formally derive a contradiction from T.

THEOREM 2.2 (COMPLETENESS THEOREM, SECOND VERSION). *T is satisfiable if and only if T is consistent. Moreover if T has infinite models then T has a model where the underlying set has cardinality κ, for all $\kappa \geq |\mathcal{L}| + \aleph_0$.*

This has an easy consequence, which is the cornerstone of model theory.

THEOREM 2.3 (COMPACTNESS THEOREM). *If every finite subset of T is satisfiable, then T is satisfiable.*

PROOF. If T is not satisfiable, then by Theorem 2.2 there is a proof of a contradiction from T. As proofs use only finitely many assumptions from T there is a finite inconsistent subset of T. □

An important question when we try to axiomatize the properties of a structure is whether we have said everything we can say. An $\mathcal{L}$-theory T is *complete* if for all $\mathcal{L}$-sentences ϕ either $T \models \phi$ or $T \models \neg\phi$. Another way to say this is that a theory is complete if any two models are elementarily equivalent.

The easiest way to get a complete theory is to take the complete theory of a structure. If $\mathcal{M}$ is a structure, let $\mathrm{Th}(\mathcal{M}) = \{\phi : \mathcal{M} \models \phi\}$.

Gödel's incompleteness theorem says that Peano axioms are not complete (and there is no reasonable way to complete them). It is easy to see that ACF is not complete as it does not decide the characteristic: For p a prime number let ψ_p be the sentence

$$\forall x \underbrace{x + \cdots + x}_{p \text{ times}} = 0.$$

Clearly neither ψ_p nor $\neg\psi_p$ is a logical consequence of ACF. But this is the only obstruction. Let ACF_p be the theory obtained by adding ψ_p to ACF and let ACF_0 be the theory obtained by adding to ACF the sentences $\{\neg\psi_p : p \text{ a prime}\}$. We show shortly that ACF_p is complete.

If κ is a cardinal, we say that a theory T is κ-*categorical* if any two models of T where the underlying set has cardinality κ are isomorphic. Since algebraically closed fields are determined up to isomorphism by their characteristic and transcendence degree it is easy to see that ACF_p is κ-categorical for all $\kappa \geq \aleph_1$.

PROPOSITION 2.4 (VAUGHT'S TEST). *If all models of T are infinite and T is κ-categorical for some infinite cardinal κ, then T is complete.*

PROOF. Suppose not. Then $T \cup \{\phi\}$ and $T \cup \{\neg\phi\}$ are satisfiable. By 2.2 we can find κ and $\mathcal{M}$ and $\mathcal{N}$ of cardinality κ such that $\mathcal{M} \models T + \phi$ and $\mathcal{N} \models T + \neg\phi$. But this is impossible, as $\mathcal{M}$ must be isomorphic to $\mathcal{N}$. □

Thus ACF_p is complete for $p \geq 0$. This can be thought of as a version of the Lefschetz principle.

COROLLARY 2.5. *Let ϕ be an $\mathcal{L}_r$-sentence. The following statements are equivalent.*

(i) *ϕ is true in the complex numbers.*

(ii) *ϕ is true in every algebraically closed field of characteristic zero.*

(iii) *ϕ is true in some algebraically closed field of characteristic zero.*

(iv) *There are arbitrarily large primes p such that ϕ is true in some algebraically closed field of characteristic p.*

(v) *There is an m such that for all $p > m$, ϕ is true in all algebraically closed fields of characteristic p.*

PROOF. The equivalence of (i)–(iii) is just the completeness of ACF_0 and (v) $\Rightarrow$ (iv) is obvious.

For (ii) $\Rightarrow$ (v) suppose $ACF_0 \models \phi$. By the completeness theorem, there is a proof of ϕ from ACF_0. That proof only uses finitely many assertions $\neg\psi_q$; thus, for large enough p, $ACF_p \models \phi$.

For (iv) $\Rightarrow$ (ii) suppose $ACF_0 \not\models \phi$. By completeness $ACF_0 \models \neg\phi$. By the above argument, $ACF_p \models \neg\phi$ for sufficiently large p; thus (iv) fails.   □

This result has a striking application.

THEOREM 2.6 (Ax). *Let $F : \mathbb{C}^n \to \mathbb{C}^n$ be an injective polynomial map. Then F is surjective.*

PROOF. Suppose not. Let $X = (X_1, \ldots, X_n)$. Let $F(X)$ be a counterexample, with coordinate functions $F_1(X), \ldots, F_n(X)$, each $F_i \in \mathbb{C}[X]$ having degree at most d. There is an $\mathcal{L}$-sentence $\Phi_{n,d}$ such that, for K a field, $K \models \Phi_{n,d}$ if and only if every injective polynomial map $G : K^n \to K^n$ whose coordinate functions have degree at most d is surjective. We can quantify over polynomials of degree at most d by quantifying over the coefficients. For example, $\Phi_{2,2}$ is the sentence

$$\forall a_{0,0} \, \forall a_{0,1} \, \forall a_{0,2} \, \forall a_{1,0} \, \forall a_{1,1} \, \forall a_{2,0} \, \forall b_{0,0} \, \forall b_{0,1} \, \forall b_{0,2} \, \forall b_{1,0} \, \forall b_{1,1} \, \forall b_{2,0}$$

$$\left(\forall x_1 \, \forall y_1 \, \forall x_2 \, \forall y_2 \right.$$

$$\sum a_{i,j} x_1^i y_1^j = \sum a_{i,j} x_2^i y_2^j \wedge \sum b_{i,j} x_1^i y_1^j = \sum b_{i,j} x_2^i y_2^j \to x_1 = x_2 \wedge y_1 = y_2 \Big)$$

$$\to \forall u \, \forall v \, \exists x \, \exists y \, \sum a_{i,j} x^i y^j = u \wedge \sum b_{i,j} x^y y^j = v.$$

If K is a finite field then $K \models \Phi_{n,d}$. It follows that $\Phi_{n,d}$ holds in any increasing union of finite fields. In particular the algebraic closure of a finite field satisfies $\Phi_{n,d}$. Hence, by Corollary 2.5, $\mathbb{C} \models \Phi_{n,d}$, a contradiction.   □

Originally logicians looked for completeness results because they lead to decidability results.

COROLLARY 2.7. *The theory ACF_p is decidable for $p \geq 0$. That is, for each p there is an algorithm which for each sentence ϕ will determine if $ACF_p \models \phi$.*

PROOF. By the completeness of ACF_p and the completeness theorem either there is a proof of ϕ or a proof of $\neg\phi$ from ACF_p. We can systematically search all finite sequences of symbols and test each one to see if it is a valid proof of either ϕ or $\neg\phi$. Eventually we will find one or the other.   □

3. Quantifier Elimination

Let F be a field. If $p(X_1, \ldots, X_n) \in F[X_1, \ldots, X_n]$, the zero set $\{x \in F^n : p(x) = 0\}$ is defined by a quantifier free $\mathcal{L}_r$-formula. We say that a subset of F^n is *constructible* if it is a boolean combination of zero sets of polynomials in $F[X_1, \ldots, X_n]$. It is easy to see that the subsets of F^n defined by quantifier free formulas are exactly the constructible subsets of F^n. If F is an algebraically closed field then Chevalley's theorem from algebraic geometry asserts that the projection of a constructible subset of F^{n+1} to F^n is constructible. Restating this model-theoretically, this says that every definable set is constructible.

This is of course not true for non-algebraically closed fields. In the reals we can define the ordering by

$$x < y \iff \exists z \; z \neq 0 \wedge x + z^2 = y,$$

but this is not a constructible subset of $\mathbb{R}^2$. Here this is the only problem. We say that a subset of an ordered field is *semialgebraic* if it is a boolean combination of zero sets of polynomials and polynomial inequalities (like $\{x : p(x) > 0\}$). It is easy to see that the semialgebraic sets are exacty the sets defined by quantifier free $\mathcal{L}_{or}$-formulas. The Tarski–Seidenberg theorem says that in the reals (or more generally in a real closed field) the projection of a semialgebraic set is semialgebraic. Thus in the real field the definable sets are exactly the semialgebraic sets.

In model theory we study the definable sets of a structure. Quantifier elimination results are very useful, as often one can show the quantifier free definable sets have good geometric properties while the definable sets have strong closure properties. For example, suppose $A \subseteq \mathbb{R}^n$ is semialgebraic. We want to show that the closure of A is also semialgebraic. Since A is definable there is an $\mathcal{L}_{or}$-formula $\phi(x_1, \ldots, x_n, a_1, \ldots, a_m)$ that defines A. Then the formula

$$\forall \varepsilon > 0 \; \exists y_1 \; \ldots \; \exists y_n \; \left(\phi(y_1, \ldots, y_n, a_1, \ldots, a_m) \wedge (x_1 - y_1)^2 + \cdots + (x_n - y_n)^2 < \varepsilon \right)$$

defines the closure of A. Since the closure of A is definable it is semialgebraic.

In the structure $(\mathbb{Q}, +, \times, 0, 1)$ we can also define the ordering by saying that the nonnegative elements are sums of four squares. Julia Robinson showed that the integers are definable in the field of rational numbers. By Gödel's incompleteness theorem this implies that the theory of the rational numbers is undecidable and the definable subsets are quite complicated.

There is a useful model-theoretic test for quantifier elimination.

THEOREM 3.1. *Let $\mathcal{L}$ be a language containing at least one constant symbol. Let T be an $\mathcal{L}$-theory and let $\phi(v_1, \ldots, v_m)$ be an $\mathcal{L}$-formula with free variables $v_1, \ldots, v_m$ (we allow the possibility that $m = 0$). The following statements are equivalent:*

(i) *There is a quantifier free $\mathcal{L}$-formula $\psi(v_1, \ldots, v_m)$ such that*

$$T \models \forall \bar{v} \; (\phi(\bar{v}) \leftrightarrow \psi(\bar{v})).$$

(ii) *If $\mathcal{A}$ and $\mathcal{B}$ are models of T, $\mathcal{C} \subseteq \mathcal{A}$ and $\mathcal{C} \subseteq \mathcal{B}$, then $\mathcal{A} \models \phi(\bar{a})$ if and only if $\mathcal{B} \models \phi(\bar{a})$ for all $\bar{a} \in \mathcal{C}$.*

PROOF. (i) $\Rightarrow$ (ii): Suppose $T \models \forall \bar{v} \, (\phi(\bar{v}) \leftrightarrow \psi(\bar{v}))$, where ψ is quantifier free. Let $\bar{a} \in \mathcal{C}$ where $\mathcal{C}$ is a substructure of $\mathcal{A}$ and $\mathcal{B}$ and the later two structures are models of T. Since quantifier free formulas are preserved under substructure and extension

$$\mathcal{A} \models \phi(\bar{a}) \Longleftrightarrow \mathcal{A} \models \psi(\bar{a})$$
$$\Longleftrightarrow \mathcal{C} \models \psi(\bar{a}) \ \ (\text{since } \mathcal{C} \subseteq \mathcal{A})$$
$$\Longleftrightarrow \mathcal{B} \models \psi(\bar{a}) \ \ (\text{since } \mathcal{C} \subseteq \mathcal{B})$$
$$\Longleftrightarrow \mathcal{B} \models \phi(\bar{a}).$$

(ii) $\Rightarrow$ (i): First, if $T \models \forall \bar{v} \, \phi(\bar{v})$, then $T \models \forall \bar{v} \, (\phi(\bar{v}) \leftrightarrow c = c)$. Second, if $T \models \forall \bar{v} \, \neg \phi(\bar{v})$, then $T \models \forall \bar{v} \, (\phi(\bar{v}) \leftrightarrow c \neq c)$. In fact, if ϕ is not a sentence we could use "$v_1 = v_1$" in place of $c = c$.

Thus we may assume that both $\phi(\bar{v})$ and $\neg \phi(\bar{v})$ are consistent with T.

Let $\Gamma(\bar{v}) = \{\psi(\bar{v}) : \psi$ is quantifier free and $T \models \forall \bar{v} \, (\phi(\bar{v}) \rightarrow \psi(\bar{v}))\}$. Let $d_1, \ldots, d_m$ be new constant symbols. We will show that $T + \Gamma(\bar{d}) \models \phi(\bar{d})$. Thus by compactness there are $\psi_1, \ldots, \psi_n \in \Gamma$ such that $T \models \forall \bar{v} \, (\bigwedge \psi_i(\bar{v}) \rightarrow \phi(\bar{v}))$. So $T \models \forall \bar{v} \, (\bigwedge \psi_i(\bar{v}) \leftrightarrow \phi(\bar{v}))$ and $\bigwedge \psi_i(\bar{v})$ is quantifier free. We need only prove the following claim.

CLAIM. $T + \Gamma(\bar{d}) \models \phi(\bar{d})$.

Suppose not. Let $\mathcal{A} \models T + \Gamma(\bar{d}) + \neg \phi(\bar{d})$. Let $\mathcal{C}$ be the substructure of $\mathcal{A}$ generated by $\bar{d}$. (Note: if $m = 0$ we need the constant symbol to ensure $\mathcal{C}$ is non-empty.)

Let $\mathrm{Diag}(\mathcal{C})$ be the set of all atomic and negated atomic formulas with parameters from $\mathcal{C}$ that are true in $\mathcal{C}$.

Let $\Sigma = T + \mathrm{Diag}(\mathcal{C}) + \phi(\bar{d})$. If Σ is inconsistent, then there are quantifier free formulas quantifier free formulas $\psi_1(\bar{d}), \ldots, \psi_n(\bar{d}) \in \mathrm{Diag}(\mathcal{C})$, such that $T \models \forall \bar{v} \, (\bigwedge \psi_i(\bar{v}) \rightarrow \neg \phi(\bar{v}))$. But then $T \models \forall \bar{v} \, (\phi(\bar{v}) \rightarrow \bigvee \neg \psi_i(\bar{v}))$. So $\bigvee \neg \psi_i(\bar{v}) \in \Gamma$ and $\mathcal{C} \models \bigvee \neg \psi_i(\bar{d})$, a contradiction. Thus Σ is consistent.

Let $\mathcal{B} \models \Sigma$. Since $\Sigma \supseteq \mathrm{Diag}(\mathcal{C})$, we may assume that $\mathcal{C} \subseteq \mathcal{B}$. But since $\mathcal{A} \models \neg \phi(\bar{d})$, $\mathcal{B} \models \neg \phi(\bar{d})$, a contradiction. $\square$

The next lemma shows that to prove quantifier elimination for a theory we need only prove quantifier elimination for formulas of a very simple form.

LEMMA 3.2. *Suppose that, for every quantifier free $\mathcal{L}$-formula $\theta(\bar{v}, w)$, there is a quantifier free $\psi(\bar{v})$ such that $T \models \forall \bar{v} \, (\exists w \, \theta(\bar{v}, w) \leftrightarrow \psi(\bar{v}))$. Then every $\mathcal{L}$-formula $\phi(\bar{v})$ is provably equivalent to a quantifier free $\mathcal{L}$-formula.*

PROOF. We prove this by induction on the complexity of ϕ. The result is clear if $\phi(\bar{v})$ is quantifier free.

For $i = 0, 1$ suppose that $T \models \forall \bar{v} \, (\theta_i(\bar{v}) \leftrightarrow \psi_i(\bar{v}))$, where ψ_i is quantifier free. If $\phi(\bar{v}) = \neg \theta_0(\bar{v})$, then $T \models \forall \bar{v} \, (\phi(\bar{v}) \leftrightarrow \neg \psi_0(\bar{v}))$.

If $\phi(\bar{v}) = \theta_0(\bar{v}) \wedge \theta_1(\bar{v})$, then $T \models \forall v \, (\phi(\bar{v}) \leftrightarrow (\psi_0(\bar{v}) \wedge \psi_1(\bar{v})))$.

In either case ϕ is provably equivalent to a quantifier free formula.

Suppose that $T \models \forall \bar{v} \left(\theta(\bar{v}, w) \leftrightarrow \psi_0(\bar{v}, w) \right)$, where ψ is quantifier free. Suppose $\phi(\bar{v}) = \exists w\, \theta(\bar{v}, w)$. Then $T \models \forall \bar{v} \left(\phi(\bar{v}) \leftrightarrow \exists w\, \psi(\bar{v}, w) \right)$. By our assumptions there is a quantifier free $\psi(\bar{v})$ such that

$$T \models \forall \bar{v}\, (\exists w\, \psi_0(\bar{v}, w) \leftrightarrow \psi(\bar{v})).$$

But then $T \models \forall \bar{v}\, (\phi(\bar{v}) \leftrightarrow \psi(\bar{v}))$. $\qquad\square$

Thus to show that T has quantifier elimination we need only verify that condition (ii) of Theorem 3.1 holds for every formula $\phi(\bar{v})$ of the form $\exists w\, \theta(\bar{v}, w)$, where $\theta(\bar{v}, w)$ is quantifier free.

THEOREM 3.3. *The theory* ACF *has quantifier elimination.*

PROOF. Let F be a field and let K and L be algebraically closed extensions of F. Suppose $\phi(v, \bar{w})$ is a quantifier free formula, $\bar{a} \in F$, $b \in K$ and $K \models \phi(b, \bar{a})$. We must show that $L \models \exists v\, \phi(v, \bar{a})$.

There are polynomials $f_{i,j}, g_{i,j} \in F[X]$ such that $\phi(v, \bar{a})$ is equivalent to

$$\bigvee_{i=1}^{l} \left(\bigwedge_{j=1}^{m} f_{i,j}(v) = 0 \wedge \bigwedge_{j=1}^{n} g_{i,j}(v) \neq 0 \right).$$

Then $K \models \bigwedge_{j=1}^{m} f_{i,j}(b) = 0 \wedge \bigwedge_{j=1}^{n} g_{i,j}(b)$ for some i.

Let $\hat{F}$ be the algebraic closure of F. We can view $\hat{F}$ as a subfield of both K and L. If any $f_{i,j}$ is not identically zero for $j = 1, \ldots, m$, then $b \in \hat{F} \subseteq L$ and we are done.

Otherwise, since

$$\bigwedge_{i=1}^{n} g_{i,j}(b) \neq 0,$$

$g_{i,j}(X) = 0$ has finitely many solutions. Let $\{c_1, \ldots, c_s\}$ be all of the elements of L where some $g_{i,j}$ vanishes for $j = 1, \ldots, m$. If we pick any element d of L with $d \notin \{c_1, \ldots, c_s\}$, then $L \models \phi(d, \bar{a})$. $\qquad\square$

The next result summarizes some simple applications.

COROLLARY 3.4. *Let K be an algebraically closed field.*

(i) *If $X \subset K$ is definable, then either X or $K \setminus X$ is finite. (This property is called strong minimality).*

(ii) *Suppose $f : K \to K$ is definable. If K has characteristic zero, there is a rational function g such that $f(x) = g(x)$ for all but finitely many x. If K has characteristic p there is a rational function g and $n \geq 0$ such that $f(x) = g(x)^{1/p^n}$ for all but finitely many x.*

PROOF. (i) X is a boolean combination of sets of the form $\{x : f(x) = 0\}$ and these sets are finite.

(ii) Assume K has characteristic 0 (the $p > 0$ case is similar). Let L be an elementary extension of K and let $a \in L \setminus K$. If σ is any automorphism of L

fixing $K(a)$, then $\sigma(f(a)) = f(a)$. But then $f(a) \in K(a)$. Thus there is a rational function $g \in K(X)$ such that $f(a) = g(a)$. The set $\{x \in K : f(x) = g(x)\}$ is either finite or cofinite by (i). If it has size N, then the fact that it has exactly N elements is expressed by a sentence true of L and K. Then L would not contain any new elements of this set. Thus $f(x) = g(x)$ for all but finitely many x. □

Indeed, (in characteristic zero) if f is definable there is a Zariski open O and a rational function g such that $f|O = g|O$.

The quantifer elimination test has many other applications. For example, consider RCF, the theory of real closed ordered fields in the language $\mathcal{L}_{or}$. The axioms for RCF consist of:

(i) the axioms for ordered fields;
(ii) $\forall x > 0 \; \exists y \; y^2 = x$;
(iii) the axiom $\forall x_0 \; \ldots \; \forall x_{n-1} \; \exists y \; y^n + x_{n-1}y^{n-1} + \cdots + x_0 = 0$ for each odd $n > 0$.

Clearly RCF is part of the $\mathcal{L}_{or}$-theory of the real field. We will see shortly that this theory is complete and hence axiomatizes the complete theory of the real field. First we show RCF has quantifier elimination. We use the algebraic facts that every ordered field has a unique real closed algebraic extension and over a real closed field any polynomial in one variable factors into a product of linear and irreducible quadratic factors (see [Lang 1984, Section XI.2], for example).

THEOREM 3.5. *The theory* RCF *has quantifier elimination in* $\mathcal{L}_{or}$.

PROOF. We apply Theorem 3.1. Let F_0 and F_1 be models of RCF and let $(R, <)$ be a common substructure. Then $(R, <)$ is an ordered domain. Let L be the real closure of the fraction field of R. By the uniqueness of real closures we can may assume that $(L, <)$ is a substructure of F_0 and F_1. Suppose $\phi(v, \bar{w})$ is quantifier free, $\bar{a} \in R$, $b \in F_0$ and $F_0 \models \phi(b, \bar{a})$. We need to show that $F_1 \models \exists v \; \phi(v, \bar{a})$. It suffices to show that $L \models \exists v \; \phi(v, \bar{a})$.

As in the proof of Theorem 3.3 (and fooling around with the order), we may assume that there are polynomials $f_1, \ldots, f_n, g_1, \ldots, g_m \in R[X]$ such that $\phi(v, \bar{a})$ is

$$\bigwedge_{i=1}^{n} f_i(v) = 0 \wedge \bigwedge_{i=1}^{m} g_i(v) > 0.$$

If any of the f_i is not zero, then since $\phi(b, \bar{a})$, b is algebraic over R and thus in L. So we may assume $\phi(v, \bar{a})$ is

$$\bigwedge_{i=1}^{m} g_i(v) > 0.$$

Since L is a real closed field, we can factor each g_i as a product of factors of the form $(X - c)$ and $(X^2 + bX + c)$, where $b^2 - 4c < 0$. The linear factors change sign at c, while the quadratic factors do not change signs. If follows that we can find $\alpha_1, \ldots, \alpha_l \in L \cup \{-\infty\}$ and $\beta_1, \ldots, \beta_l \in L \cup \{+\infty\}$ such that, for $v \in F_0$, $\phi(v, \bar{a})$ if and only if

$$\bigvee_{i=1}^{l} \alpha_i < v < \beta_i.$$

Since $F_0 \models \phi(b, \bar{a})$, we have $\alpha_i < b < \beta_i$ for some i. Then $L \models \phi(\frac{1}{2}(\alpha+\beta), \bar{a})$. $\square$

COROLLARY 3.6. *RCF is complete and decidable.*

PROOF. Let ϕ be a sentence. By quantifier elimination there is a quantifier free sentence ψ such that RCF $\models \phi \leftrightarrow \psi$. We can embed the rational numbers in any real closed field F and $F \models \psi$ if and only if $\mathbb{Q} \models \psi$. Thus $F \models \phi$ if and only if $\mathbb{Q} \models \psi$. In particular if F_1 and F_2 are real closed fields then $F_1 \models \phi$ if and only if $F_2 \models \phi$.

Hence RCF is complete. Decidability follows as in 2.7. $\square$

We also have an analog of Corollary 3.4.

COROLLARY 3.7. *If R is real closed and $X \subset R$ is definable, then X is a finite union of points and intervals (this is called o-minimality).*

PROOF. Definable subsets of R are boolean combinations of $\{x : f(x) > 0\}$ which are finite unions of intervals. $\square$

4. Model Completeness

We say that a theory T is *model complete* if whenever $\mathcal{M}$ and $\mathcal{N}$ are models of T and $\mathcal{M} \subset \mathcal{N}$, then $\mathcal{M}$ is an elementary substructure of $\mathcal{N}$.

PROPOSITION 4.1. *If T has quantifier elimination, then T is model complete.*

PROOF. Let $\mathcal{M} \subset \mathcal{N}$. Suppose $\phi(\bar{v})$ is a formula and $\bar{a} \in M^n$. There is a quantifier free formula $\psi(\bar{v})$ such that

$$T \models \forall \bar{v} \, (\phi(\bar{v}) \leftrightarrow \psi(\bar{v})).$$

Since ψ is quantifier free, $\mathcal{M} \models \psi(\bar{a}) \Leftrightarrow \mathcal{N} \models \psi(\bar{a})$. Thus $\mathcal{M} \models \psi(\bar{a}) \Leftrightarrow \mathcal{N} \models \psi(\bar{a})$. $\square$

Model completeness can arise in cases where quantifier elimination fails. For example, let T be the $\mathcal{L}_r$-theory of the real numbers (without a symbol for the order). The formula $\exists y \, y^2 = x$ is not equivalent to a quantifier free formula (recall that quantifier free definable sets in $\mathcal{L}_{or}$ are constructible), so T does not have quantifier elimination. On the other hand the ordering of a real closed field is definable in the field language, thus if F and K are real closed fields and F is a subfield of K, then the ordering on F agrees with the ordering inherited from K (that is, F is an $\mathcal{L}_{or}$-substructure of K). Thus, by quantifier elimination in $\mathcal{L}_{or}$, F is an elementary substructure of K.

This example shows that "quantifier elimination" is sensitive to the exact choice of language. In general we can always enrich the language so that we have quantifier elimination. If T_0 is an $\mathcal{L}$-theory, we can add to our language an n-ary relation symbol R_ϕ for each formula ϕ with n free variables and we could let T be the theory where we add to T_0 axioms $\forall \bar{x} \ (\phi(\bar{x}) \leftrightarrow R_\phi(\bar{x}))$ for each formula ϕ. The theory T will have quantifier elimination, but this would be useless as we would not be able to say anything sensible about the quantifier free formulas. The goal is to show we have quantifer elimination in a language where the quantifier free formulas are simple.

Wilkie showed that the theory of $(R, +, -, \times, <, \exp)$ is model complete, but

$$y > 0 \wedge \exists w \ (wy = x \wedge z = y \exp(w))$$

is not equivalent to a quantifier free formula in the language

$$\{+, -, \times, <, 0, 1, \exp\}$$

(or any expansion by total real analytic functions). Van den Dries, Macintyre and I showed that you can eliminate quantifiers in a much more expressive language, but we do not know the simplest language for quantifier elimination.

Model completeness itself has useful consequences. For example, the model completeness of ACF leads to an easy proof of a version of the Nullstellensatz.

THEOREM 4.2. *Let F be an algebraically closed field and let $I \subset F[X_1, \ldots, X_n]$ be a prime ideal. Then there is $a \in F^n$ such that $f(a) = 0$ for all $f \in I$.*

PROOF. Let K be the algebraic closure of the fraction field of $F[X_1, \ldots, X_n]/I$. If $x_i \in K$ is X_i/I, then $f(x_1, \ldots, x_n) = 0$ for all $f \in I$. Choose $f_1, \ldots, f_m$ generating I. Then

$$K \models \exists y_1, \ldots, y_n \bigwedge_{i=1}^{m} f_i(y_1, \ldots, y_n) = 0.$$

As this is a sentence with parameters from F, by model completeness this sentence is also true in F. Thus there is $a \in F^n$ such that $f_i(a) = 0$ for $i = 1, \ldots, m$ and hence $f(a) = 0$ for all $f \in I$. $\square$

A very similar argument can be used to reprove Artin's solution to Hilbert's seventeenth problem.

THEOREM 4.3. *Let F be a real closed field. Suppose that $f(X_1, \ldots, X_n) \in F(X_1, \ldots, X_n)$ and that $f(a) \geq 0$ for all $a \in F^n$ (we call f positive semi-definite). Then f is a sum of squares of rational functions.*

PROOF. If not, we can extend the order of F to $F(X_1, \ldots, X_n)$ such that $f < 0$ (see [Lang 1984, Section XI.2]). Let K be the real closure of $F(X_1, \ldots, X_n)$ with this ordering. Then

$$K \models \exists y_1 \ldots \exists y_n \ f(y_1, \ldots, y_n) < 0,$$

since we can use $X_1, \ldots, X_n$ as witnesses. By model completeness

$$F \models \exists y_1 \ldots \exists y_n \; f(y_1, \ldots, y_n) < 0. \qquad \square$$

DEFINITION. We say that a theory T is the *model companion* of a universal theory T_0 if:

(i) every model of T is a model of T_0,
(ii) every model of T_0 can be extended to a model of T, and
(iii) T is model complete.

For example, the theory of algebraically closed fields is the model companion of the theory of integral domains and the theory of real closed fields is the model companion of the theory of ordered domains. More interesting examples can be found in [Chatzidakis 2000].

Model theoretic methods can sometimes be used to obtain effective bounds. Compactness arguments alone can lead to crude bounds.

PROPOSITION 4.4. *There is a computable function $\tau(n, d)$ such if F is a real closed field and $f = g/h \in F(X_1, \ldots, X_n)$ where f and g are polynomials of degree at most d and f is positive semidefinite then f is the sum of squares of at most $\tau(n, d)$ rational functions with numerator and denominator of degree at most $\tau(n, d)$.*

PROOF. Fix n, d. We first claim that there is an M such that any positive semidefinite rational function in n variables with numerator and denominator of degree at most d is a sum of at most M squares of rational functions with numerator and denominator of degree at most M. Suppose not. Let $c_1, \ldots, c_N$ be new constants which will be coefficients of a rational function f in n-variables with numerator and denominator of degree at most d. Let Φ_M be a sentence asserting "f is not a sum of at most M squares of functions of degree at most M". Then RCF + "f is positive semidefinite" + $\{\neg \Phi_M : M \geq 1\}$ is satisfiable, contradicting Hilbert's 17th problem.

Given n and d, let $\tau(n, d)$ be the least M as above. Since RCF is decidable, we can compute $\tau(n, d)$. $\qquad \square$

5. Types

Suppose $\mathcal{M}$ is an $\mathcal{L}$-structure and $A \subseteq M$. Let $\mathcal{L}_A$ be the language obtained by adding to $\mathcal{L}$ constant symbols for all elements of A. Let $\mathrm{Th}_A(M)$ be the set of all $\mathcal{L}_A$-sentences true in $\mathcal{M}$.

DEFINITION. An *n-type* over A is a set of $\mathcal{L}_A$-formulas in free variables $x_1, \ldots, x_n$ that is consistent with $\mathrm{Th}_A(\mathcal{M})$. A *complete n-type* over A is a maximal n-type. In other words a complete type is a set p of $\mathcal{L}_A$-formulas consistent with $\mathrm{Th}_A(\mathcal{M})$ in the free variables $x_1, \ldots, x_n$ such that for any $\mathcal{L}_A$-formula $\phi(\bar{x})$ either $\phi(\bar{x}) \in p$ or $\neg \phi(\bar{x}) \in p$. Let $S_n(A)$ be the set of all complete n-types over A.

We sometimes refer to incomplete types as *partial types*.

By compactness every n-type over A is realized in some elementary extension of $\mathcal{M}$.

There is one easy way to get complete types. Suppose $\mathcal{N}$ is an elementary extension of $\mathcal{M}$ and $\bar{b} \in N^n$. Let $\operatorname{tp}(\bar{b}/A) = \{\phi(\bar{x}) \in \mathcal{L}_A : \mathcal{N} \models \phi(\bar{b})\}$. It is easy to see that $\operatorname{tp}(\bar{b}/A)$ is a complete type.

If $p \in S_n(A)$ we say that $\bar{b}$ *realizes* p if $\operatorname{tp}(\bar{b}/A) = p$.

What do types tell us?

PROPOSITION 5.1. *Suppose* $\bar{a}, \bar{b} \in M^n$ *and* $\operatorname{tp}(\bar{a}/A) = \operatorname{tp}(\bar{b}/A)$. *Then there is an elementary extension* $\mathcal{N}$ *of* $\mathcal{M}$ *and an* $\mathcal{L}$-*automorphism of* $\mathcal{N}$ *which fixes* A *and maps* $\bar{a}$ *to* $\bar{b}$.

PROOF. We carefully iterate the following lemma.

LEMMA 5.2. *Suppose* $\mathcal{M}$ *is an* $\mathcal{L}$-*structure,* $A \subset M$ *and* $f : A \to M$ *is a partial elementary map (i.e.,* $M \models \phi(a_1, \ldots, a_n) \Leftrightarrow \mathcal{M} \models \phi(f(a_1), \ldots, f(a_n))$). *If* $b \in M$, *we can find* $\mathcal{N}$ *an elementary extension of* $\mathcal{N}$ *and extend* f *to a partial elementary map from* $A \cup \{b\}$ *into* N.

PROOF. Let c be a new constant symbol. Let

$$\Gamma = \{\phi(c, f(a_1), \ldots, f(a_n)) : \mathcal{M} \models \phi(b, a_1, \ldots, a_n), a_1, \ldots, a_n \in A\} \cup \operatorname{Th}_M(\mathcal{M}).$$

Suppose we find a structure $\mathcal{N}$ and an element $c \in N$ satisfying all of the formulas in Γ. Since $\mathcal{N} \models \operatorname{Th}_M(\mathcal{M})$, $\mathcal{N}$ is an elementary extension of $\mathcal{M}$. It is also easy to see that we can extend f to an elementary map by $b \mapsto c$.

So it suffices to show that Γ is satisfiable. By compactness it suffices to show that every finite subset of Γ is satisfiable. Taking conjunctions it is enough to show that if $\mathcal{M} \models \phi(b, a_1, \ldots, a_n)$ then $\mathcal{M} \models \exists v \, \phi(v, f(a_1), \ldots, f(a_n))$. But this is clear since $\mathcal{M} \models \exists v \, \phi(v, a_1, \ldots, a_n)$ and f is elementary. $\square$

The type space $S_n(A)$ can be topologized as follows. For each $\mathcal{L}_A$-formula $\phi(x_1, \ldots, x_n)$ let $B_\phi = \{p \in S_n(A) : \phi \in p\}$. The *Stone topology* on $S_n(A)$ is the topology generated by using the sets B_ϕ as basic open sets.[1]

PROPOSITION 5.3. $S_n(A)$ *is compact and totally disconnected.*

PROOF. Suppose $\{B_{\phi_i} : i \in I\}$ is a cover of $S_n(A)$ by basic open sets. Suppose there is no finite subcover. Let $\Gamma = \{\neg\phi_i(x_1, \ldots, x_n) : i \in I\}$. Since there in no finite subcover every finite subset of Γ is satisfiable. By compactness Γ is satisfiable and this yields a type that is not contained in any B_{ϕ_i}.

Since $S_n(A) \setminus B_\phi = B_{\neg\phi}$, each B_ϕ is open and closed. Thus $S_n(A)$ is totally disconnected. $\square$

---

[1] $S_n(A)$ can be thought of as the set of ultrafilters on the Boolean algebra of A-definable subsets of M^n so it is in fact the Stone space of a Boolean algebra.

Suppose K is an algebraically closed field and F is a subfield of K. What are the complete n-types over F? Suppose $p \in S_n(F)$. Let $I_p = \{f \in F[X_1, \ldots, X_n] : "f(x_1, \ldots, x_n) = 0" \in p\}$. Let $\operatorname{Spec}(F[X_1, \ldots, X_n])$ be the set of prime ideals of $F[X_1, \ldots, X_n]$. We topologize this space by taking sub-basic open sets $\{P : f \notin P\}$, for $f \in F[\bar{X}]$.

PROPOSITION 5.4. $p \mapsto I_p$ is a continuous bijection between $S_n(F)$ and
$$\operatorname{Spec}(F[\bar{X}]).$$

PROOF. If $fg \in I_p$, then $"f(\bar{x})g(\bar{x}) = 0" \in p$. Since p is complete either $"f(\bar{x}) = 0" \in p$ or $"g(\bar{x}) = 0" \in p$. Thus I_p is prime. It is just as easy to see that it is an ideal.

If P is a prime ideal, then we can find a prime ideal $P_1 \in K[\bar{X}]$ such that $P_1 \cap F[\bar{X}] = P$. Let K_1 be the algebraic closure of $K[\bar{X}]/P_1$ and let $a_i = X_i/P$. For $f \in K[\bar{X}]$ $f(\bar{a}) = 0$ if and only if $f \in P_1$, thus $I_{\operatorname{tp}(\bar{a}/F)} = P$. Thus the map is surjective.

By quantifier elimination if $I_p \neq I_q$, then $p \neq q$.

Continuity is clear. $\qquad\square$

This shows that the Zariski topology on $\operatorname{Spec}(F[\bar{X}])$ is compact.

We can identify (as objects) $S_n(F)$ and $\operatorname{Spec}(F[\bar{X}])$, but the Stone topology is much finer that the Zariski topology. The Stone topology corresponds to the topology generated by the constructible sets.

If $p \in S_n(F)$, let $V = \{x \in K^n : f(x) = 0 \text{ for all } f \in I_p\}$. Then the type p asserts that $\bar{x} \in V$ and $\bar{x} \notin W$ for any $W \subset V$ defined over F. Thus realizations of p are points of V generic over F.

What about real closed fields? If F is an ordered subfield of a real closed field and p is an n-type, let I_p be as above and let $C_p = \{f/I_p : f \in F[\bar{X}]$ and $"f(\bar{x}) > 0 \in p\}$. Then $p \mapsto (I_p, C_p)$ is a bijection onto the set of pairs of real prime ideals P (prime ideals where $-1 \neq \sum a_i b_i^2$ where $a_i > 0$, $a_i \in F[\bar{X}]$, $b_i \in P$) and orderings of $F[\bar{X}]/P$. This is the *real spectrum* of $F[\bar{X}]$.

In particular if R is a real closed field, then elements of $S_1(R)$ correspond to either elements of R or cuts in the ordering of R.

6. Saturation

It is often useful to work in a very rich model of a theory. For example, it is sometimes easier to prove things in an algebraically closed field of infinite transcendence degree. Or when dealing with the reals it is useful to use nonstandard methods by assuming there are infinite elements. In model theory we make this precise in the following way.

DEFINITION 6.1. Let κ be an infinite cardinal. We say that a structure $\mathcal{M}$ is κ-*saturated* if for every $A \subset M$ with $|A| < \kappa$ if $p \in S_1(A)$, then there is b in M

such that b realizes p. An easy induction shows that in this case every n-type over A is also realized in $\mathcal{M}$.

We say that $\mathcal{M}$ is *saturated* if it is $|M|$-saturated.

LEMMA 6.2. *If $\mathcal{M}$ is saturated, $A \subset M$ and $|A| < |M|$, then $\operatorname{tp}(\bar{a}/A) = \operatorname{tp}(\bar{b}/A)$ if and only if there is an automorphism of M fixing A mapping $\bar{a}$ to $\bar{b}$.*

PROOF. The argument from 5.1 can be done completely inside M.   □

PROPOSITION 6.3. *An algebraically closed field K is saturated if and only if it has infinite transcendence degree.*

PROOF. Suppose $A \subset K$ is finite and F is the field generated by A. Let p be the 1-type over A which says that x is transcendental over F. If K is $\aleph_0$-saturated, then p must be realized in K. Thus every $\aleph_0$-saturated algebraically closed field has infinite transcendence degree.

On the other hand suppose K has infinite transcendence degree and $F \subset K$ is a field generated by fewer that $|K|$ elements. Let $p \in S_1(F)$ and let I_p be as in Proposition 5.4. If $I_p = \{0\}$, then p simply says "x is transcendental over F", and we can find a realization in K. If I_p is generated by $f(X)$, then any zero of f realizes p and we can find a realization in K.   □

Unfortunately saturated models are not so easy to come by in general. In general $|S_n(A)|$ can be as large as $2^{|A|+|\mathcal{L}|+\aleph_0}$. For example, 1-types over $\mathbb{Q}$ in the theory of real closed fields, correspond to cuts in the rationals so $|S_1(\mathbb{Q})| = 2^{\aleph_0}$. Thus set theoretic problems arise. Under assumptions like the generalized continuum hypothesis or the existence of inaccessible cardinals we can find saturated models, but it is also possible that there are no saturated real closed fields.

Suppose $|\mathcal{L}| \le \aleph_0$ and λ is an infinite cardinal. We say that an $\mathcal{L}$-theory T is *λ-stable* if and only for all $\mathcal{M} \models T$ and all $A \subset M$, if $|A| = \lambda$, then $|S_n(A)| = \lambda$. It is easy to see that algebraically closed fields are λ-stable for all infinite λ.

PROPOSITION 6.4. *If T is λ-stable, then T has a saturated model of size λ^+.*

PROOF. We build a saturated model of size λ^+ as a union of an elementary chain of models $(\mathcal{M}_\alpha : \alpha < \lambda^+)$ where each $\mathcal{M}_\alpha$ has size λ. Let $\mathcal{M}_0$ be any model of size λ. For α a limit let $\mathcal{M}_\alpha$ be the union of the $\mathcal{M}_\beta$, for $\beta < \alpha$.

Given $\mathcal{M}_\alpha$. Let $(p_\beta : \beta < \lambda)$ list all 1-types over $\mathcal{M}_\alpha$. Build a chain of elementary extensions $(\mathcal{N}_\beta, \beta < \lambda)$ where $\mathcal{N}_0 = \mathcal{M}_\alpha$ and where $\mathcal{N}_\beta$ contains a realization of p_β. Let $\mathcal{M}_{\alpha+1}$ be the union of the $\mathcal{N}_\beta$.

Let $\mathcal{M} = \bigcup_{\alpha < \lambda^+} \mathcal{M}_\alpha$. Then $|M| = \lambda^+$. If $A \subset M$ and $|A| = \lambda$, then $A \subset M_\alpha$ for some α. Thus any 1-type over A is already realized in $\mathcal{M}_{\alpha+1}$.   □

7. Interpretability and Imaginaries

It is often very useful to study the structures which can be defined inside a give structure. For example, let K be a field and let G be the group $GL_2(K)$. Let $X = \{(a, b, c, d) \in K^4 : ad - bc \neq 0\}$. Let $f : X^2 \to X$ by

$$f((a_1, b_1, c_1, d_1), (a_2, b_2, c_2, d_2))$$
$$= (a_1 a_2 + b_1 c_2, \ a_1 b_2 + b_1 c_2, \ c_1 a_2 + d_1 c_2, \ c_1 b_2 + d_1 d_2).$$

Clearly X and f are definable and the set X with the operation f is isomorphic to $GL_2(K)$.

We say that an $\mathcal{L}_0$-structure $\mathcal{N}$ is *definable* in an $\mathcal{L}$-structure $\mathcal{M}$ if and only if we can find a definable (in $\mathcal{L}$) subset X of M^n for some n and we can interpret the symbols of $\mathcal{L}_0$ as definable subsets and functions on X so that the resulting $\mathcal{L}_0$-structure is isomorphic to $\mathcal{N}$.

The example above shows that $GL_n(K)$ is definable in K. It is also easy to see that any linear algebraic group is definable in K.

We give a more interesting example. Let F be a field and let G be the group of matrices of the form

$$\begin{pmatrix} a & b \\ 0 & 1 \end{pmatrix}$$

where $a, b \in K, a \neq 0$. We will show that F is definable in the group G. Let

$$\alpha = \begin{pmatrix} 1 & 1 \\ 0 & 1 \end{pmatrix} \text{ and } \beta = \begin{pmatrix} \tau & 0 \\ 0 & 1 \end{pmatrix}$$

where $\tau \neq 0, 1$. Let

$$A = \{g \in G : g\alpha = \alpha g\} = \left\{ \begin{pmatrix} 1 & x \\ 0 & 1 \end{pmatrix} : x \in F \right\}$$

and

$$B = \{g \in G : g\beta = \beta g\} = \left\{ \begin{pmatrix} x & 0 \\ 0 & 1 \end{pmatrix} : x \neq 0 \right\}.$$

Clearly A and B are definable.

B acts on A by conjugation:

$$\begin{pmatrix} x & 0 \\ 0 & 1 \end{pmatrix}^{-1} \begin{pmatrix} 1 & y \\ 0 & 1 \end{pmatrix} \begin{pmatrix} x & 0 \\ 0 & 1 \end{pmatrix} = \begin{pmatrix} 1 & y/x \\ 0 & 1 \end{pmatrix}.$$

Clearly the action $(a, b) \mapsto b^{-1}ab$ is definable. We can define the map $i : A \backslash \{1\} \to B$ by $i(a) = b$ if $b^{-1}ab = 1$, i.e.,

$$i \begin{pmatrix} 1 & x \\ 0 & 1 \end{pmatrix} = \begin{pmatrix} x & 0 \\ 0 & 1 \end{pmatrix}.$$

Define an operation $*$ on A by

$$a * b = \begin{cases} i(b)^{-1}ab & \text{if } b \neq 1 \\ 1 & \text{if } b = 1. \end{cases}$$

It is now easy to see that $(F, +, \times, 0, 1)$ is isomorphic to $(A, \cdot, *, 1, \alpha)$. Thus the field is definable in G.

This will not be true for all algebraic groups. For example, if E is an elliptic curve and $\oplus$ is the addition law on E then we cannot interpret a field in the group $(E, \oplus)$.

Often we want to do more general constructions. For example, suppose we have a definable group G and a definable normal subgroup H. We might want to look at the group G/H. It is possible that G/H does not correspond to a definable group in our structure. But it does correspond to the cosets of a definable equivalence relation.

We say that an $\mathcal{L}_0$-structure $\mathcal{N}$ is *interpretable* in an $\mathcal{L}$-structure M if there is a definable set X, a definable equivalence relation E on X, and for each symbol of $\mathcal{L}$ we can find definable E-invariant sets on X, such that X/E with the induced structure is isomorphic to $\mathcal{N}$.

As an example let us show that we can interpret the additive group of integers in the field $\mathbb{Q}_p$. First note that we can define $\mathbb{Z}_p = \{x \in \mathbb{Q}_p : \exists y \ y^2 = px^2 + 1\}$ (at least for $p \neq 2$). Let $U = \{x \in \mathbb{Z}_p : \exists y \in \mathbb{Z}_p : xy = 1\}$ be the units of $\mathbb{Z}_p$. Then $(\mathbb{Z}, +)$ is isomorphic to the multiplicative group $\mathbb{Q}_p^*/U$. We can define the ordering on $\mathbb{Q}_p^*/U$ by

$$x/U \geq y/U \iff \frac{x}{y} \in \mathbb{Z}_p.$$

Quotient constructions are so useful that we often enrich our structure so that we can deal with all quotients as elements of the structure. Let $\mathcal{M}$ be an $\mathcal{L}$-structure. If E is a $\varnothing$-definable equivalene relation on M^n, let $S_E = M^n/E$ and let $\pi_E : M^n \to M^n/E$ be the quotient map. Let $\mathcal{M}^{\mathrm{eq}}$ be the structure whose underlying set is the disjoint union of $\mathcal{M}$ and all of the S_E for E a $\varnothing$-definable equivalence relation. In addition to the relations and functions of $\mathcal{L}$, we add function symbols for each map π_E. We call the new elements of $\mathcal{M}^{\mathrm{eq}}$ *imaginary* elements.

If a structure $\mathcal{N}$ is interpretable in $\mathcal{M}$, then $\mathcal{N}$ is definable in $\mathcal{M}^{\mathrm{eq}}$. On other hand, not much has changed, if $X \subseteq M^n$ is definable in $\mathcal{M}^{\mathrm{eq}}$ then X is already definable in $\mathcal{M}$.

An important property of many of the theories of fields that we will consider is that the passage from K to K^{eq} is unnecessary.

Van den Dries showed that in real closed fields any definable equivalence relation has a definable set of representatives. Nothing this strong could be true in algebraically closed fields as a set of represntatives for the equivalence relation $xEy \leftrightarrow x^2 = y^2$ would be infinite and coinfinite.

We say that $\mathcal{M}$ has *elimination of imaginaries* if, whenever E is a definable equivalence relation on M^n, there is for some m a definable function $f : M^n \to M^m$ such that $xEy \leftrightarrow f(x) = f(y)$.

THEOREM 7.1 (POIZAT). *Algebraically closed fields have elimination of imagi-naries.*

In other words, if X is a constructible set and E is a constructible equivalence relation on X, then X/E can be viewed as a constructible set.

The proof of Theorem 7.1 proceeds by first showing that a theory has elimi-nation of imaginaries if and only if for any saturated model $\mathcal{M}$ and any $X \subset M^n$ definable there is $\bar{a} \in M^m$ for some m such that for all automorphisms σ of $\mathcal{M}$, σ fixes X setwise if and only if $\sigma(\bar{a}) = \bar{a}$. We call $\bar{a}$ *canonical parameter for X.* If X is defined by $\phi(\bar{x}, \bar{b})$, we could define an equivalence relation $\bar{b}_1 E \bar{b}_2$ if and only if $\phi(\bar{x}, \bar{b}_1) \leftrightarrow \phi(\bar{x}, \bar{b}_2)$. Then $\bar{b}/E$ is a canonical parameter for X. In general canonical parameters will only be found in $\mathcal{M}^{\text{eq}}$.

Suppose K is an algebraically closed field and X is an irreducible variety. There is a smallest subfield $k \subset K$ such that X is fixed by an automorphism if and only if k is fixed pointwise (k is called the *field of definition* of X). The subfield k must be finitely generated and if $\bar{a}$ generates k, then $\bar{a}$ is a canonical parameter. From this observation and quantifier elimination one can derive elimination of imaginaries.

The field of p-adics is a natural example where elimination of imaginaries fails. We saw above that we can interpret the integers in $\mathbb{Q}_p$. Analysis using quantifier elimination for the p-adics, shows that any definable set is either finite or uncountable, so the integers cannot be isomorphic to a definable set.

Here is another instructive example where elimination of imaginaries fails. Let K be an algebraically closed field of characteristic zero. Let C be curve of genus at least one and let $\mathcal{C}$ be the structure with underlying set C and relation symbols for all constructible subsets of C^n. Since there is a rational map $\pi : C \to K$, we can intepret the field on C using the equivalence relation $xEy \Longleftrightarrow \pi(x) = \pi(y)$. If C/E was definably isomorphic to a definable set $X \subset C^n$, this would give rise to a definable map $f : K \to C$. But (by Corollary 3.4) there is a rational map $g : K \to C$ which agrees with f on all but a finite set. By genus considerations g is constant.

It is often very important to understand the groups and fields interpretable in a structure. For algebraically closed fields we get a very satisfying answer. It is easy to see that if K is an algebraically closed field any algebraic group defined over K is interpretable in K and hence, by elimination of imaginaries, isomorphic to a definable group. The following theorem is related to Weil's theorems on group chunks.

THEOREM 7.2. (i) (van den Dries and Hrushovski) *If a group G is definable in an algebraically closed field K, then G is definably isomorphic to the K-rational points of an algebraic group defined over K.*

(ii) (Poizat) *If F is an infinite field definable in an algebraically closed field, then F is definably isomorphic to K.*

Appendix: Formulas

Here I give a precise definition of formulas. Let $\mathcal{L}$ be a language.

DEFINITION. (i) The set of $\mathcal{L}$-*terms* is the smallest set $\mathcal{T}$ such that all constant symbols of $\mathcal{L}$ are in $\mathcal{T}$,

(ii) all variables are in $\mathcal{T}$, and

(iii) if $t_1, \ldots, t_n$ are in $\mathcal{T}$ and $\hat{f}$ is an n-ary function symbol of $\mathcal{L}$, then

$$\hat{f}(t_1, \ldots, t_n) \in \mathcal{T}.$$

The set of *atomic $\mathcal{L}$-formulas* is the smallest set $\mathcal{A}$ such that

(i) if t_1 and t_2 are terms, then $t_1 = t_2$ is in $\mathcal{A}$, and

(ii) if $t_1, \ldots, t_n$, are terms and $\hat{R}$ is an n-ary function symbol, then $\hat{R}(t_1, \ldots, t_n)$ is in $\mathcal{A}$.

The set of $\mathcal{L}$-formulas is the smallest set $\mathcal{F}$ such that

(i) every atomic $\mathcal{L}$-formula is in $\mathcal{F}$;

(ii) if $\phi \in \mathcal{F}$, then $\neg \phi \in \mathcal{F}$;

(iii) if ϕ and ψ are in $\mathcal{F}$, then $(\phi \wedge \psi)$, $(\phi \vee \psi)$, $(\phi \rightarrow \psi)$ and $(\phi \leftrightarrow \psi)$ are in $\mathcal{F}$;

(iv) if ϕ is in $\mathcal{F}$ and x_i is a variable, then $\exists x_i\ \phi$ and $\forall x_i\ \phi$ are in $\mathcal{F}$.

For example, $x_1 + (x_2 \times (x_1 + 1))$ is an $\mathcal{L}_{or}$-term, $x_1 \times (x_2 + x_3) = x_1 + 1$ and $x_1 < x_3 + x_7$ are atomic $\mathcal{L}_{or}$-formulas, and $\exists x_1\ (x_1 \times (x_2 + x_3) = x_1 + 1 \wedge x_2 < x_1)$ is an $\mathcal{L}_{or}$-formula.

Bibliographical Notes

The following are good basic texts: [Chang and Keisler 1990; Hodges 1993; 1997; Poizat 1985; Sacks 1972]. For an introduction to the model theory of algebraically closed, real closed, differentially closed and separably closed fields, see [Marker et al. 1996]. Proofs of Theorem 7.1 can be found in the same volume or in [Poizat 1989]. A proof of 7.2 can be found in [Poizat 1987].

References

[Chang and Keisler 1990] C. C. Chang and H. J. Keisler, *Model theory*, Third ed., Studies in Logic and the Foundations of Math. **73**, North-Holland, Amsterdam, 1990.

[Chatzidakis 2000] Z. Chatzidakis, "A survey on the model theory of difference fields", pp. ??–?? in *Model theory, algebra and geometry*, edited by D. Haskell et al., Math. Sci. Res. Inst. Publ. **39**, Cambridge Univ. Press, New York, 2000.

[Hodges 1993] W. Hodges, *Model theory*, Encyclopedia of Mathematics and its Applications **42**, Cambridge University Press, Cambridge, 1993.

[Hodges 1997] W. Hodges, *A shorter model theory*, Cambridge University Press, Cambridge, 1997.

[Lang 1984] S. Lang, *Algebra*, Second ed., Addison-Wesley, Reading, MA, 1984.

[Marker et al. 1996] D. Marker, M. Messmer, and A. Pillay, *Model theory of fields*, Lecture Notes in Logic **5**, Springer, Berlin, 1996.

[Poizat 1985] B. Poizat, *Cours de théorie des modèles: une introduction à la logique mathématique contemporaine*, Nur al-Mantiq wal Ma'rifah **1**, Bruno Poizat, Villeurbanne, 1985.

[Poizat 1987] B. Poizat, *Groupes stables: une tentative de conciliation entre la géométrie algébrique et la logique mathématique*, Nur al-Mantiq wal Ma'rifah **2**, Bruno Poizat, Lyon, 1987.

[Poizat 1989] B. Poizat, "An introduction to algebraically closed fields and varieties", pp. 41–67 in *The model theory of groups* (Notre Dame, IN, 1985–1987), edited by A. Nesin and A. Pillay, Univ. Notre Dame Press, Notre Dame, IN, 1989.

[Sacks 1972] G. E. Sacks, *Saturated model theory*, Math. Lecture Note Series, W. A. Benjamin, Reading, MA, 1972.

DAVID MARKER
UNIVERSITY OF ILLINOIS AT CHICAGO
DEPARTMENT OF MATHEMATICS, STATISTICS, AND COMPUTER SCIENCE (M/C 249)
851 S. MORGAN ST.
CHICAGO, IL 60613
UNITED STATES
marker@math.uic.edu

Model Theory, Algebra, and Geometry
MSRI Publications
Volume **39**, 2000

Classical Model Theory of Fields

LOU VAN DEN DRIES

ABSTRACT. We begin with some thoughts on how model theory relates to
other parts of mathematics, and on the indirect role of Gödel's incompleteness theorem in this connection. With this in mind we consider in Section 2
the fields of real and p-adic numbers and show how these algebraic objects
are understood model-theoretically: theorems of Tarski, Kochen, and Macintyre. This leads naturally to a discussion of the famous work by Ax,
Kochen and Ershov in the mid sixties on henselian fields and its number-
theoretic implications.

In Section 3 we add analytic structure to the real and p-adic fields, and
indicate how results such as the Weierstrass preparation theorem can be
used to extend much of Section 2 to this setting. Here we make contact
with the theory of subanalytic sets developed by analytic geometers in the
real case.

In Section 4 we focus on o-minimal expansions of the real field that are
not subanalytic, such as the real exponential field (Wilkie's theorem). We
indicate in a diagram the main known o-minimal expansions of the real
field. We also provide a translation into the coordinate-free language of
manifolds via "analytic-geometric categories". (This has been found useful
by geometers.)

CONTENTS

1. Introduction

In model theory we associate to a structure $\mathcal{M}$ invariants of a logical nature like $\mathrm{Th}(\mathcal{M})$, the set of first-order sentences which are true in $\mathcal{M}$. Other invariants of this kind are the category of definable sets and maps over $\mathcal{M}$ or over $\mathcal{M}^{\mathrm{eq}}$ and the category of definable groups and definable homomorphisms over $\mathcal{M}$ or over $\mathcal{M}^{\mathrm{eq}}$. If we are lucky we can find a well-behaved notion of dimension for the objects in these categories, which make these objects behave more or less like algebraic varieties and algebraic groups.

We consider a little more closely the simplest of the above invariants, namely $\mathrm{Th}(\mathcal{M})$. To use it for gaining a better understanding of $\mathcal{M}$, it is desirable that $\mathrm{Th}(\mathcal{M})$ can be *effectively* described. In practice we want $\mathrm{Th}(\mathcal{M})$ to be axiomatizable by *finitely* many axiom *schemes*.

EXAMPLE. $\mathrm{Th}(\mathbb{C}, +, \cdot, 0, 1)$ is axiomatized by:

- field axioms (finite in number)
- $\forall x_1 \ldots \forall x_n \, \exists y \, (y^n + x_1 y^{n-1} + \cdots + x_n = 0)$, for $n = 1, 2, 3, \ldots$
- $\underbrace{1 + \cdots + 1}_{n \text{ times}} \neq 0$, for $n = 1, 2, 3, \ldots$

EXAMPLE (GÖDEL). $\mathrm{Th}(\mathbb{Z}, +, \cdot)$ cannot be effectively described in any reasonable way, so in contrast to the field of complex numbers, the ring of integers is "wild". (But $\mathbb{Z}$ as ordered additive group is tame again!)

We use here "tame" and "wild" very *informally*, to suggest the distinction between good and bad model-theoretic behaviour.

The requirement of effective axiomatizability of $\mathrm{Th}(\mathcal{M})$ has been known since Gödel to be a serious constraint on $\mathcal{M}$. It implies some highly intrinsic model-theoretic properties in the *tame* direction, such as non-interpretability of the ring of integers. Though these properties are far weaker than stability, simplicity, o-minimality, etcetera, this axiomatizability demand can serve as a useful guide in initial model-theoretic explorations of certain mathematical structures.

Ironically, Gödel's work is often characterized as saying that only for "uninteresting" $\mathcal{M}$ can $\mathrm{Th}(\mathcal{M})$ be effectively axiomatizable. This attitude overlooks the fact that even in ostensibly *nontame* subjects like number theory, the solution of problems frequently involves ingenious moves into tame territory! Thus the relevance of the slogan (proposed by Hrushovski):

model theory = geography of tame mathematics

EXAMPLE. The field $\mathbb{Q}$ of rational numbers is not tame, but its completions $\mathbb{R}$, $\mathbb{Q}_2$, $\mathbb{Q}_3$, $\mathbb{Q}_5$, ... are all tame (J. Robinson, Tarski, Ax, Kochen, Ershov). It is not known if the field $\mathbb{F}_p((t))$ is tame.

2. Elimination Theory and Henselian Fields

How does one prove that a given structure $\mathcal{M}$ is *tame*? First, choose a set T of axioms such that $\mathcal{M} \models T$, and then try to show that

1. T admits QE (quantifier elimination), or
2. T is model complete, or

$\vdots$

If this works, one typically obtains a complete description of $\mathrm{Th}(\mathcal{M})$, and in the bargain a lot of positive information about (the category of) definable sets and maps as well, for example a notion of dimension for definable sets. It should be mentioned that, especially for QE, the right choice of primitives (language) is important.

This general scheme is perhaps best illustrated by the field of real numbers.

2.1. The field of real numbers. For $\mathcal{M} := (\mathbb{R}, <, 0, 1, +, -, \cdot)$ we choose $T := \mathrm{RCF}$ (the axioms for Real Closed ordered Fields):

- axioms for ordered fields
- $\forall x \, \exists y \, (x > 0 \rightarrow x = y^2)$
- $\forall x_1 \ldots \forall x_{2n+1} \, \exists y \, (y^{2n+1} + x_1 y^{2n} + \cdots + x_{2n+1} = 0)$, for $n = 1, 2, 3, \ldots$

How do we show that T admits QE? There are several model-theoretic criteria that can be helpful. (We usually associate the names of A. Robinson, J. Shoenfield and L. Blum with these tests.) Here is one that we shall apply to $T = \mathrm{RCF}$.

PROPOSITION (QE-TEST). *An $\mathcal{L}$-theory T admits QE*

for any models $\mathcal{M}$ and $\mathcal{N}$ of T, each $\mathcal{L}$-embedding $\mathcal{A} \rightarrow \mathcal{N}$ where $\mathcal{A} \subseteq \mathcal{M}$ and $\mathcal{A} \neq \mathcal{M}$ can be extended to an $\mathcal{L}$-embedding $\mathcal{A}' \rightarrow \mathcal{N}'$ from some strictly larger $\mathcal{L}$-substructure $\mathcal{A}'$ of $\mathcal{M}$ into some elementary extension $\mathcal{N}'$ of $\mathcal{N}$.

To apply this test to RCF we only need to know the following about ordered domains [Artin and Schreier 1926]:

1. Each ordered domain A has a real closure A^{rc}, that is, A^{rc} is a real closed ordered field extending A and algebraic over the fraction field of A.
2. Each embedding $A \rightarrow L$ of an ordered domain A into a real closed ordered field L extends to an embedding $A^{\mathrm{rc}} \rightarrow L$.
3. If A is a real closed field and $A(b)$, $A(c)$ are two ordered field extensions with $b, c \notin A$ such that b and c determine the same cut in A, then there is an A-isomorphism of ordered fields from $A(b)$ onto $A(c)$ sending b to c.

These facts easily imply that RCF admits QE: Let K and L be real closed ordered fields, $A \subseteq K$ an ordered subring and $i : A \longrightarrow L$ an embedding (of ordered rings). Assuming $A \neq K$, we want to show that i can be extended as required in the QE-test. By fact 2. above we can reduce to the case that A itself is a real closed ordered field. Take any $b \in K \setminus A$. Then b determines a cut in A: $U < b < V$, where $U := \{x \in A : x < b\}$ and $V := \{x \in A : b < x\}$. Thus $i(U) < i(V)$ in L. Replacing L by a suitable elementary extension we can take an element $c \in L$ such that $i(U) < c < i(V)$. Then by fact 3 above we can extend i to an ordered field embedding from $A(b)$ into L.   □

That RCF admits QE was first proved by Tarski [1951] by other means. Some routine but noteworthy consequences are:

1. $\mathrm{Th}(\mathbb{R}, <, 0, 1, +, -, \cdot) = \{\text{logical consequences of RCF}\}$, and thus the theory $\mathrm{Th}(\mathbb{R}, <, 0, 1, +, -, \cdot)$ is decidable.
2. The field $\mathbb{Q}^{\mathrm{rc}}$ of real *algebraic* numbers is an elementary substructure of the field of real numbers.
3. Definable = Semialgebraic (for any real closed field).
4. If $S \subseteq \mathbb{R}^{m+n}$ is semialgebraic, there is a semialgebraic map $f : \pi S \longrightarrow \mathbb{R}^n$ such that $\Gamma(f) \subseteq S$:

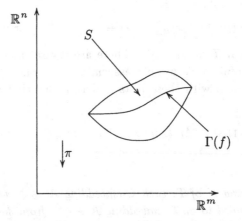

This last result can be read off directly from the axioms of RCF, since the existentially quantified variables in these axioms can be witnessed *definably*, by choosing for each positive element its *positive* square root, and for each odd degree polynomial its *least* zero. This trick can also be used to show that each definable function is piecewise built up from the field operations and the Skolem functions we just indicated. This way of arguing is generally available for other structures whose elementary theory we manage to axiomatize.

2.2. The field of p-adic numbers. Let p be a prime number, and equip $\mathbb{Q}$ with the (nonarchimedean) absolute value defined by

$$|a|_p := p^{-e} \quad \text{for } a = p^e \frac{b}{c} \text{ with } e, b, c \in \mathbb{Z}, \ p \nmid bc.$$

The completion of $(\mathbb{Q}, |\ |_p)$ is called the field of p-adic numbers and is denoted by $(\mathbb{Q}_p, |\ |_p)$. Its elements can be represented uniquely as absolutely convergent series $\sum_{k \in \mathbb{Z}} a_k p^k$ with all $a_k \in \{0, 1, \ldots, p-1\}$, and $a_k = 0$ for all $k < k_0$ for some $k_0 \in \mathbb{Z}$. The ring $\mathbb{Z}_p$ of p-adic *integers* is given by

$$\mathbb{Z}_p := \text{closure of } \mathbb{Z} \text{ in } \mathbb{Q}_p = \{x \in \mathbb{Q}_p : |x|_p \leq 1\}.$$

It is a compact subring of the locally compact field $\mathbb{Q}_p$.

The pair $(\mathbb{Q}_p, \mathbb{Z}_p)$ is an example of a valued field: A *valued field* is a pair (K, V) with K a field and V a *valuation ring* of K (a subring of K such that $x \in K^\times \Longrightarrow x \in V$ or $x^{-1} \in V$). We remark that a valuation ring V has only one maximal ideal $\mathfrak{m}(V) = V \setminus U$, where U is the multiplicative group of units of V.

To a valued field (K, V) we associate:

- its residue field $k := V/\mathfrak{m}(V)$, and

- its value group $\Gamma := K^\times/U$, viewed as an ordered abelian group with

$$aU \leq bU \Longleftrightarrow \frac{b}{a} \in V.$$

(By convention the group operation of Γ is written additively.)

DEFINITION. A valuation ring V is said to be **henselian** if each polynomial $X^n + a_1 X^{n-1} + \cdots + a_{n-1} X + a_n \in V[X]$ with $a_{n-1} \notin \mathfrak{m}(V)$ and $a_n \in \mathfrak{m}(V)$ has a (necessarily unique) zero in $\mathfrak{m}(V)$.

EXAMPLE. $\mathbb{Z}_p$ is henselian; $k[\![t]\!]$ is henselian for any field k.

DEFINITION. A **p-adically closed field** is a valued field (K, V) such that $\text{char}(K) = 0$, V is henselian with $\mathfrak{m}(V) = pV$, $k \simeq \mathbb{F}_p$, and $[\Gamma : n\Gamma] = n$ for $n = 1, 2, 3, \ldots$.

So $(\mathbb{Q}_p, \mathbb{Z}_p)$ is a p-adically closed field. Note that the definition of *p-adically closed field* basically describes a set of axioms in the language of valued fields whose models are exactly the p-adically closed fields.

THEOREM [Ax and Kochen 1965; 1966; Kochen 1969]. *The theory of p-adically closed fields is complete and model complete.*

Kochen used this to characterize the p-adic rational functions in any number of variables that take only values in $\mathbb{Z}_p$ for arguments in $\mathbb{Q}_p$. A one-variable example, which in some sense generates them all, is

$$\frac{1}{p(X^p - X) - p(X^p - X)^{-1}}$$

(this is "the p-adic version of Hilbert's seventeen problem").

For elimination of quantifiers we need to add the right (definable) predicates, just as in the case of the real field where we have to single out the set of squares:

THEOREM [Macintyre 1976]. *The theory of p-adically closed fields admits QE when we extend the language of valued fields with unary relation symbols P_n (for $n = 2, 3, 4, \ldots$) and add their "defining axioms":*

$$\forall x \, (P_n(x) \leftrightarrow \exists y \, (x = y^n)).$$

For a nice treatment of the theorems of Kochen and Macintyre, as well as their generalizations, see [Prestel and Roquette 1984]. Macintyre's theorem has the same kind of consequences for $\mathbb{Q}_p$ as Tarski's theorem for $\mathbb{R}$: it leads to a theory of semialgebraic sets over $\mathbb{Q}_p$ (with dimension theory for such sets, curve selection, ...); see also [Macintyre 1986] for more on this.

Denef discovered new ways to exploit this, namely for the study of various kinds of Poincaré series and local zeta functions associated to p-adic (semi-) algebraic sets. We refer to Denef's paper in this volume for more details.

We did not follow here the chronological order: the above p-adic developments came after the material to be discussed next.

2.3. Henselian valued fields of equicharacteristic 0. Consider a *henselian* valued field (K, V) of equicharacteristic 0, i.e., $\mathrm{char}(k) = 0$ (hence $\mathrm{char}(K) = 0$). Then Ax and Kochen [1965; 1966] and Ershov [1965] proved:

THEOREM. $\mathrm{Th}(K, V)$ *is determined by* $\mathrm{Th}(k)$ *and* $\mathrm{Th}(\Gamma)$, *where k is the residue field and Γ is the ordered value group.*

This extends in some sense work by Mac Lane and Kaplansky [Kaplansky 1942], who showed that under mild assumptions on Γ we can embed (K, V) into the generalized formal power series field $k((t^\Gamma))$ consisting of all formal power series $\sum_{\gamma \in \Gamma} a_\gamma t^\gamma$ with coefficients $a_\gamma \in k$, and with well ordered support

$$\{\gamma \in \Gamma : a_\gamma \neq 0\}.$$

(The valuation ring of $k((t^\Gamma))$ consists of the series with support in $\Gamma^{\geq 0}$.)

By suitably adapting Kaplansky's embedding technique, Ax and Kochen, and independently Ershov, showed that if (K_1, V_1) and (K_2, V_2) are sufficiently saturated henselian valued fields of equicharacteristic 0 with $\mathrm{Th}(k_1) = \mathrm{Th}(k_2)$ and $\mathrm{Th}(\Gamma_1) = \mathrm{Th}(\Gamma_2)$, then (K_1, V_1) and (K_2, V_2) are "back-and-forth" equivalent, and thus $\mathrm{Th}(K_1, V_1) = \mathrm{Th}(K_2, V_2)$.

The following is a routine consequence of the preceding theorem, although it is not mentioned explicitly in [Ax and Kochen 1965; 1966] or [Ershov 1965].

COROLLARY. Given an elementary statement σ about valued fields, there are elementary statements $\sigma_1, \ldots, \sigma_k$ about fields and elementary statements $\tau_1, \ldots, \tau_k$ about ordered groups such that for *all* henselian valued fields (K, V) of equicharacteristic 0 we have

$$(K, V) \models \sigma \Leftrightarrow \text{there exists } i \in \{1, \ldots, k\} \text{ such that } k \models \sigma_i \text{ and } \Gamma \models \tau_i.$$

The valued field $(\mathbb{Q}_p, \mathbb{Z}_p)$ is of mixed characteristic, and the valued field

$$\left(\mathbb{F}_p((t)), \mathbb{F}_p[[t]]\right)$$

is of equicharacteristic p. While neither of these valued fields is of equicharacteristic 0, the *uniformity* in the equivalence above implies a surprising connection between them:

COROLLARY. Let σ be an elementary statement about valued fields. Then

$$(\mathbb{Q}_p, \mathbb{Z}_p) \models \sigma \iff \left(\mathbb{F}_p((t)), \mathbb{F}_p[[t]]\right) \models \sigma$$

for all but finitely many primes p.

PROOF. Take σ_i and τ_i $(i = 1, \ldots, k)$ as in the previous corollary. Then by Gödel's completeness theorem there must be a formal proof of

$$\sigma \leftrightarrow (\sigma_1 \wedge \tau_1) \vee \cdots \vee (\sigma_k \wedge \tau_k)$$

from the axioms for henselian valued fields of equicharacteristic 0. But in such a proof we use only finitely many of the axioms saying that the residue field has characteristic 0. Thus this equivalence also holds in $(\mathbb{Q}_p, \mathbb{Z}_p)$ and in $\left(\mathbb{F}_p((t)), \mathbb{F}_p[[t]]\right)$, for all but finitely many p. Now use the fact that $(\mathbb{Q}_p, \mathbb{Z}_p)$ and $\left(\mathbb{F}_p((t)), \mathbb{F}_p[[t]]\right)$ have the same residue field $\mathbb{F}_p$ and the same value group $\mathbb{Z}$. □

APPLICATION. S. Lang showed in the early 1950s that each homogeneous polynomial of degree $d \geq 1$ in more than d^2 variables over $\mathbb{F}_p((t))$ has a non-trivial zero in that field. Hence by the last corollary, *given any $d \geq 1$*, this statement remains true when we replace $\mathbb{F}_p((t))$ by $\mathbb{Q}_p$, for all but finitely many p. This establishes an asymptotic form of a conjecture by E. Artin. Exceptions indeed occur [Terjanian 1966], and the finite set of exceptional primes depends on d.

What about QE for henselian valued fields? There are several results, by P. J. Cohen, V. Weispfenning, F. Delon, J. Denef, J. Pas, and others, that take the following general form:

Henselian valued fields of equicharacteristic 0 have (uniformly) relative QE: field quantifiers can be eliminated at the cost of introducing quantifiers over the residue field, and over the value group.

The exact language used here can make a difference for the applications. The next example is due to Pas, and is useful in motivic integration; see Denef's paper in this volume.

EXAMPLE. For the valued field $\mathbb{C}((t))$ we have (full) QE in the language with three sorts of variables: variables ranging over the field itself, variables ranging over the residue field $\mathbb{C}$, and variables ranging over the value group $\mathbb{Z}$ (viewed as ordered abelian group with unary predicates for the sets $n\mathbb{Z}$, with $n = 2, 3, \ldots$). Moreover, these sorts are related in the usual way, except that instead of the residue class map $\mathbb{C}[[t]] \to \mathbb{C}$ we consider the *leading coefficient map* $\mathbb{C}((t)) \to \mathbb{C}$ associating to each series its leading coefficient.

3. Expanding by Restricted Analytic Functions

Here things are easier in the p-adics than in the reals! Write $|a| := |a|_p$ for $a \in \mathbb{Q}_p$. Let $X = (X_1, \ldots, X_m)$ and put

$$\mathbb{Z}_p\langle X \rangle := \left\{ f = \sum_{\alpha \in \mathbb{N}^m} c_\alpha X^\alpha \in \mathbb{Z}_p[[X]] : |c_\alpha| \to 0 \text{ as } |\alpha| \to \infty \right\},$$

where $|\alpha| := \alpha_1 + \cdots + \alpha_m$. Each $f \in \mathbb{Z}_p\langle X \rangle$ defines a function

$$x \mapsto f(x) = \sum_{\alpha \in \mathbb{N}^m} c_\alpha x^\alpha : \mathbb{Z}_p^m \to \mathbb{Z}_p.$$

We extend the p-adic absolute value on $\mathbb{Z}_p$ to a norm on the ring $\mathbb{Z}_p\langle X \rangle$ by putting $|f| := \max_{\alpha \in \mathbb{N}^m} |c_\alpha|$.

We construe $\mathbb{Z}_p$ as an $\mathcal{L}_{\mathrm{an}}^D$-structure where the language $\mathcal{L}_{\mathrm{an}}^D$ has the following symbols:

- $0, 1, +, -, \cdot$ (ring operation symbols)
- P_n, for $n = 2, 3, 4, \ldots$ (to denote the set of n-th powers in $\mathbb{Z}_p$)
- an m-ary function symbol f for each $f \in \mathbb{Z}_p\langle X \rangle$
- a binary function symbol D to be interpreted as restricted division:

$$D(x, y) = \begin{cases} \dfrac{x}{y} & \text{if } |x| \leq |y| \neq 0, \\ 0 & \text{otherwise.} \end{cases}$$

Removal of D gives the language $\mathcal{L}_{an}$. The next result is in [Denef and van den Dries 1988], where it is applied to prove rationality of Poincaré series of p-adic analytic varieties.

THEOREM. *The $\mathcal{L}_{\mathrm{an}}^D$-structure $\mathbb{Z}_p$ admits QE.*

This amounts to a theory of p-adic subanalytic sets. (A subset of $\mathbb{Z}_p^m$ is *subanalytic* if it is the projection of a subset of $\mathbb{Z}_p^{m+n}$ defined by a quantifier free formula in $\mathcal{L}_{an}$. By the theorem this is the same as a subset of $\mathbb{Z}_p^m$ defined by a quantifier free formula in $\mathcal{L}_{an}^D$.)

IDEA OF PROOF. Consider for example a formula $\exists y \ f(x, y) = 0$, where $f \in \mathbb{Z}_p\langle X, Y \rangle$, $X = (X_1, \ldots, X_m)$, $Y = (Y_1, \ldots, Y_n)$ with $x = (x_1, \ldots, x_m)$ and $y = (y_1, \ldots, y_n)$ ranging over $\mathbb{Z}_p^m$ (the parameter space) and $\mathbb{Z}_p^n$ respectively. We have to find the (quantifier free) conditions on the parameter x for solvability in y of the equation $f(x, y) = 0$. The case $n = 0$ being trivial, assume $n > 0$. Below we use the lexicographic ordering of $\mathbb{N}^n$.

STEP 1: Write $f(X,Y) = \sum_{i \in \mathbb{N}^n} a_i Y^i$ with $a_i = a_i(X) \in \mathbb{Z}_p\langle X \rangle$. Using the noetherianity of $\mathbb{Z}_p\langle X \rangle$, we can take $d \in \mathbb{N}$ such that each a_i is of the form

$$a_i = \sum_{|j| < d} c_{ij} a_j$$

with $c_{ij} \in \mathbb{Z}_p\langle X \rangle$, such that for each j with $|j| < d$ we have $|c_{ij}| \to 0$ as $|i| \to \infty$.

STEP 2: Partition the x-space $\mathbb{Z}_p^m$ into the subsets Z and S_j for $|j| < d$, each quantifier free definable in the language $\mathcal{L}_{\mathrm{an}}$, such that

- if $x \in Z$, then $f(x, Y) = 0$ (so $f(x, y) = 0$ for all $y \in \mathbb{Z}_p^n$);
- if $x \in S_j$, then $f(x, Y) \neq 0$, $|a_j(x)| = \max_{i \in \mathbb{N}^n} |a_i(x)|$, and j is lexicographically maximal with this property.

(Note that Z is just the set defined by the formula $\bigwedge_{|j| < d} a_j(x) = 0$.)

STEP 3: Fix $j \in \mathbb{N}^n$ with $|j| < d$, and let x range over S_j. Put

$$v_{ij}(x) := \begin{cases} \dfrac{a_i(x)}{a_j(x)} & \text{if } i < j,\ |i| < d, \\[2ex] \dfrac{a_i(x)}{pa_j(x)} & \text{if } i > j,\ |i| < d. \end{cases}$$

Then $v_{ij}(x) \in \mathbb{Z}_p$. Put $v_j(x) := (v_{ij}(x))_{|i| < d, i \neq j}$. We now carry out a standard change of variables, by substituting

$$T_d(Y) := (Y_1 + Y_n^{d^{n-1}}, \ldots, Y_{n-1} + Y_n^d, Y_n)$$

for Y. We also factor out the "last" dominating coefficient $a_j(x)$. The combined result is an identity (for $x \in S_j$):

$$f(x, T_d(Y)) = a_j(x) F_j(x, v_j(x), Y)$$

where $F_j(X, V_j, Y) \in \mathbb{Z}_p\langle X, V_j, Y \rangle$, $V_j = (V_{ij})_{|i| < d, i \neq j}$, and

$$F_j \bmod p \in \mathbb{F}_p[X, V_j, Y]$$

is monic in Y_n of degree $j_1 d^{n-1} + \cdots + j_n$.

STEP 4: By p-adic Weierstrass Preparation we have

$$F_j = U \cdot (Y_n^e + c_1 Y_n^{e-1} + \cdots + c_e),$$

where U is a unit of $\mathbb{Z}_p\langle X, V_j, Y \rangle$, $c_1, \ldots, c_e \in \mathbb{Z}_p\langle X, V_j, Y' \rangle$, $Y' = (Y_1, \ldots, Y_{n-1})$, and $e = j_1 d^{n-1} + \cdots + j_n$. Thus for $x \in S_j$ and $y \in \mathbb{Z}_p^n$ we have

$$f(x, T_d(y)) = a_j(x) U(x, v_j(x), y) g(x, y', y_n),$$

where $y' := (y_1, \ldots, y_{n-1})$ and

$$g(x, y', y_n) := y_n^e + c_1(x, v_j(x), y') y_n^{e-1} + \cdots + c_e(x, v_j(x), y').$$

STEP 5: Note the equivalences

$$x \in S_j \ \wedge \ \exists y \ f(x,y) = 0 \Longleftrightarrow x \in S_j \ \wedge \ \exists y \ f(x, T_d(y)) = 0$$
$$\Longleftrightarrow x \in S_j \ \wedge \ \exists y \ g(x,y) = 0$$
$$\Longleftrightarrow x \in S_j \ \wedge \ \exists y' \ \phi(x, v_j(x), y'),$$

where ϕ is quantifier free in $\mathcal{L}_{an}$. For the last equivalence we used that y_n occurs only polynomially in g, so that Macintyre's theorem can be applied. We eliminated $\exists y_n$ at the cost of introducing fractions $v_{ij}(x)$, but these fractions only involve x and not any of the y-variables.

While we were dealing here with a rather special kind of formula, the above reduction of $\exists y$ to $\exists y'$ does contain the main ideas for a general quantifier elimination. □

REMARKS. 1. The appeal to noetherianity of $\mathbb{Z}_p\langle X \rangle$ may seem to make this proof non-constructive. But this appeal is made only for convenience. At the cost of complicating the exposition we could replace it by finitely many applications of p-adic Weierstrass division. This should not be too surprising, since one way to prove the noetherianity of $\mathbb{Z}_p\langle X \rangle$ goes via the p-adic Weierstrass division theorem.

2. One can indicate a few simple schemes of universal axioms in $\mathcal{L}_{an}^D$ and true in $\mathbb{Z}_p$, that together with the axioms for p-adically closed valuation rings formally imply the QE above. It follows that the definable functions are piecewise superpositions of semialgebraic functions, functions given by the power series in $\mathbb{Z}_p\langle X \rangle$, and D.

Does this method also work for $\mathbb{R}$? Yes, *except* that in Weierstrass preparing a convergent power series over $\mathbb{R}$ (or $\mathbb{C}$) the domain of convergence may decrease drastically. Thus we have to work more locally, and exploit the compactness of $[-1,1]^m$ (whereas we didn't need the compactness of $\mathbb{Z}_p^m$ in the arguments above).

Details: The language $\mathcal{L}_{an}^D$ (real version) has the following symbols:

- $0, 1, +, -, \cdot, <$
- for each power series $f \in \mathbb{R}[\![X_1, \ldots, X_m]\!]$ converging in a neighborhood of $[-1,1]^m$ a corresponding function symbol, also denoted f, to be interpreted as the "restricted analytic" function on $\mathbb{R}^m$ given by

$$x \mapsto \begin{cases} f(x) & \text{if } x \in [-1,1]^m, \\ 0 & \text{otherwise;} \end{cases}$$

- a binary function symbol D for restricted division:

$$D(x,y) = \begin{cases} \dfrac{x}{y} & \text{if } |x| \le |y| \le 1, \ y \ne 0, \\ 0 & \text{otherwise.} \end{cases}$$

The proof in [Denef and van den Dries 1988] that $\mathbb{R}$ admits QE in this language $\mathcal{L}_{an}^D$ gives an alternative approach to the theory of subanalytic sets. This subject was originally developed using other tools by Gabrielov, Hironaka and other analytic geometers. Indeed, a subset of $\mathbb{R}^n$ is definable using the language $\mathcal{L}_{an}^D$ if and only if it is subanalytic in its projective completion $\mathbb{P}^n(\mathbb{R})$.

An explicit axiomatization of the theory of $\mathbb{R}$ in the language $\mathcal{L}_{an}^D$ by finitely many schemes appears in [van den Dries et al. 1994]. This axiomatization is used to show that each field of power series over $\mathbb{R}$ with exponents in a divisible ordered abelian group carries a natural extra structure making it an elementary extension of $\mathbb{R}$ in the language $\mathcal{L}_{an}^D$. These power series fields with their extra analytic structure turn out to be important in the model theoretic study of the real field with restricted analytic functions and the unrestricted exponential function.

4. o-Minimal Expansions of the Real Field

For general background on o-minimality, see Macpherson's article in this volume, or the book [van den Dries 1998]. One direction in o-minimal studies that is close to classical model theory of fields involves constructing new o-minimal expansions of the field of reals. This activity received a big boost from Wilkie's theorem [1996a] that the real exponential field is model-complete. (Its o-minimality then follows from earlier work by Khovanskii.)

In this section we focus on some recent (post 1994) examples and constructions of o-minimal expansions of the real field. The main such expansions known at present are then indicated in an inclusion diagram, following Macintyre's lead.

Geometers often need the setting of manifolds rather than being tied to the particular coordinate systems of cartesian spaces $\mathbb{R}^n$. In Section 4.3 below we indicate how to accomplish this by means of the *analytic-geometric categories* of [van den Dries and Miller 1996]. This was used in [Schmid and Vilonen 1996].

4.1. Special constructions (see [van den Dries and Speissegger 1998a; 1998b]).

1. The expansion $\mathbb{R}_{an^*}$, the field of reals with functions given by generalized convergent power series. In this structure one has among the basic functions those of the form $x \mapsto \sum_{n=0}^{\infty} a_n x^{r_n} : [0,1] \to \mathbb{R}$ with real coefficients a_n and real exponents r_n such that $r_n \uparrow +\infty$ and $\sum |a_n|(1+\varepsilon)^{r_n} < \infty$ for some $\varepsilon > 0$. In particular, $\mathbb{R}_{an^*}$ defines the function

$$\zeta(-\log x) = \sum_{n=1}^{\infty} x^{\log n}$$

on $[0, e^{-1})$. The structure $\mathbb{R}_{an^*}$ is model complete in its "natural" language, o-minimal, and polynomially bounded.

2. The expansion $\mathbb{R}_{\mathcal{G}}$, the field of reals with functions given by multisummable real power series. Among the basic operations of $\mathbb{R}_{\mathcal{G}}$ are the C^∞ functions

$f : [0,1] \longrightarrow \mathbb{R}$ whose restriction to $(0,1]$ extends to a holomorphic function on a sector

$$S(R, \phi) := \{ z \in \mathbb{C} : |z| < R, |\arg z| < \phi \}$$

for some $R > 1$ and $\phi > \frac{\pi}{2}$, such that there exist positive constants A, B with $|f^{(n)}(z)| \leq AB^n (n!)^2$ for all $z \in S(R, \phi)$, and $\lim_{z \to 0, z \in S(R,\phi)} f^{(n)}(z) = f^{(n)}(0)$.

Two examples of such functions:

-
$$f(x) = \int_0^\infty \frac{e^{-t}}{1 + xt} \, dt, \quad \text{for } 0 \leq x \leq 1.$$

Its Taylor expansion at 0 is the *divergent* series $\sum_{n=0}^\infty (-1)^n n! \, x^n$.

- The continuous function ψ on $[0,1]$ given by Stirling's expansion

$$\log \Gamma(x) = (x - \tfrac{1}{2}) \log x - x + \tfrac{1}{2} \log \pi + \psi\left(\frac{1}{x}\right), \quad \text{for } x \geq 1.$$

The structure $\mathbb{R}_{\mathcal{G}}$ is model complete in its natural language and polynomially bounded. Note that the Gamma function on $(0, +\infty)$ is definable in $\mathbb{R}_{\mathcal{G},\exp}$.

4.2. General constructions

I. If $\tilde{\mathbb{R}}$ is a polynomially bounded o-minimal expansion of the real field in which $\exp|_{[0,1]}$ is definable, then $(\tilde{\mathbb{R}}, \exp)$ is an exponentially bounded o-minimal expansion of the real field, and $\mathrm{Th}(\tilde{\mathbb{R}}, \exp, \log)$ admits QE relative to $\mathrm{Th}(\tilde{\mathbb{R}})$, see [van den Dries and Speissegger 1998b].

II. Suppose $\tilde{\mathbb{R}}$ is an o-minimal expansion of the real field by (total) C^∞ functions. Then Wilkie [1996b] proved that $\tilde{\mathbb{R}}$ remains o-minimal when expanded by the $\tilde{\mathbb{R}}$-Pfaffian functions. Here a C^∞ function $f : \mathbb{R}^n \to \mathbb{R}$ is $\tilde{\mathbb{R}}$-Pfaffian if there are C^∞ functions $f_1, \ldots, f_k : \mathbb{R}^n \to \mathbb{R}$ (not necessarily definable in $\tilde{\mathbb{R}}$) such that $f = f_k$, and there are C^∞ functions $F_{ij} : \mathbb{R}^{n+i} \to \mathbb{R}$, definable in $\tilde{\mathbb{R}}$, such that

$$\frac{\partial f_i}{\partial x_j}(x) = F_{ij}(x, f_1(x), \ldots, f_i(x))$$

for $i = 1, \ldots, k$, $j = 1, \ldots, n$, $x \in \mathbb{R}^n$.

This inspired further work along this line by Lion and Rolin [1998]. This in turn led Speissegger [1999] to prove that any o-minimal expansion $\tilde{\mathbb{R}}$ of the real field remains o-minimal when further expanded by the so-called *Rolle leaves* of 1-forms of class C^1 definable in $\tilde{\mathbb{R}}$. This expansion operation can then be iterated infinitely often to produce an o-minimal "Pfaffian closure" of $\tilde{\mathbb{R}}$.

The diagram on the next page lists the main o-minimal expansions of the real field that can be obtained by the methods above. An arrow $\mathbb{R}_{\mathcal{A}} \to \mathbb{R}_{\mathcal{B}}$ means that the definable sets of $\mathbb{R}_{\mathcal{A}}$ are also definable in $\mathbb{R}_{\mathcal{B}}$. The bottom-left corner $\mathbb{R}_{\mathrm{alg}}$ is just the ordered field of real numbers with no further structure, and $\mathbb{R}_{\mathrm{an}}$ is the ordered real field expanded by the restricted analytic functions. The bottom arrows connect the polynomially bounded expansions, the upward pointing ones go to the expansions that can be built on top of the polynomially bounded ones by adding exp, and taking the "Pfaffian" closure.

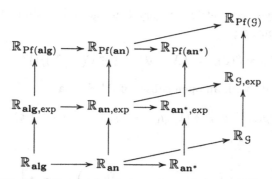

The known o-minimal expansions of the real field.

4.3. Analytic-geometric categories. In this subsection "manifold" means "real analytic manifold whose topology is Hausdorff with a countable basis". An **analytic-geometric category** $\mathcal{C}$ is given if each manifold M is equipped with a collection $\mathcal{C}(M)$ of distinguished subsets, called the $\mathcal{C}$-**subsets** of M, such that for all manifolds M and N the following holds:

1. $M \in \mathcal{C}(M)$ and $\mathcal{C}(M)$ is a boolean algebra.
2. If $A \in \mathcal{C}(M)$ then $A \times \mathbb{R} \in \mathcal{C}(M \times \mathbb{R})$.
3. If $f : M \to N$ is a proper analytic map and $A \in \mathcal{C}(M)$, then $f(A) \in \mathcal{C}(N)$.
4. If $A \subseteq M$ and $\{U_i\}_{i \in I}$ is an open covering of M, then $A \in \mathcal{C}(M)$ iff $A \cap U_i \in \mathcal{C}(U_i)$ for all $i \in I$.
5. The boundary of every bounded $\mathcal{C}$-subset of $\mathbb{R}$ is finite.

We make a category $\mathcal{C}$ out of this by letting the objects of $\mathcal{C}$ be the pairs (A, M) with M a manifold and $A \in \mathcal{C}(M)$. The morphisms $(A, M) \to (B, N)$ of objects (A, M) and (B, N) are continuous maps $f : A \to B$ such that

$$\Gamma(f) := \{(a, f(a)) : a \in A\} \subseteq A \times B$$

belongs to $\mathcal{C}(M \times N)$. The composition of morphisms is given by composition of maps.

An object (A, M) of $\mathcal{C}$ is called $\mathcal{C}$-set A in M (or just the $\mathcal{C}$-set A). Call a morphism $f : (A, M) \to (B, N)$ a $\mathcal{C}$-**map** $f : A \to B$ if M and N are clear from the context.

The subanalytic subsets of a manifold M are necessarily $\mathcal{C}$-sets in M. Conversely we define an analytic-geometric category by taking just the subanalytic subsets of each manifold as its $\mathcal{C}$-sets.

By "o-minimal structure on $\mathbb{R}_{\mathbf{an}}$" we mean the system of definable sets of an o-minimal expansion of $\mathbb{R}_{\mathbf{an}}$. There is a one-to-one correspondence between o-minimal structures on $\mathbb{R}_{\mathbf{an}}$ in this sense, and analytic-geometric categories, as we now explain.

Given an analytic-geometric category $\mathcal{C}$ we get an o-minimal structure $\mathcal{S} = \mathcal{S}(\mathcal{C})$ on $\mathbb{R}_{\mathbf{an}}$ by putting

$$\mathcal{S}_n = \mathcal{S}(\mathcal{C})_n := \{X \subseteq \mathbb{R}^n : X \in \mathcal{C}(\mathbb{P}^n(\mathbb{R})))\} \quad .$$

Here we identify $\mathbb{R}^n$ with an open subset of $\mathbb{P}^n(\mathbb{R})$ via

$$(y_1, \ldots, y_n) \mapsto (1 : y_1 : \cdots : y_n) : \mathbb{R}^n \to \mathbb{P}^n(\mathbb{R}).$$

Equivalently, for $A \subseteq \mathbb{R}^n$:

$$A \in \mathcal{S}(\mathcal{C})_n \iff \left\{ \left(\frac{x_1}{\sqrt{1 + x_1^2}}, \ldots, \frac{x_n}{\sqrt{1 + x_n^2}} \right) : x \in A \right\} \in \mathcal{C}(\mathbb{R}^n).$$

From an o-minimal structure $\mathcal{S} = (\mathcal{S}_n)$ on $\mathbb{R}_{\mathbf{an}}$ we get an analytic-geometric category $\mathcal{C} = \mathcal{C}(\mathcal{S})$ by defining the $\mathcal{C}$-subsets of an m-dimensional manifold M to be those $A \subseteq M$ such that for each $x \in M$ there is an open neighborhood U of x, an open $V \subseteq \mathbb{R}^m$ and an analytic isomorphism $h : U \to V$ with $h(U \cap A) \in \mathcal{S}_m$.

For each analytic-geometric category $\mathcal{C}$ and each o-minimal structure $\mathcal{S}$ on $\mathbb{R}_{\mathbf{an}}$ we have $\mathcal{C}(\mathcal{S}(\mathcal{C})) = \mathcal{C}$ and $\mathcal{S}(\mathcal{C}(\mathcal{S})) = \mathcal{S}$.

Let $\mathcal{C}$ be an analytic-geometric category, M, N manifolds of dimension m, n respectively and $A \in \mathcal{C}(M)$. We can now state a number of basic facts directly in terms of this category. They follow easily from corresponding o-minimal results, using charts and partitions of unity.

1. Every analytic map $f : M \to N$ is a $\mathcal{C}$-map.
2. $\mathrm{cl}(A), \mathrm{int}(A) \in \mathcal{C}(M)$.
3. $\mathrm{Reg}_k^p(A) \in \mathcal{C}(M)$ for each $k \in \mathbb{N}$ and positive $p \in \mathbb{N}$.
4. If A is also a C^1 submanifold of M, then its tangent, cotangent, and conormal bundles are $\mathcal{C}$-sets in their corresponding ambient manifolds: $TA \in \mathcal{C}(TM)$, $T^*A \in \mathcal{C}(T^*M)$, and $T_A^*M \in \mathcal{C}(T^*M)$.
5. A is locally connected, and has locally a finite number of components. If C is a component of A, then $C \in \mathcal{C}(M)$.
6. Every connected $\mathcal{C}$-set is path connected. The set of components of A is a locally finite subcollection of $\mathcal{C}(M)$.
7. If $\mathcal{F} \subseteq \mathcal{C}(M)$ is locally finite, then $\bigcap \mathcal{F} \in \mathcal{C}(M)$ and $\bigcup \mathcal{F} \in \mathcal{C}(M)$.
8. If $\varnothing \neq \mathcal{A} \subseteq \mathcal{C}(M)$ is locally finite, then

$$\dim\left(\bigcup \mathcal{A}\right) = \max\{\dim(A) : A \in \mathcal{A}\}.$$

9. If $f : A \to N$ is a proper $\mathcal{C}$-map, then $\dim(C) \geq \dim(f(C))$ for all $\mathcal{C}$-sets $C \subseteq A$.
10. If $A \neq \varnothing$, then $\dim(\mathrm{closure}(A) \setminus A) < \dim(A)$.

The following results require more effort. The first one is due to Bierstone, Milman and Pawłucki in the subanalytic case.

1. If A is closed and $p \in \mathbb{N}$, there is a $\mathcal{C}$-map $f : M \to \mathbb{R}$ of class C^p with $A = f^{-1}(0)$.

2. (Whitney stratification) Let $S \in \mathcal{C}(M)$ be closed and p be a positive integer. For every locally finite $\mathcal{A} \subseteq \mathcal{C}(M)$ there is a C^p Whitney stratification $\mathcal{P} \subseteq \mathcal{C}(M)$ of S, compatible with $\mathcal{A}$, with each stratum connected and relatively compact.

Let $f : S \to N$ be a proper $\mathcal{C}$-map and $\mathcal{F} \subseteq \mathcal{C}(M)$, $\mathcal{G} \subseteq \mathcal{C}(N)$ be locally finite. Then there is a C^p Whitney stratification $(\mathcal{S}, \mathcal{T})$ of f with connected strata such that $\mathcal{S} \subseteq \mathcal{C}(M)$ is compatible with $\mathcal{F}$ and $\mathcal{T} \subseteq \mathcal{C}(N)$ is compatible with $\mathcal{G}$.

References

[Artin and Schreier 1926] E. Artin and O. Schreier, "Algebraische Konstruktion reeller Körper", *Hamb. Abh.* **5** (1926), 85–99.

[Ax and Kochen 1965] J. Ax and S. Kochen, "Diophantine problems over local fields, I, II", *Amer. J. Math.* **87** (1965), 605–630, 631–648.

[Ax and Kochen 1966] J. Ax and S. Kochen, "Diophantine problems over local fields, III", *Ann. of Math.* (2) **83** (1966), 437–456.

[Denef and van den Dries 1988] J. Denef and L. van den Dries, "*p*-adic and real subanalytic sets", *Ann. of Math.* (2) **128**:1 (1988), 79–138.

[van den Dries 1998] L. van den Dries, *Tame topology and o-minimal structures*, London Math. Soc. Lecture Note Series **248**, Cambridge Univ. Press, Cambridge, 1998.

[van den Dries and Miller 1996] L. van den Dries and C. Miller, "Geometric categories and o-minimal structures", *Duke Math. J.* **84**:2 (1996), 497–540.

[van den Dries and Speissegger 1998a] L. van den Dries and P. Speissegger, "The real field with convergent generalized power series", *Trans. Amer. Math. Soc.* **350**:11 (1998), 4377–4421.

[van den Dries and Speissegger 1998b] L. van den Dries and P. Speissegger, "The field of reals with multisummable series and the exponential function", preprint, 1998. Available at http://www.math.utoronto.ca/~speisseg/preprints.html. To appear in *Proc. London Math. Soc.*

[van den Dries et al. 1994] L. van den Dries, A. Macintyre, and D. Marker, "The elementary theory of restricted analytic fields with exponentiation", *Ann. of Math.* (2) **140**:1 (1994), 183–205.

[Ershov 1965] Y. L. Ershov, "On elementary theories of local fields", *Algebra i Logika (Seminar)* **4**:2 (1965), 5–30.

[Kaplansky 1942] I. Kaplansky, "Maximal fields with valuations", *Duke Math. J.* **9** (1942), 303–321.

[Kochen 1969] S. Kochen, "Integer valued rational functions over the *p*-adic numbers: A *p*-adic analogue of the theory of real fields", pp. 57–73 in *Number Theory* (Houston, 1967), edited by W. J. Leveque and E. G. Straus, Proc. Sympos. Pure Math. **12**, Amer. Math. Soc., Providence, 1969.

[Lion and Rolin 1998] J.-M. Lion and J.-P. Rolin, "Volumes, feuilles de Rolle de feuilletages analytiques et théorème de Wilkie", *Ann. Fac. Sci. Toulouse Math.* (6) **7**:1 (1998), 93–112.

[Macintyre 1976] A. Macintyre, "On definable subsets of p-adic fields", *J. Symbolic Logic* **41**:3 (1976), 605–610.

[Macintyre 1986] A. Macintyre, "Twenty years of p-adic model theory", pp. 121–153 in *Logic colloquium '84* (Manchester, 1984), edited by J. B. Paris et al., Stud. Logic Found. Math. **120**, North-Holland, Amsterdam, 1986.

[Prestel and Roquette 1984] A. Prestel and P. Roquette, *Formally p-adic fields*, Lecture Notes in Math. **1050**, Springer, Berlin, 1984.

[Schmid and Vilonen 1996] W. Schmid and K. Vilonen, "Characteristic cycles of constructible sheaves", *Invent. Math.* **124**:1-3 (1996), 451–502.

[Speissegger 1999] P. Speissegger, "The Pfaffian closure of an o-minimal structure", *J. Reine Angew. Math.* **508** (1999), 189–211.

[Tarski 1951] A. Tarski, *A decision method for elementary algebra and geometry*, 2nd ed., Univ. of California Press, Berkeley and Los Angeles, 1951.

[Terjanian 1966] G. Terjanian, "Un contre-exemple à une conjecture d'Artin", *C. R. Acad. Sci. Paris Sér. A-B* **262** (1966), A612.

[Wilkie 1996a] A. J. Wilkie, "Model completeness results for expansions of the ordered field of real numbers by restricted Pfaffian functions and the exponential function", *J. Amer. Math. Soc.* **9** (1996), 1051–1094.

[Wilkie 1996b] A. J. Wilkie, "A general theorem of the complement and some new o-minimal structures", preprint, 1996. To appear in *Selecta Math.*

LOU VAN DEN DRIES
UNIVERSITY OF ILLINOIS
DEPARTMENT OF MATHEMATICS
1409 W GREEN ST.
URBANA, IL 61801-2917
UNITED STATES
vddries@math.uiuc.edu

Model Theory, Algebra, and Geometry
MSRI Publications
Volume **39**, 2000

Model Theory of Differential Fields

DAVID MARKER

ABSTRACT. This article surveys the model theory of differentially closed fields, an interesting setting where one can use model-theoretic methods to obtain algebraic information. The article concludes with one example showing how this information can be used in diophantine applications.

A *differential field* is a field K equipped with a derivation $\delta : K \to K$; recall that this means that, for $x, y \in K$, we have $\delta(x + y) = \delta(x) + \delta(y)$ and $\delta(xy) = x\,\delta(y) + y\,\delta(x)$. Roughly speaking, such a field is called differentially closed when it contains enough solutions of ordinary differential equations. This setting allows one to use model-theoretic methods, and particularly dimension-theoretic ideas, to obtain interesting algebraic information.

In this lecture I give a survey of the model theory of differentially closed fields, concluding with an example — Hrushovski's proof of the Mordell–Lang conjecture in characteristic zero — showing how model-theoretic methods in this area can be used in diophantine applications. I will not give the proofs of the main theorems. Most of the material in Sections 1–3 can be found in [Marker et al. 1996], while the material in Section 4 can be found in [Hrushovski and Sokolovic $\geq$ 2000; Pillay 1996]. The primary reference on differential algebra is [Kolchin 1973], though the very readable [Kaplansky 1957] contains most of the basics needed here, as does the more recent [Magid 1994]. The book [Buium 1994] also contains an introduction to differential algebra and its connections to diophantine geometry. We refer the reader to these sources for references to the original literature.

1. Differentially Closed Fields

Throughout this article all fields will have characteristic zero. A *differential field* is a field K equipped with a derivation $\delta : K \to K$. The field of *constants* is $C = \{x \in K : \delta(x) = 0\}$.

We will study differential fields using the language $\mathcal{L} = \{+, -, \cdot, \delta, 0, 1\}$, the language of rings augmented by a unary function symbol δ. The theory of

differential fields, DF, is given by the axioms for fields of characteristic zero and
the axioms

$$\forall x \, \forall y \, \delta(x+y) = \delta(x) + \delta(y),$$

$$\forall x \, \forall y \, \delta(xy) = x\delta(y) + y\delta(x),$$

which assert that δ is a derivation.

If K is a differential field, we define $K\{X_1, \ldots, X_n\}$, the ring of *differential polynomials* over K, to be the following polynomial ring in infinitely many
variables:

$$K\left[X_1, \ldots, X_n, \delta(X_1), \ldots, \delta(X_n), \ldots, \delta^m(X_1), \ldots, \delta^m(X_n), \ldots\right].$$

We extend δ to a derivation on $K\{X_1, \ldots, X_n\}$ by setting $\delta(\delta^n(X_i)) = \delta^{n+1}(X_i)$.

We say that K is an *existentially closed* differential field if, whenever $f_1, \ldots,$
$f_m \in K\{X_1, \ldots, X_n\}$ and there is a differential field L extending K containing
a solution to the system of differential equations $f_1 = \cdots = f_m = 0$, there is
already a solution in K. Robinson gave an axiomatization of the existentially
closed differential fields. Blum gave a simple axiomatization that refers only to
differential polynomials in one variable.

If $f \in K\{X_1, \ldots, X_n\} \setminus K$, the *order* of f is the largest m such that $\delta^m(X_i)$
occurs in f for some i. If f is a constant, we say f has order -1.

DEFINITION. A differential field K is *differentially closed* if, whenever $f, g \in$
$K\{X\}$, g is nonzero and the order of f is greater than the order of g, there is
$a \in K$ such that $f(a) = 0$ and $g(a) \neq 0$.

In particular, any differentially closed field is algebraically closed.

For each m and d_0 and d_1 we can write down an $\mathcal{L}$-sentence ϕ_{m,d_0,d_1} that
asserts that if f is a differential polynomial of order m and degree at most d_0
and g is a nonzero differential polynomial of order less than m and degree at
most d_1, then there is a solution to $f(X) = 0$ and $g(X) \neq 0$. For example, $\phi_{2,1,1}$
is the formula

$\forall a_0 \, \forall a_1 \, \forall a_2 \, \forall a_3 \, \forall b_0 \, \forall b_1 \, \forall b_2$

$\quad (a_3 \neq 0 \land (b_0 \neq 0 \lor b_1 \neq 0 \lor b_2 \neq 0)$

$\qquad \rightarrow \exists x \, (a_3 \delta(\delta(x)) + a_2 \delta(x) + a_1 x + a_0 = 0 \land b_2 \delta(x) + b_1 x + b_0 \neq 0)).$

The $\mathcal{L}$-theory DCF is axiomatized by DF and the set of axioms ϕ_{m,d_0,d_1}, for
all m, d_0 and d_1. The models of DCF are exactly the differentially closed fields.

It is not hard to show that if $f, g \in K\{X\}$ are as above, then there is $L \supseteq K$
containing a solution to the system $f(X) = 0$ and $Yg(X) - 1 = 0$. Indeed we
could take L to be the fraction field of $K\{X\}/P$, where P is a minimal differential
prime ideal with $f \in P$. Iterating this construction shows that any differential
field can be extended to a differentially closed field. Thus any existentially closed
field is differentially closed.

The next theorem of Blum shows that the converse holds (see [Marker et al. 1996] for the proof).

THEOREM 1.1. *The theory* DCF *has quantifier elimination and hence is model complete.*

COROLLARY 1.2. (i) DCF *is a complete theory.*

(ii) *A differential field is existentially closed if and only if it is differentially closed.*

PROOF. (i) The rational numbers with the trivial derivation form a differential subfield of any differentially closed field. If K_0 and K_1 are models of DCF and ϕ is a quantifier free sentence, then there is a quantifier free sentence ψ such that DCF $\models \phi \leftrightarrow \psi$. But $K_i \models \psi$ if and only if $\mathbb{Q} \models \psi$. Hence $K_0 \models \phi$ if and only if $K_1 \models \phi$ and DCF is complete.

(ii) We already remarked that every existentially closed field is differentially closed. Suppose K is differentially closed. Suppose $f_1 = \cdots = f_m = 0$ is a system of polynomial differential equations solvable in an extension L of K. We can find K_1 an extension of L which is differentially closed. By model completeness K is an elementary submodel of K_1. Since there is a solution in K_1 there is a solution in K. □

Pierce and Pillay [1998] have given a more geometric axiomatization of DCF. Suppose K is a differential field and $V \subseteq K^n$ is an irreducible algebraic variety defined over K. Let $I(V) \subset K[X_1, \ldots, X_n]$ be the ideal of polynomials vanishing on V and let $f_1, \ldots, f_m$ generate $I(V)$. If $f = \sum a_{i_1,\ldots,i_m} X_1^{i_1} \cdots X_m^{i_m}$, let $f^\delta = \sum \delta(a_{i_1,\ldots,i_m}) X_1^{i_1} \cdots X_m^{i_m}$. The tangent bundle $T(V)$ can be identified with the variety

$$T(V) = \left\{ (x,y) \in K^{2n} : x \in V \wedge \sum_{j=1}^n y_j \frac{\partial f_i}{\partial X_j}(x) = 0 \text{ for } i = 1, \ldots, m \right\}$$

We define the *first prolongation* of V to be the algebraic variety

$$V^{(1)} = \left\{ (x,y) \in K^{2n} : x \in V \wedge \sum_{j=1}^n y_j \frac{\partial f_i}{\partial X_j}(x) + f_i^\delta(x) = 0 \text{ for } i = 1, \ldots, m \right\}.$$

If V is defined over the constant field C, then each f_i^δ vanishes, and $V^{(1)}$ is $T(V)$. In general, for $a \in V$, the vector space $T_a(V) = \{b : (a,b) \in T(V)\}$ acts regularly on $V_a^{(1)} = \{b : (a,b) \in V^{(1)}\}$, making $V^{(1)}$ a torsor under $T(V)$. It is easy to see that $(x, \delta(x)) \in V^{(1)}$ for all $x \in V$. Thus the derivation is a section of the first prolongation.

THEOREM 1.3. *For K be a differential field, the following statements are equivalent:*

(i) K *is differentially closed.*

(ii) *K is existentially closed.*

(iii) *K is algebraically closed and for every irreducible algebraic variety $V \subseteq K^n$ if W is an irreducible subvariety of $V^{(1)}$ such that the projection of W onto V is Zariski dense in V and U is a Zariski open subset of V, then $(x, \delta(x)) \in U$ for some $x \in V$.*

PROOF. We know (i) $\Longleftrightarrow$ (ii).

(iii) $\Rightarrow$ (i) Suppose $f(X) \in K\{X\}$ has order n and $g(X)$ has lower order. Say $f(X) = p(X, \delta(X), \dots, \delta^n(X))$ and $g = q(X, \delta(X), \dots \delta^{n-1}(X))$, where p and q are polynomials. Without loss of generality p is irreducible. Set $V = K^n$ and

$$W = \{(x, y) \in K^{2n} : y_1 = x_2, y_2 = x_3, \dots, y_{n-1} = x_n, p(x_1, \dots, x_n, y_n) = 0\}.$$

It is easy to see that W is irreducible and W projects generically onto K^n. Let $U = \{(x, y) \in W : q(x) \neq 0\}$. By (iii) there is $x \in K^n$ such that $(x, \delta(x)) \in U$. Then $f(x_1) = 0$ and $g(x_1) \neq 0$.

(i) $\Rightarrow$ (iii) Let V, W and U be as in (iii). Let (x, y) be a generic point of U over K. One can show that there is a differential field L extending $K(x, y)$ with $\delta(x) = y$ (indeed we can extend δ to $K(x, y)$). We may assume L is differentially closed. Then $(x, \delta(x)) \in U$ and by model completeness we can find a solution in K. $\qquad \square$

2. The Kolchin Topology

We say that an ideal I in $K\{X_1, \dots, X_n\}$ is a δ-ideal if $\delta(f) \in I$ whenever $f \in I$. If $I \subset K\{X_1, \dots, X_n\}$, let $V_\delta(I) = \{x \in K^n : f(x) = 0 \text{ for all } f \in I\}$. We can topologize K^n by taking the $V_\delta(I)$ as basic closed sets. This topology is referred to as the *Kolchin topology* or the *δ-topology*.

There are infinite ascending sequences of δ-ideals. For example,

$$\langle X^2 \rangle \subset \langle X^2, \delta(X)^2 \rangle \subset \cdots \subset \langle X^2, \delta(X)^2, \dots, \delta^m(X)^2 \rangle \subset \cdots,$$

where $\langle f_1, \dots, f_n \rangle$ is the δ-ideal generated by $f_1, \dots, f_n$. But radical δ-ideals are well behaved (for a proof see [Kaplansky 1957] or [Marker et al. 1996]).

THEOREM 2.1 (RITT–RAUDENBUSH BASIS THEOREM). (i) *There are no infinite ascending chains of radical differential ideals in $K\{X_1, \dots, X_n\}$. For any radical differential ideal I there are $f_1, \dots, f_m$ such that $I = \sqrt{\langle f_1, \dots, f_m \rangle}$.*
(ii) *If $I \subset K\{X_1, \dots, X_n\}$ is a radical δ-ideal, there are distinct prime δ-ideals $P_1, \dots, P_m$ such that $I = P_1 \cap \cdots \cap P_m$ and $P_1, \dots, P_m$ are unique up to permutation.*

Thus the δ-topology is Noetherian and any δ-closed set is a finite union of irreducible δ-closed sets.

THEOREM 2.2 (SEIDENBERG'S DIFFERENTIAL NULLSTELLENSATZ). *Let K be a differentially closed field. $I \mapsto V_\delta(I)$ is a one to one correspondence between radical δ-ideals and δ-closed sets.*

PROOF. It is easy to see that $I_\delta(Y)$ is a radical δ-ideal for all $Y \subseteq K^n$. Suppose I and J are radical δ-ideals and $g \in J \setminus I$. By Theorem 2.1 there is a prime δ-ideal $P \supseteq I$ with $g \notin P$. It suffices to show there is $x \in V_\delta(P)$ with $g(x) \neq 0$. Let $P = \sqrt{\langle f_1, \dots, f_m \rangle}$, and let L be a differentially closed field containing $K\{X_1, \dots, X_n\}/P$. Let $x = (X_1/P, \dots, X_n/P)$. Clearly $f(x) = 0$ for $f \in P$ and $g(x) \neq 0$. In particular,

$$L \models \exists v_1 \dots \exists v_n \ f_1(v_1, \dots, v_n) = \dots = f_m(v_1, \dots, v_n) = 0 \wedge g(v_1, \dots, v_n) \neq 0.$$

By model completeness the same sentence is true in K. Thus there is $x \in K^n$ such that $x \in V_\delta(P) \setminus V_\delta(J)$. $\qquad \square$

By the basis theorem every δ-closed set is definable. We say that a subset of K^n is δ-constructible if it is a finite boolean combination of δ-closed sets. The δ-constructible sets are exactly those defined by quantifier free $\mathcal{L}$-formulas. Quantifier elimination implies that the δ-constructible sets are exactly the definable sets. Thus the projection of a δ-constructible set is δ-constructible.

3. ω-Stability and Dimension

Let K be a differentially closed fields and let F be a differential subfield of K. If $p \in S_n(F)$, let $I_\delta(p) = \{f \in F\{X_1, \dots, X_n\} : \text{``} f(x_1, \dots, x_n) = 0\text{''} \in p\}$. The arguments for types in algebraically closed fields in [Marker 2000] work here to show that $p \mapsto I_p$ is a bijection from $S_n(F)$ onto the space of prime δ-ideals.

COROLLARY 3.1. *DCF is ω-stable.*

PROOF. Let K and F be as above. We must show that $|S_n(F)| = |F|$. But for all p, we can find $f_1, \dots, f_m$ such that $I_\delta(p) = \sqrt{\langle f_1, \dots, f_m \rangle}$. Thus the number of complete n-types is equal to $|F\{X_1, \dots, X_n\}| = |F|$. $\qquad \square$

There is an important algebraic application of ω-stability. If F is a differential field, we say that a differentially closed $K \supseteq F$ is a *differential closure* of F if for any differentially closed $L \supseteq F$ there is a differential embedding of K into L fixing F.

This is related to a general model-theoretic notion mentioned in [Hart 2000]. A *prime model* of T over A is a model $\mathcal{M} \models T$ with $A \subseteq M$, such that if $\mathcal{N} \models T$ and $A \subset N$, then there is an elementary embedding $j : \mathcal{M} \to \mathcal{N}$ such that $j|A$ is the identity. For DCF, prime model extensions are exactly differential closures (recall that, by model completeness, all embeddings are elementary).

THEOREM 3.2. *Let T be an ω-stable theory, $\mathcal{M} \models T$ and $A \subseteq M$. There is a prime model of T over A. If $\mathcal{N}_0$ and $\mathcal{N}_1$ are prime models of T over A, then $\mathcal{N}_0$ and $\mathcal{N}_1$ are isomorphic over A.*

The existence of prime models was proved by Morley and uniqueness (under less restrictive assumptions) is due to Shelah. The following corollary was later given a slightly more algebraic proof by Kolchin.

COROLLARY 3.3. *Every differential field F has a differential closure and any two differential closures of F are unique up to isomorphism over F.*

Differential closures need not be minimal. Let F be the differential closure of $\mathbb{Q}$. Independent results of Kolchin, Rosenlicht and Shelah show there is a nontrivial differential embedding $j : F \to F$ with $j(F)$ a proper subfield of F.

Since DCF is ω-stable, we can assign Morley rank to types and definable sets. This gives us a potentially useful notion of dimension. It is interesting to see how this corresponds to more algebraic notions of dimension.

There are two natural cardinal dimensions. Suppose $V \subseteq K^n$ is an irreducible δ-closed set. Let $K[V]$ be the *differential coordinate ring* $K\{X_1, \ldots, X_n\}/I_\delta(V)$. Let td$(V)$ be the transcendence degree of $K[V]$ over K. Often td(V) is infinite. We say that elements of a differential ring are *differentially dependent* over K if they satisfy a differential polynomial over K. Let td$_\delta(V)$ be the size of a maximal differentially independent subset of $K[V]$ over V. Note that td(V) is finite if and only if td$_\delta(V) = 0$. If V is an algebraic variety of dimension d, then td$_\delta(V) = d$. Suppose W and V are proper irreducible δ-closed subsets of K with $W \subset V$. Then td$_\delta(V) =$ td$_\delta(W) = 0$ and td$(W) <$ td(V).

There is a natural ordinal dimension that arises from the Noetherian topology. This is the analog of Krull dimension in Noetherian rings. If V is a non-empty irreducible δ-closed set, we define dim$_\delta(V)$ as follows. If V is a point, then dim$_\delta(V) = 0$. Otherwise

$$\dim_\delta(V) = \sup\{\dim_\delta(W) + 1 : W \subset V \text{ is irreducible, } \delta\text{-closed and nonempty}\}.$$

Since $V_\delta(X) \subset V_\delta(\delta(X)) \subset \cdots \subset V_\delta(\delta^n(X)) \subset \cdots K$, we have dim$_\delta(K) \geq \omega$. The remarks above imply that if $V \subset K$ is δ-closed, then dim$_\delta(V) \leq$ td(V). Hence dim$_\delta(K) = \omega$.

There are two model-theoretic notions of dimension, Morley rank and U-rank. We refer to [Hart 2000] for the definition of Morley rank but will describe U-rank in this context. If A, B and C are subsets of a differentially closed field, we say that B and C are *independent* over A if the differential field generated by $B \cup A$ and the differential field generated by $C \cup A$ are linearly disjoint over the algebraic closure of the differential field generated by A. If $B \supset A$, $p \in S_n(A)$, $q \in S_n(B)$ and $p \subset q$, we say that q is a *forking* extension of p, if for any a realizing q, a and B are dependent over A.

We define the *U-rank* of a type $p \in S_n(A)$ inductively as follows:

(i) U$(p) \geq 0$.
(ii) If α is a limit ordinal, then U$(p) \geq \alpha$ if and only if U$(p) \geq \beta$ for all $\beta < \alpha$.

(iii) $U(p) \geq \alpha + 1$ if and only if there is $B \supset A$ and $q \in S_n(B)$, q a forking extension of p and $U(q) \geq \alpha$.

We write $U(b/A)$ for $U(\mathrm{tp}(b/A))$. If X is definable over A, we let $U(X)$ be the maximum $U(b/A)$ for $b \in X$.

In algebraically closed fields we also have four notions of dimension (transcendence degree, Krull dimension, Morley rank and U-rank), all of which agree. In DCF the situation is different.

THEOREM 3.4. *We have*

$$U(V) \leq \mathrm{RM}(V) \leq \dim_\delta(V) \leq \omega \cdot \mathrm{td}_\delta(V)$$

and if $\mathrm{td}_\delta(V) = 0$, *then* $\dim_\delta(V) \leq \mathrm{td}(V)$.

Theorem 3.4 is a combination of results of Poizat, Johnson and Pong (see [Pong ≥ 2000] for details). There are examples due to Kolchin, Poizat, and Hrushovski and Scanlon showing that any of these inequalities may be strict. Although these notions may disagree, it is easy to see that U-rank is finite if and only if transcendence degree is finite. Thus finite dimensionality does not depend on which notion of dimension we choose. It is also easy to see that $U(K^n) = \omega n$ so the notions of dimension agree on K^n (and on all algebraic varieties).

The following result of Pong [≥ 2000] shows the usefulness of U-rank. It is part of his proof that any finite rank δ-closed set is affine.

THEOREM 3.5. *Suppose* $V \subset \mathbb{P}^n$ *is* δ-*closed and* $U(V) < \omega$. *If* $H \subset \mathbb{P}^n$ *is a generic hyperplane, then* $H \cap V = \varnothing$.

PROOF. Let $\mathcal{H}$ be the set of all hyperplanes in $\mathbb{P}^n$. Since $\mathcal{H}$ is isomorphic to $\mathbb{P}^n$, $U(\mathcal{H}) = \omega n$. Similarly for any point x, the set of hyperplanes through x has U-rank $\omega(n-1)$ over x. Let $I = \{(v, H) : H \in \mathcal{H}, v \in V \cap H\}$. Suppose $(v, H) \in I$ and $U((v, H))$ is maximal. The Lascar inequality (valid in any superstable theory) asserts that

$$U(v, H) \leq U(H/v) \oplus U(v),$$

where $\oplus$ is the symmetric sum of ordinals. In this case H is in the set of hyperplanes through v, so $U(H/v) \leq \omega(n-1)$. Since $U(V)$ is finite, $U(v) < \omega$. Thus $U(v, H) < \omega n$. But if H were a generic hyperplane $U(v, H) \geq U(H) = \omega n$, a contradiction. $\square$

We conclude this section by summarizing some important results about interpretability in DCF.

THEOREM 3.6 (POIZAT). *DCF has elimination of imaginaries. In particular the quotient of a* δ-*constructible set by a* δ-*constructible equivalence relation is* δ-*constructible.*

THEOREM 3.7. (i) (PILLAY) *Any group interpretable in a differentially closed field K is definably isomorphic to the K-rational points of a differential algebraic group defined over K.*

(ii) (SOKOLOVIC) *Any infinite field of finite rank interpretable in a differentially closed field is definably isomorphic to the field of constants.*

(iii) (PILLAY) *Any field of infinite rank interpretable in $K \models$ DCF is definably isomorphic to K.*

4. Strongly Minimal Sets in Differentially Closed Fields

Let K be a $\aleph_0$-saturated differentially closed field. Let $X \subset K^n$ be definable. By adding parameters to the language we assume that X is defined over $\varnothing$. Recall that X is *strongly minimal* if whenever $Y \subseteq X$ is definable then either Y or $X \setminus Y$ is finite. For $A \subseteq K$ let $\mathrm{acl}^\delta(A)$ be the algebraic closure of the differential field generated by A. In DCF this is exactly the model-theoretic notion of algebraic closure. If X is strongly minimal let $\mathrm{acl}^\delta_X(A) = \mathrm{acl}^\delta(A) \cap X$. For $A \subseteq X$, let $\dim(A)$ be the maximum cardinality of an acl^δ-independent subset of $\mathrm{acl}^\delta(X)$.

We say that a strongly minimal set X is *trivial* if

$$\mathrm{acl}^\delta_X(A) = \bigcup_{a \in A} \mathrm{acl}^\delta_X(\{a\})$$

for all $A \subseteq X$. Examples of trivial strongly minimal sets are a set with no structure or the natural numbers with the successor function.

We say that a strongly minimal set X is *locally modular* if

$$\dim(A \cup B) = \dim(A) + \dim(B) - \dim(A \cap B)$$

whenever A and B are finite dimensional acl^δ_X-closed subsets of X and $A \cap B \neq \varnothing$. Vector spaces are good examples of locally modular strongly minimal sets. Indeed general results of Hrushovski show that any nontrivial locally modular strongly minimal set is essentially a vector space.

To fully understand the model theory of any ω-stable theory it is essential to understand the strongly minimal sets. In differentially closed fields this is particularly fascinating because there are trivial, nontrivial locally modular and non-locally modular strongly minimal sets.

Trivial strongly minimal sets first arose in the proofs that differential closures need not be minimal. For example the solution sets to the differential equations $\delta(X) = \dfrac{X}{X+1}$ and $\delta(X) = X^3 - X^2$ are (after throwing out 0 and, in the second case, 1) sets of total indiscernibles (that is, sets with no additional structure).

There is one obvious non-locally modular strongly minimal set, C the field of constants. Hrushovski and Sokolovic proved that this is essentially the only one.

THEOREM 4.1. *If $X \subset K^n$ is strongly minimal and non-locally modular, then X is non-orthogonal to the constants.*

The proof proceeds by first showing that strongly minimal sets in differentially closed field are Zariski geometries, in the sense of [Hrushovski and Zilber 1996]. The main theorem on Zariski geometries says that non-locally modular Zariski geometries are non-orthogonal to definable fields, but the only finite rank definable field is, by Theorem 3.7, the constants.

Hrushovski and Sokolovic also showed that nontrivial locally modular strongly minimal sets arise naturally in studying abelian varieties as differential algebraic groups. In his proof of the Mordell conjecture for function fields, Manin proved the following result.

THEOREM 4.2. *If A is an abelian variety, then there is a nontrivial differential algebraic group homomorphism $\mu : A \to K^n$.*

For example, if A is the elliptic curve $y^2 = x(x-1)(x-\lambda)$, where $\delta(\lambda) = 0$, then

$$\mu(x, y) = \frac{\delta(x)}{y}.$$

If $\delta(\lambda) \neq 0$, then μ is a second order differential operator. Let $A^{\#}$ be the δ-closure of the torsion points of A. We can choose μ so that $A^{\#}$ is the kernel of μ. Building on 4.2 and further work of Buium, Hrushovski and Sokolovic showed:

THEOREM 4.3. *Suppose A is a simple abelian variety defined over K. Either*

(i) *A is isomorphic to an abelian variety B defined over the constants, or*
(ii) *$A^{\#}$ is locally modular and strongly minimal.*

Moreover, any nontrivial locally modular strongly minimal set is non-orthogonal to $A^{\#}$ for some such A, and $A^{\#}$ and $B^{\#}$ are non-orthogonal if and only if A and B are isogenous.

Thus Hrushovski and Sokolovic have completely characterized the nontrivial strongly minimal sets. Understanding the trivial ones is still a difficult open problem. We say that a strongly minimal set X is $\aleph_0$-categorical if in any model the dimension of the elements of the model in X is infinite. One open question is: In DCF is every trivial strongly minimal set $\aleph_0$-categorical? Hrushovski has proved this for strongly minimal sets of transcendence degree one.

5. Diophantine Applications

Hrushovski used Theorem 4.3 in his proof of the Mordell–Lang conjecture for function fields in characteristic zero.

THEOREM 5.1. *Suppose $K \supset k$ are algebraically closed fields of characteristic zero, A is an abelian variety defined over K such that no infinite subabelian* *variety of A is isomorphic to an abelian variety defined over k, Γ is a finite rank subgroup of A, and X is a subvariety of A such that $X \cap \Gamma$ is Zariski dense in X. Then X is a finite union of cosets of abelian subvarieties of A.*

I will sketch the ideas of the proof; for full details see [Hrushovski 1996], [Bouscaren 1998] or [Pillay 1997]. First I give the full proof in one easy case.

THEOREM 5.2. *Suppose $K \supset k$ are algebraically closed fields of characteristic zero, A is a simple abelian variety defined over K that is not isomorphic to an abelian variety defined over k, Γ is the torsion points of A, and X is a proper subvariety of A. Then $X \cap \Gamma$ is finite.*

PROOF. The main idea is to move to a differential field setting where we may apply model-theoretic tools. In doing so we will replace the group Γ by $A^\#$, a small (that is, finite-dimensional) differential algebraic group.

The first step is to define a derivation $\delta : K \to K$ such that $k = \{x \in K : \delta(x) = 0\}$. One can show that if $\hat{K}$ is the differential closure of K then the field of constants of $\hat{K}$ is still k. Thus without loss of generality we may assume that K is a differentially closed field and k is the constant field of K.

Let $A^\#$ be the δ-closure of the torsion points of A. It suffices to show that $A^\# \cap X$ is finite. Suppose not. Since $A^\#$ is strongly minimal, $A^\# \setminus X$ is finite. But then the Zariski closure of $A^\#$ is contained in $X \cup A^\# \setminus X$, which is properly contained in A. This is a contradiction since the torsion points are Zariski dense. $\square$

The proof above does not explicitly use the fact that $A^\#$ is locally modular (though this does come into the proof that it is strongly minimal). Local modularity plays more of a role if we consider larger groups Γ. Suppose K, k, A and X are as above and Γ is a finite rank subgroup of A. Let $\mu : A \to K^n$ be a definable homomorphism with kernel $A^\#$. Since Γ has finite rank, the image of Γ under μ is contained in $V \subset K^n$ which is a finite dimensional k-vector space. Consider $G = \mu^{-1}(V)$. This is a definable finite Morley rank subgroup of A. Some analysis of this group allows us to conclude that it is 1-based (see [Hart 2000] for a definition: basically this means that all of the strongly minimal sets are locally modular). Hrushovski and Pillay showed that in an 1-based group G any definable subset of G^n is a boolean combination of cosets of definable subgroups. In particular $X \cap G$ will be a finite union of cosets of definable subgroups of G. If any of these subgroups is infinite, then its Zariski closure is an algebraic subgroup and must hence be the whole group. This would contradict the fact that the Zariski closure would be contained in X.

To prove Theorem 5.1 in general, we use the fact that every abelian variety is isogenous to a direct sum of simple abelian subvarieties together with a number of techniques from finite Morley rank group theory.

References

[Bouscaren 1998] E. Bouscaren, "Proof of the Mordell–Lang conjecture for function fields", pp. 177–196 in *Model theory and algebraic geometry*, edited by E. Bouscaren, Lecture Notes in Math. **1696**, Springer, Berlin, 1998.

[Buium 1994] A. Buium, *Differential algebra and Diophantine geometry*, Actualités math., Hermann, Paris, 1994.

[Hart 2000] B. Hart, "Stability theory and its variants", pp. **??**–**??** in *Model theory, algebra and geometry*, edited by D. Haskell et al., Math. Sci. Res. Inst. Publ. **39**, Cambridge Univ. Press, New York, 2000.

[Hrushovski 1996] E. Hrushovski, "The Mordell–Lang conjecture for function fields", *J. Amer. Math. Soc.* **9**:3 (1996), 667–690.

[Hrushovski and Sokolovic ≥ 2000] E. Hrushovski and Z. Sokolovic, "Minimal subsets of differentially closed fields". To appear in *Trans. Amer. Math. Soc.*

[Hrushovski and Zilber 1996] E. Hrushovski and B. Zilber, "Zariski geometries", *J. Amer. Math. Soc.* **9**:1 (1996), 1–56.

[Kaplansky 1957] I. Kaplansky, *An introduction to differential algebra*, Actualités Sci. Ind. **1251**, Hermann, Paris, 1957.

[Kolchin 1973] E. R. Kolchin, *Differential algebra and algebraic groups*, Pure and Applied Mathematics **54**, Academic Press, New York, 1973.

[Magid 1994] A. R. Magid, *Lectures on differential Galois theory*, Amer. Math. Soc., Providence, RI, 1994.

[Marker 2000] D. Marker, "Introduction to model theory", pp. 15–35 in *Model theory, algebra and geometry*, edited by D. Haskell et al., Math. Sci. Res. Inst. Publ. **39**, Cambridge Univ. Press, New York, 2000.

[Marker et al. 1996] D. Marker, M. Messmer, and A. Pillay, *Model theory of differentiable fields*, Lecture Notes in Logic **5**, Springer, Berlin, 1996.

[Pierce and Pillay 1998] D. Pierce and A. Pillay, "A note on the axioms for differentially closed fields of characteristic zero", *J. Algebra* **204**:1 (1998), 108–115.

[Pillay 1996] A. Pillay, *Differential algebraic groups and the number of countable differentially closed fields*, Lecture Notes in Logic **5**, Springer, Berlin, 1996.

[Pillay 1997] A. Pillay, "Model theory and Diophantine geometry", *Bull. Amer. Math. Soc.* (*N.S.*) **34**:4 (1997), 405–422. Erratum in **35**:1 (1998), 67.

[Pong ≥ 2000] W. Y. Pong, "Some applications of ordinal dimensions to the theory of differentially closed fields". To appear in *J. Symb. Logic*.

DAVID MARKER
UNIVERSITY OF ILLINOIS AT CHICAGO
DEPARTMENT OF MATHEMATICS, STATISTICS, AND COMPUTER SCIENCE (M/C 249)
851 S. MORGAN ST.
CHICAGO, IL 60613
UNITED STATES
marker@math.uic.edu

Model Theory, Algebra, and Geometry
MSRI Publications
Volume **39**, 2000

A Survey on the Model Theory of Difference Fields

ZOÉ CHATZIDAKIS

ABSTRACT. We survey the model theory of difference fields, that is, fields
with a distinguished automorphism σ. After introducing the theory ACFA
and stating elementary results, we discuss independence and the various
concepts of rank, the dichotomy theorems, and, as an application, the
Manin–Mumford conjecture over a number field. We conclude with some
other applications.

Difference field are fields with a distinguished automorphism σ. They were
first studied by Ritt in the 1930s. A good reference for the algebraic results is
[Cohn 1965]. Interest in the model theory of difference fields started at the end of
the eighties, particularly during the MSRI logic year, because of two questions.

The first question stemmed from the failure of Zil'ber's conjecture: there is
a strongly minimal theory extending the theory of algebraically closed fields of
any given characteristic. People were looking at the possibility of finding a non-
definable automorphism σ of $\mathbb{F}_p^{\mathrm{alg}}$ (the algebraic closure of the field $\mathbb{F}_p$ with p
elements), such that $\mathrm{Th}(\mathbb{F}_p^{\mathrm{alg}}, +, \cdot, \sigma)$ is strongly minimal. This question so far
remains open.

The second problem had to do with the difference fields $\mathcal{F}_q = (\mathbb{F}_p^{\mathrm{alg}}, +, \cdot, \phi_q)$,
where q is a power of p and $\phi_q : x \mapsto x^q$ is a power of the Frobenius automor-
phism $x \mapsto x^p$. The hope was to generalise the work of Ax on finite fields to
these structures, and in particular to describe the theory of the non-principal
ultraproducts of the difference fields $\mathcal{F}_q$.

These questions led Macintyre, van den Dries and Wood to look for a model
companion of the theory of difference fields, and to prove various results (decidability, description of the completions, etc ...) for this theory, henceforth called
ACFA. For details and attribution of results, see [Macintyre 1997]. I should also
mention that the second problem was solved recently, by Hrushovski [1996b] and

---

Notes based on lectures given at MSRI, January 98.

Macintyre [≥ 2000], showing that non-principal ultraproducts of $\mathcal{F}_q$'s are models of ACFA.

In 1994, Hrushovski and I started looking at stability-type properties of the theory ACFA. Our main result is a dichotomy result for types of rank 1 for models of characteristic 0, which was later partially extended to the case of positive characteristic with the help of Peterzil [Chatzidakis and Hrushovski 1999; Chatzidakis et al. 1999]. It has some applications to the description of types of finite rank, and to groups definable in models of ACFA. These results were used by Hrushovski [1995] to find explicit bounds in the Manin–Mumford conjecture.

The first four sections of the paper give a survey of the results obtained to-date for difference fields. In Section 5 we state the results used by Hrushovski in his proof of the Manin–Mumford conjecture over a number field, and show how he effectively derives from them the bounds. In the last section we conclude with the statements of some other applications due to Hrushovski and Scanlon.

Acknowledgements. I take this opportunity to thank MSRI for their support during the special semester on the Model Theory of Fields, and to show my appreciation for the congenial atmosphere. I would also like to thank D. Haskell and E. Hrushovski for many improvements to this paper.

1. Description and Elementary Results on the Theory ACFA

We work in the language $\mathcal{L} = \{+, -, \cdot, 0, 1, \sigma\}$, where $+, -, \cdot$ are the usual ring operations, 0 and 1 are constants, and σ is a unary function.

1.1. Some examples

(1) The shift operator. Consider the field $K = \mathbb{C}(t)$, and define σ by

$$\sigma|_{\mathbb{C}} = \mathrm{id}, \qquad \sigma(t) = t + 1.$$

The name "difference field" originated from this example: an equation of the form $P(f(t), f(t+1), \ldots, f(t+n)) = 0$, where f is the unknown function to be found and P is a polynomial over K, is called an algebraic difference equation. One can replace K by other fields, e.g., the field of meromorphic functions on $\mathbb{C}$ or on $\mathbb{R}$.

(2) Let K be a field, K^s its separable closure and $\sigma \in \mathrm{Gal}(K^s/K)$. Then (K^s, σ) is a difference field. Note that because the algebraic closure K^{alg} of K is purely inseparable over K^s, σ extends uniquely to an automorphism of K^{alg}. One often identifies $\mathrm{Gal}(K^s/K)$ and $\mathrm{Aut}(K^{\mathrm{alg}}/K)$.

The structures $\mathcal{F}_q$ described above are a particular example. More generally, we have:

(3) Let K be a perfect field of characteristic $p > 0$, and q a power of p. Then (K, ϕ_q) is a difference field. If the field K is algebraically closed then

$(\mathbb{F}_p^{\text{alg}}, \phi_q) \prec (K, \phi_q)$. This is because for fixed q the map $x \mapsto x^q$ is definable in the language of fields, and because the theory of algebraically closed fields is model complete.

1.2. Definitions, notation and some basic algebraic results.

In the literature, a difference field is a field K with a distinguished monomorphism σ. If σ is onto, then (K, σ) is called an *inversive* difference field. However, a simple inductive limit argument shows that every difference field has a unique (up to isomorphism) inversive closure. We will assume in what follows, that *all our difference fields are inversive*. The references are to [Cohn 1965].

Let K be a difference field, and let $\bar{X} = (X_1, \dots, X_n)$ be indeterminates. A difference polynomial over K in $X_1, \dots, X_n$ is an ordinary polynomial with coefficients in K, in the variables $X_1, \dots, X_n, \sigma(X_1), \dots, \sigma^i(X_j), \dots$. The ring of those difference polynomials is denoted $K[X_1, \dots, X_n]_\sigma$, and σ extends naturally to $K[X_1, \dots, X_n]_\sigma$, in the way suggested by the names of the variables.

NOTE. As defined, σ is not onto. It is sometimes convenient to consider the inversive closure of this ring, namely $K[\sigma^i(X_1), \dots, \sigma^i(X_n)]_{i \in \mathbb{Z}}$, but we will not do this here.

There is a natural notion of σ-*ideal*, i.e., an ideal closed under σ, and of *reflexive* σ-ideal ($a \in I \iff \sigma(a) \in I$). The analog of a radical ideal is called a perfect σ-ideal: a σ-ideal I is *perfect* if $a \in I$ whenever $a^j \sigma^i(a) \in I$ for some $i, j \in \mathbb{N}$. A *prime* σ-*ideal* is a reflexive σ-ideal which is prime. Note that a prime σ-ideal is perfect. $K[X_1, \dots, X_n]_\sigma$ does not satisfy the ascending chain condition on σ-ideals; however it satisfies it for perfect σ-ideals, and therefore for prime σ-ideals. This allows one to define σ-closed sets and σ-varieties (also called irreducible σ-closed sets) in affine n-spaces. They correspond dually to perfect σ-ideals and prime σ-ideals, and are the basic closed sets of a noetherian topology.

Let K be a difference field, a a tuple of elements (in some difference field extending K). We denote by $K(a)_\sigma$ the difference field generated by a over K, by $\text{acl}_\sigma(Ka)$ its algebraic closure, and by $\deg_\sigma(a/K)$ the transcendence degree of $K(a)_\sigma$ over K. If a is a single element and $\deg_\sigma(a/K)$ is infinite, then a is called *transformally transcendental*. The elements $\sigma^j(a)$, $j \in \mathbb{Z}$, are then algebraically independent over K. If $\deg_\sigma(a/K)$ is finite, then a is called *transformally algebraic*. There are natural notions of transformal transcendence basis and transformal dimension.

1.3. An axiomatisation of the theory ACFA.

Consider the theory ACFA, whose models are the $\mathcal{L}$-structures K satisfying these conditions:

(i) K is an algebraically closed field.

(ii) $\sigma \in \text{Aut}(K)$.

(iii) If U and V are (affine) varieties defined over K, with $V \subseteq U \times \sigma(U)$ projecting generically onto U and $\sigma(U)$, then there is a tuple a in K such that $(a, \sigma(a)) \in V$.

Here, by a variety, we mean an absolutely irreducible Zariski closed set, i.e., a set defined by polynomial equations, and which is not the proper union of two smaller Zariski closed sets. The set $\sigma(U)$ is the variety obtained from U by applying σ to the coefficients of the defining polynomials of U. When we say that V projects generically onto U, we mean that the image of V under the natural projection $U \times \sigma(U) \to U$ is Zariski dense in U (i.e., not contained in any proper Zariski closed subset). Note that (iii) is indeed a conjunction of first-order sentences, since (by classical results on polynomial rings over fields) the fact that polynomials $f_1(\bar{X}), \ldots, f_n(\bar{X})$ generate a prime ideal of $K[\bar{X}]$ is an elementary condition on the coefficients of $f_1, \ldots, f_n$. Similarly for the inclusion of ideals in $K[\bar{X}]$.

THEOREM. ACFA *is the model companion of the theory of difference fields.*

SKETCH OF PROOF. We first need to show that every difference field embeds in a model of ACFA. Axioms (i) and (ii) pose no problem, as every automorphism of a field extends to its algebraic closure. Let U and V be as in (iii). Choose a generic point (a, b) of V over K (i.e., the ideal of polynomials over K vanishing at (a, b) is exactly the ideal of polynomials vanishing at all points of V), in some field containing K. Then a is a generic of U, and b is a generic of $\sigma(U)$. By elementary properties of algebraically closed fields, the isomorphism $\tau : K(a) \to K(b)$ that extends σ and sends a to b extends to an automorphism of the algebraic closure of $K(a, b)$.

This shows that every difference field embeds in a model of ACFA. It remains to show that the models of ACFA are existentially closed. Let $(K, \sigma) \models$ ACFA, let $\varphi(x)$, x a tuple of variables, be a quantifier-free formula with parameters in K, and assume that $\varphi(x)$ has a solution in some difference field (L, σ) extending K. The usual trick of replacing the inequality $y \neq 0$ by $\exists z \ yz - 1 = 0$, shows that one can assume that $\varphi(x)$ is a conjunction of σ-equations. Let $a \in L$ satisfy φ. For n large enough, the σ-ideal I generated by the set

$$\{f(X, \sigma(X), \ldots, \sigma^n(X)) \mid$$
$$f(Y, Y_1, \ldots, Y_n) \in K[Y, Y_1, \ldots, Y_n], \ f(a, \sigma(a), \ldots, \sigma^n(a)) = 0\}$$

is precisely the prime σ-ideal of difference polynomials over K annulled by a. Thus any point satisfying these equations will satisfy $\varphi(x)$.

Let U be the variety defined over K with generic $(a, \sigma(a), \ldots, \sigma^{n-1}(a))$, and V the variety defined over K with generic $(a, \sigma(a), \ldots, \sigma^{n-1}(a), \sigma(a), \ldots, \sigma^n(a))$. Then U and V satisfy the hypotheses of axiom (iii), and therefore there is a tuple b in K such that $(b, \sigma(b)) \in V$. Then $b = (c, \sigma(c), \ldots, \sigma^{n-1}(c))$ for some c, and $K \models \varphi(c)$. □

1.4. The Frobenius automorphisms. Before continuing with the elementary properties of ACFA, we will state precisely the result of Hrushovski, from which follows that non-principal ultraproducts of $\mathcal{F}_q$'s are models of ACFA. It is then a consequence of Tchebotarev's theorem on the distribution of primes that ACFA is exactly the theory of all non-principal ultraproducts of $\mathcal{F}_q$'s, see [Macintyre 1997].

THEOREM [Hrushovski 1996b]. *Let U, V be varieties with $V \subseteq U \times \sigma(U)$, and assume that the projections are onto and have finite fibers. Let $d_1 = [K(V) : K(U)]$, $d_2 = [K(V) : K(\sigma(U))]_i$ (purely inseparable degree); let $c = d_1/d_2$ and $d = \dim(V)$. Then for some constant $C > 0$, depending on the two varieties U and V, and which remains bounded when U and V move inside an algebraic family of varieties,*

$$\left|\mathrm{Card}(\{a \in (\mathbb{F}_p^{\mathrm{alg}})^n \mid (a, a^q) \in V\}) - cq^d\right| \leq Cq^{d-1/2}.$$

1.5. PROPOSITION. *If $(K, \sigma) \models$ ACFA, then the subfield $\mathrm{Fix}(\sigma)$ of K fixed by σ is a pseudo-finite field.*

PROOF. By [Ax 1968], one needs to show that: $\mathrm{Fix}(\sigma)$ is perfect; $\mathrm{Fix}(\sigma)$ has exactly one algebraic extension of each degree; every (absolutely irreducible) variety defined over $\mathrm{Fix}(\sigma)$ has an $\mathrm{Fix}(\sigma)$-rational point.

The first assertion is obvious, and the third one follows easily from axiom (iii). For the second assertion, it suffices to show that for each $n > 1$, the system

$$\sigma^n(x) = x, \qquad \sigma^j(x) \neq x \quad \text{for } j = 1, \ldots, n-1,$$

has a solution in K. Since K is existentially closed, it suffices to find a difference field extending K in which this system has a solution. Consider the field $K(X_1, \ldots, X_n)$ in n indeterminates, and extend σ by defining $\sigma(X_j) = X_{j+1}$ for $j < n$ and $\sigma(X_n) = X_1$. Then X_1 is a solution of the system. $\square$

In characteristic $p > 0$ one shows similarly that if $m \neq 0$ and n are integers, then the set of elements of K satisfying $\sigma^m(x) = x^{p^n}$ is a pseudo-finite field.

1.6. It turns out that many of the proofs given in [Ax 1968] for pseudo-finite fields generalise to models of ACFA. Parts (1)–(5) of the following result appear in [Macintyre 1997].

PROPOSITION. (1) *Let (K_1, σ_1) and (K_2, σ_2) be models of ACFA, and let E be a common difference subfield. Then*

$$(K_1, \sigma_1) \equiv_E (K_2, \sigma_2) \iff (E^{\mathrm{alg}}, \sigma_1|_{E^{\mathrm{alg}}}) \simeq_E (E^{\mathrm{alg}}, \sigma_2|_{E^{\mathrm{alg}}}).$$

(2) *From this one deduces immediately that the completions of ACFA are obtained by describing the action of σ on the algebraic closure of the prime field ($\mathbb{Q}^{\mathrm{alg}}$ or $\mathbb{F}_p^{\mathrm{alg}}$). This then entails the decidability of the theory ACFA, as well as of its extensions ACFA$_0$ and ACFA$_p$ obtained by specifying the characteristic of the field.*

(3) *It also gives a description of the types. Let E be a difference field, a and b*
   two tuples from a model K of ACFA *containing E. Then* $\mathrm{tp}(a/E) = \mathrm{tp}(b/E)$
   if and only if there is an isomorphism φ from the difference field $\mathrm{acl}_\sigma(Ea) =_{\mathrm{def}}$
   $E(a)_\sigma^{\mathrm{alg}}$ *onto the difference field* $\mathrm{acl}_\sigma(Eb)$ *which is the identity on E and sends*
   a to b.

(4) *If E is an algebraically closed difference field, then*

$$\mathrm{ACFA} \cup \mathrm{qftp}(E) \vdash \mathrm{tp}(E),$$

   where $\mathrm{qftp}(E)$ *denotes the quantifier-free type of E.*

(5) *The algebraic closure (in the model-theoretic sense) of a set A coincides*
   with the algebraic closure (in the ordinary field sense) of the difference field
   generated by A (which we denote by $\mathrm{acl}_\sigma(A)$*).*

(6) *Let $K \models$* ACFA, *let U be a variety, $l \geq 1$, and V a subvariety of $U \times \sigma(U) \times$*
   $\cdots \times \sigma^l(U)$. *Let* $\pi_1 : U \times \sigma(U) \times \cdots \times \sigma^l(U) \to U \times \sigma(U) \times \cdots \times \sigma^{l-1}(U)$
   and $\pi_2 : U \times \sigma(U) \times \cdots \times \sigma^l(U) \to \sigma(U) \times \cdots \times \sigma^l(U)$ *be the two canonical*
   projections, and assume that $\sigma\pi_1(V)$ and $\pi_2(V)$ have the same generics. Then
   the set of points $\tilde{V} = \{x \in U(K) \mid (x, \sigma(x), \ldots, \sigma^l(x)) \in V\}$ *is Zariski dense*
   in U.

(7) *If $(K, \sigma) \models$* ACFA *and $m \geq 1$, then $(K, \sigma^m) \models$* ACFA.

PROOF. (1) The left to right implication is almost immediate. For the other one,
moving K_2 by some E-isomorphism, we may assume that $E = E^{\mathrm{alg}}$ and that K_1
and K_2 are linearly disjoint over E. This implies that the ring $K_1 \otimes_E K_2$ is a
domain. Define $\sigma(a \otimes b) = \sigma_1(a) \otimes \sigma_2(b)$ for $a \in K_1$ and $b \in K_2$; then σ extends
to an automorphism of the quotient field L of $K_1 \otimes_E K_2$, which agrees with σ_1
on K_1 and σ_2 on K_2. Now, (L, σ) embeds in a model (M, σ) of ACFA, and by
model-completeness we have $(K_1, \sigma_1) \prec (M, \sigma)$ and $(K_2, \sigma_2) \prec (M, \sigma)$.

The first part of (2), (3) and (4) are immediate, applying compactness to (1).
The decidability follows from the recursive axiomatisation of ACFA, together
with the effective computability of Galois groups of the splitting fields over $\mathbb{Q}$
and $\mathbb{F}_p$ of polynomials of $\mathbb{Z}[T]$.

(5) Let $A = \mathrm{acl}_\sigma(A) \subseteq K \models$ ACFA and $b \in K \setminus A$, $B = \mathrm{acl}_\sigma(Ab)$; let B_1 be an
A-isomorphic copy of B, linearly disjoint over A. As in (1), there is a model of
ACFA containing the difference fields B and B_1. By (3), $\mathrm{tp}(B_1/A) = \mathrm{tp}(B/A)$,
which shows that $\mathrm{tp}(b/A)$ is not algebraic.

(6) We may assume that U and V are affine. Let

$$W \subseteq U \times \sigma(U) \times \cdots \times \sigma^{l-1}(U)$$

be the Zariski closure of $\pi_1(V)$. By assumption, $\pi_2(V)$ is Zariski dense in $\sigma(W)$,
and we may therefore assume that $l = 1$. The proof that every difference field
embeds in a model of ACFA shows that if K is sufficiently saturated, then K
contains a point a such that $(a, \sigma(a))$ is a generic of V. This shows that $\tilde{V}$ is
dense in U. $\qquad\square$

2. Independence and Rank

2.1. Definition of independence. Let A, B and C be subsets of a model K of ACFA. We say that A and B are *independent over* C, and write $A \downarrow_C B$, if $\mathrm{acl}_\sigma(CA)$ and $\mathrm{acl}_\sigma(CB)$ are linearly disjoint over $\mathrm{acl}_\sigma(C)$. This notion has all the usual properties of independence in algebraically closed fields. Recall that by Proposition 1.6(5), $\mathrm{acl}_\sigma(A)$ is the model-theoretic algebraic closure of the set A in the model K.

2.2. Definition of the SU-rank. We define a rank based on independence in the usual way, that is, for p a type over E, realised by a tuple a:

- $\mathrm{SU}(p) = \mathrm{SU}(a/E) \geq 0$,
- $\mathrm{SU}(p) \geq \alpha$ for α a limit ordinal, if and only if $\mathrm{SU}(p) \geq \beta$ for every $\beta < \alpha$,
- $\mathrm{SU}(p) \geq \alpha+1$ if and only if there is $F \supseteq E$ such that $a \not\downarrow_E F$ and $\mathrm{SU}(a/F) \geq \alpha$.

Then $\mathrm{SU}(p)$ is the least ordinal α such that $\mathrm{SU}(p) \not\geq \alpha + 1$. If $\varphi(x)$ is a formula with parameters in $E = \mathrm{acl}_\sigma(E)$, one also defines $\mathrm{SU}(\varphi) = \max\{\mathrm{SU}(a/E) \mid a \text{ satisfies } \varphi\}$.

2.3. The SU-rank shares the properties of the usual U-rank, and in particular, the Lascar rank inequality: if a, b are tuples and E a set, then $\mathrm{SU}(a/Eb) + \mathrm{SU}(b/E) \leq \mathrm{SU}(a, b/E) \leq \mathrm{SU}(a/Eb) \oplus \mathrm{SU}(b/E)$, where $\oplus$ denotes the natural sum on ordinal numbers. (Recall that $1 + \omega = \omega$, while $1 \oplus \omega = \omega + 1$.)

2.4. Some examples. Let E be a difference subfield of a model K of ACFA and a a tuple in K. From the definition of the SU-rank, it is clear that:

- $\mathrm{SU}(a/E) = 0 \iff a \in \mathrm{acl}_\sigma(E)$.
- $\mathrm{SU}(a/E) = 1 \iff a \notin \mathrm{acl}_\sigma(E)$, and for every $F \supseteq E$, either $a \downarrow_E F$ or $a \in \mathrm{acl}_\sigma(F)$.

Earlier we defined $\deg_\sigma(a/E)$, which is also an invariant of $\mathrm{tp}(a/E)$. It has some relation with SU-rank, since independence is defined in terms of non-forking in algebraically closed fields. For instance, one has, for $E \subseteq F$ difference fields and a a tuple with $\deg_\sigma(a/E) < \infty$,

$$a \downarrow_E F \iff \deg_\sigma(a/F) < \deg_\sigma(a/E),$$

and this implies

$$\mathrm{SU}(a/E) \leq \deg_\sigma(a/E).$$

Thus in particular, every non-algebraic type containing the equation $\sigma(x) = x^2 + 1$ has SU-rank 1. This inequality can be strict; see the example in 2.6 below.

2.5. One can also show that the SU-rank of an element transformally transcendental over the difference field E is ω: let a be such an element, and consider the sequence (b_i), $i \in \mathbb{N}$, defined by $b_0 = a$, $b_{i+1} = \sigma(b_i) - b_i$. Then the fields $L_i = E(b_i)_\sigma$ form a decreasing sequence of subfields of $E(a)_\sigma$, with tr $\deg(L_i/L_{i+1}) = 1$. By additivity of rank, we obtain $\mathrm{SU}(a/L_i) = i$, which implies that $\mathrm{SU}(a/E) \geq \omega$. On the other hand, $\mathrm{SU}(a/E) \not\geq \omega + 1$: if $a \underset{E}{\not\smile} F$ then $\deg_\sigma(a/F) < \infty$, which implies that $\mathrm{SU}(a/F) < \omega$. Hence $\mathrm{SU}(a/E) = \omega$.

Note that this gives an example of the left-hand equality in 2.3: $\mathrm{SU}(a/Eb_1) = 1$, $\mathrm{SU}(b_1/E) = \omega$, and $\mathrm{SU}(a/E) = \omega$. For a tuple b in K this also yields: $\mathrm{SU}(b/E) < \omega \iff \deg_\sigma(b/E) < \infty$.

2.6. EXAMPLE. *Consider the formula* $\varphi(x) : \sigma^{-2}(x) = x^2 + 1$ (*in characteristic* $\neq 2$). *Then* $\mathrm{SU}(\varphi) = 1$.

PROOF. By 2.4, we want to show that if E is any difference field and a is any solution of $\sigma^{-2}(x) = x^2 + 1$, then either $a \in \mathrm{acl}_\sigma(E)$, or $a \underset{}{\smile} E$, i.e.: $\deg_\sigma(a/E) = 2$. Let $E = \mathrm{acl}_\sigma(E)$ be a field, and a a realisation of φ, $a \notin E$. We need to show that $\deg_\sigma(a/E) = 2$. Since E is an arbitrary algebraically closed difference field, this will imply: if $F = \mathrm{acl}_\sigma(F)$ contains E and $\mathrm{tp}(a/F)$ forks over E, then $a \in F$, and therefore that $\mathrm{SU}(\varphi) = 1$.

Suppose by way of contradiction that $\deg_\sigma(a/E) = 1$, and let $K = E(a, \sigma(a))$, $m = [K : E(a)]$ and $n = [K : E(\sigma(a))]$. Observe that K contains all $\sigma^{-j}(a)$ for $j \geq 0$ (because $\sigma^{-2j}(a) \in E(a)$). Since $E(\sigma^2(a))$ is a Galois extension of $E(a)$, we have that $[K(\sigma^2(a)) : K]$ divides $[E(\sigma^2(a)) : E(a)] = 2$.

Assume first that $[K(\sigma^2(a)) : K] = 1$. Then $\sigma(K) = K$, which implies that $K = E(a)_\sigma$. On the other hand, $E(a)_\sigma$ contains the infinite algebraic extension $E(\sigma^{2j}(a))_{j \in \mathbb{N}}$ of $E(a)$, which gives us a contradiction.

Thus $[K(\sigma^2(a)) : K] = 2$, and therefore $E(\sigma^2(a)) \cap K = E(a)$. So we have:

$$[E(\sigma(a), \sigma^2(a)) : E(a)] = [E(\sigma(a), \sigma^2(a)) : K][K : E(a)] = 2m$$
$$= [E(\sigma(a), \sigma^2(a)) : E(\sigma^2(a))][E(\sigma^2(a)) : E(a)] = 2n$$

since $[E(\sigma(a), \sigma^2(a)) : E(\sigma^2(a))] = [K : E(\sigma(a))]$. This implies $m = n$. On the other hand,

$$[E(\sigma(a), \sigma^2(a)) : E(\sigma(a))] = [K : E(a)] = m$$
$$= [E(\sigma(a), \sigma^2(a)) : K][K : E(\sigma(a))] = 2n$$

which gives $m = 2n$ and the desired contradiction.   □

2.7. The independence theorem. *Let* $E = \mathrm{acl}_\sigma(E) \subseteq K$, *let* a, b, c_1 *and* c_2 *be tuples from* K *such that* a, b, c_1 *and* c_2 *are independent over* E *and* $\mathrm{tp}(c_1/E) = \mathrm{tp}(c_2/E)$. *Then there is* c (*in some elementary extension of* K) *independent from* (a, b) *over* K, *and realising* $\mathrm{tp}(c_1/\mathrm{acl}_\sigma(Ea)) \cup \mathrm{tp}(c_2/\mathrm{acl}_\sigma(Eb))$.

A generalised version of this theorem holds: let $n \geq 3$, let $x_1, \ldots, x_n$ be tuples of variables, and let W be a set of proper subsets of $\{1, \ldots, n\}$ closed under intersection. Assume that for each $w \in W$ we are given a complete type $p_w(x_w)$ over $E = \mathrm{acl}_\sigma(E)$, in the variables $x_w = \{x_i \mid i \in w\}$, which can be realised by some $(a_i \mid i \in w)$ such that the elements $a_i, i \in w$, are independent over E (i.e., for each $j \in w$, the tuple a_j is independent from the set $\{a_i \mid i \in w, \ i \neq j\}$ over E). Assume moreover that if $v \subset w$ are in W then $p_v(x_v) \subset p_w(x_w)$. Then the type

$$\bigcup_{w \in W} p_w(x_w)$$

can be realised by some tuple $a_1, \ldots, a_n$, with $a_1, \ldots, a_n$ independent over E. The independence theorem corresponds to the case

$$n = 3, \quad W = \{\{1,2\}, \{1,3\}, \{2,3\}\}.$$

2.8. Independence and non-forking. Using the independence theorem, one proves that independence as defined above coincides with the usual notion of non-forking. Namely, assume that a and F are independent over $E = \mathrm{acl}_\sigma(E)$, and let $p(x) = \mathrm{tp}(a/F)$. Assume that $(F_i)_{i \in \mathbb{N}}$ is an E-indiscernible sequence of realisations of $\mathrm{tp}(F/E)$, and let $p_i(x)$ be the type over F_i which is the image of $p(x)$ by an E-automorphism mapping F to F_i. Then $\cup_i p_i(x)$ is consistent. Thus any completion of ACFA is simple in the sense of [Shelah 1980].

The connections between the independence theorem and simplicity were first observed by Hrushovski in the context of pseudofinite fields (and more generally bounded PAC fields); see [Hrushovski 1991; Hrushovski and Pillay 1994]. The case $n > 3$ of the generalised independence theorem goes beyond simplicity, and its model theoretic meaning remains to be clarified.

Recall that any PAC field which is not separably closed is unstable by a result of Duret [Duret 1980]. Hrushovski recognised the usefulness of the independence theorem for studying definable groups and generalising the techniques of stability theory to the context of pseudofinite fields, and more generally, to models of S_1-theories (an S_1-theory has finite SU-rank and some definability property of the SU-rank).

The independence theorem is indeed a good substitute for "uniqueness of non-forking extension" which is true in stable theories, and allows one to generalise the concepts of generic type of a group and of stabilisers of types to groups definable in finite fields, and later, to groups definable in models of ACFA.

The independence theorem was later generalised by B. Kim and A. Pillay [Kim and Pillay 1997] to Lascar types in simple theories. Moreover their result gives a nice characterisation of non-forking. The results on definable groups were also generalised to the context of simple theories; see [Pillay 1998; Wagner 1997].

2.9. The independence theorem is also used in the proof of these two statements:

PROPOSITION. *Let K be a model of* ACFA.

(1) $\mathrm{Th}(K)$ *has elimination of imaginaries.*
(2) *Let $S \subseteq \mathrm{Fix}(\sigma)^n$ be definable in K. Then S is definable in the pure field* $\mathrm{Fix}(\sigma)$ *(maybe with additional parameters from* $\mathrm{Fix}(\sigma)$).

2.10. Groups of finite SU-rank. Using the techniques developed in [Hrushovski and Pillay 1994; 1995], one obtains for instance a generalisation of a well-known result of algebraic geometry:

PROPOSITION. *Let G be a group of finite SU-rank defined over a model K of ACFA, and let $\{X(i) \mid i \in I\}$ be a family of definable subsets of G. There is a definable group H contained in the subgroup of G generated by the $X(i)$, $i \in I$, such that $X(i)H/H$ is finite for every $i \in I$.*

Note that no uniformity is assumed in the family $X(i)$, just that each of them is definable by some formula. The proof gives more information. Without loss of generality we will assume that for every $i \in I$ there is $j \in I$ such that $X(j) = X(i)^{-1}$.

(1) There are elements $i_1, \ldots, i_n \in I$ such that $H \subseteq X(i_1) \cdots X(i_n)$ and $n \leq 2\,\mathrm{SU}(G)$.

(2) Assume that G is a subgroup of an algebraic group defined over K, that G and the sets $X(i)$ are irreducible σ-closed sets, and that the identity element of the group belongs to all $X(i)$'s. Then H is the subgroup generated by all the $X(i)$'s, and the number n in (1) is $\leq \mathrm{SU}(G)$.

2.11. Finite simple groups. One can then use Hrushovski's result on the Frobenius automorphisms ϕ_q of Section 1.4 to get information about certain classes of finite simple groups. With the exception of the sporadic groups and the alternating groups, finite simple groups are defined in terms of algebraic groups, and form families (e.g., $\mathrm{PSL}_n(\mathbb{F}_q)$ for fixed n and q ranging over prime powers). All but the Suzuki and Ree families are already definable over finite fields in the language of fields $\{+, -, \cdot, 0, 1\}$. The Suzuki and Ree families become uniformly definable in the structures $\mathcal{F}_{p^{m+1}}$, for $p = 2$ or 3, as m varies over the positive integers. Indeed, these groups are defined as follows: we have some algebraic group G (in the family B_2, G_2 or F_4), and an algebraic automorphism φ of G whose square induces the Frobenius map ϕ_p on $G(\mathbb{F}_p^{\mathrm{alg}})$. Then the subgroup ${}^2G(p^{2m+1})$ is the subgroup of $G(\mathbb{F}_{p^{2m+1}})$ left fixed by $\varphi\phi_{p^{m+1}}^{-1}$ (see [Suzuki 1982, p. 388]). This implies that ${}^2G(p^{2m+1})$ is the subgroup of $G(\mathbb{F}_p^{\mathrm{alg}})$ defined by the equation $\sigma(g) = \varphi(g)$ in the structure $\mathcal{F}_{p^{m+1}}$.

The results in the previous subsection apply and give for instance: if m is large enough, then any non-trivial conjugacy class of ${}^2G(p^{2m})$ generates the whole group in at most $\dim(G)+1$ steps ($\dim(B_2) = 10$, $\dim(G_2) = 14$, $\dim(F_4) = 52$).

2.12. PROPOSITION. *Let G be a group of finite SU-rank defined over a model K of ACFA. There is an algebraic group H defined over K, and a definable group homomorphism f from some definable subgroup G_0 of G of finite index in G into $H(K)$, with $\mathrm{Ker}(f)$ finite central.*

Note that $f(G_0)$ has infinite index in $H(K)$, since $H(K)$ has SU-rank $\omega \dim(H)$. However, if H_0 is the smallest quantifier-free definable subgroup of H containing $f(G_0)$, then $f(G_0)$ has finite index in H_0, and $\mathrm{SU}(G) = \mathrm{SU}(H_0)$.

3. Study of Types of Finite Rank

In this section we will study types of finite SU-rank. First a reduction to types of SU-rank 1:

3.1. PROPOSITION. *Let $E = \mathrm{acl}_\sigma(E)$ and a a tuple with $0 < \mathrm{SU}(a/E) < \omega$. Then there is a tuple b independent from a over E, and an element*

$$c \in \mathrm{acl}_\sigma(Eab) \setminus \mathrm{acl}_\sigma(Eb)$$

such that $\mathrm{SU}(c/Eb) = 1$.

3.2. Orthogonality. Recall that two types p and q are *orthogonal* (denoted by $\perp$), if for every set E containing the sets over which p and q are defined, if a and b realise non-forking extensions of p and q respectively to E, then $a \downharpoonleft_E b$. A type is *orthogonal to a formula* if it is orthogonal to any type containing this formula. Two formulas $\varphi(x)$ and $\psi(y)$ are *orthogonal* if and only if, for every $F = \mathrm{acl}_\sigma(F)$ containing the parameters needed to define φ and ψ, and any tuples a and b satisfying φ and ψ respectively, a and b are independent over F.

Rephrased in terms of orthogonality, Proposition 3.1 says that every type of finite SU-rank is non-orthogonal to a type of SU-rank 1.

3.3. Modularity. Let $E = \mathrm{acl}_\sigma(E)$ be a subset of the model K of ACFA, and let $R \subseteq K^n$ be the set of realisations of a set of types over E (so, for instance, a subset of K^n which is definable over E). We say that R is *modular* (*over E*) if and only if any two subsets A and B of R are independent over $\mathrm{acl}_\sigma(EA) \cap \mathrm{acl}_\sigma(EB)$. We say that a (possibly incomplete) type over E, or a formula, is *modular* (*over E*) if its set of realisations is modular over E.

REMARKS. (1) The definition of modularity first appears in an unstable context in [Cherlin and Hrushovski 1998], where it is given in terms of $\mathrm{acl}^{\mathrm{eq}}$. This agrees with our definition because ACFA eliminates imaginaries. This notion of modularity generalises several notions introduced in the eighties: locally modular, one-based, module-like. All three were defined in a stable context, and some required the underlying set to be a set of realisations of rank 1 types, or of regular types.

(2) It suffices to check modularity for finite sets A and B.

(3) A modular set satisfies the stronger property: if $A \subseteq R$ and $B \subseteq K$, then A and B are independent over $\mathrm{acl}_\sigma(EA) \cap \mathrm{acl}_\sigma(EB)$.

(4) The set of realisations of (a set of) modular types of SU-rank 1 is modular. A subset of a modular set is modular. Any trivial type of SU-rank 1 is modular (a

type p over E is *trivial* if $\mathrm{acl}_\sigma(EA) = \bigcup_{a \in A} \mathrm{acl}_\sigma(Ea)$ for any set A of realisations of p).

(5) Assume that the elements of R have SU-rank 1 over E. Then the modularity of R can be rephrased as follows: there is no rank-2 family of definable curves on R^2.

(6) If p and q are non-orthogonal types of SU-rank 1, and if p is modular then so is q.

(7) Assume that R is the set of realisations of a type of SU-rank 1 over E, and that R is modular and stable, stably embedded. Then R satisfies the stronger property: any two subsets A and B of R are independent over $\mathrm{acl}_\sigma(EA) \cap \mathrm{acl}_\sigma(EB) \cap R$, provided this intersection is non-empty. This coincides with the classical notion of local modularity known to model theorists.

3.4. Additional remarks on modularity

(1) Modularity is a very strong property. In particular it implies that no field is interpretable. As we will see below, if a stable group G is modular, then there is essentially only one possible group law on G (see Proposition 4.2 below). Modular stable groups are abelian by finite.

Let me show by an example that an algebraically closed field k cannot be modular (we work in the language of rings $\{+, -, \cdot, 0, 1\}$). Indeed, consider three (algebraically) independent elements a, b, c in some algebraically closed field containing k, and let $d = ac + b$. Then the algebraic closures of the fields $k(a, b)$ and $k(c, d)$ intersect in k; but clearly (a, b) and (c, d) are not independent over k since e.g., $\mathrm{tr}\deg(k(a, b, c, d)/k) = 3 < \mathrm{tr}\deg(k(a, b)/k) + \mathrm{tr}\deg(k(c, d)/k) = 4$. The failure of modularity is of course due to the existence of the two-dimensional family $C_{a,b}$ of curves $y = ax + b$ in the plane.

(2) Let $K \models \mathrm{ACFA}$, let $R \subseteq K^n$ be definable over $E = \mathrm{acl}_\sigma(E)$, and assume that R is modular. This gives us information about the field of definition of the σ-closure $\bar{R}$ of R: if $a \in R$ is a generic of an irreducible component Z of $\bar{R}$, then Z is defined over $\mathrm{acl}_\sigma(Ea)$. When R is quantifier-free definable, then $\bar{R}$ is what we could call a "good approximation" of R, because $\deg_\sigma(\bar{R} \setminus R) < \deg_\sigma(R)$. When R is not quantifier-free definable, then usually $\deg_\sigma(\bar{R} \setminus R) = \deg_\sigma(R)$, and in an unstable context it may happen that any set S containing R and satisfying $\deg_\sigma(S \setminus R) < \deg_\sigma(R)$ "needs" parameters from outside the algebraic closure of the field of definition of $\bar{R}$.

3.5. PROPOSITION. *Let $K \models \mathrm{ACFA}$, let $E = \mathrm{acl}_\sigma(E) \subseteq K$ and let p be a non-trivial modular type over E, of SU-rank 1. Then p is non-orthogonal to the generic of a definable subgroup of some (simple) commutative algebraic group, i.e., a simple abelian variety, or the multiplicative group $\mathbb{G}_m$, or the additive group $\mathbb{G}_a$; the latter case can only occur in positive characteristic.*

3.6. Zil'ber's conjecture and the dichotomy. Zil'ber's conjecture states: Let T be a strongly minimal theory. Then either all types of T are modular, or T interprets a (pure) algebraically closed field.

This conjecture was disproved by Hrushovski. However, the philosophy behind Zil'ber's conjecture remains true: in most natural situations, the conjecture should be valid. An axiomatic system of such "natural situations" is given in [Hrushovski and Zilber 1996] (Zariski geometries).

The dichotomy "modular/field" was proved for strongly minimal types in differentially closed fields (see [Hrushovski and Ž. Sokolović 1994]) and for minimal types in separably closed fields (see [Hrushovski 1996c; Delon 1998]). Its interest lies in the fact that there is a complete characterisation of the fields of rank 1 interpretable in the theory of differentially closed fields or in the theory of separably closed fields of positive degree of imperfection: they are definably isomorphic to, respectively, the field of constants and the field of elements which are q-th powers for all power q of the characteristic. This is an important tool in Hrushovski's proof of the Geometric Mordell–Lang conjecture.

4. The Dichotomy Theorems

As explained in the previous section, our goal was to prove the following dichotomy: a type of SU-rank 1 is either non-orthogonal to one of the fixed fields, or it is modular. We first proved the characteristic 0 case, in a stronger form. The proof in that case is very algebraic and uses ramification theory. We were then able to establish the dichotomy in positive characteristic by completely different methods, see 4.4 for some details.

The dichotomy result allows us to get a good description of certain definable sets in the modular case (see Sections 4.2 and 4.7 below) and a semi-minimal analysis of types of finite rank 4.6.

4.1. THEOREM (THE DICHOTOMY IN CHARACTERISTIC 0). *Let p be a type of SU-rank 1 over $E = \mathrm{acl}_\sigma(E)$. Then either $p \not\perp (\sigma(x) = x)$, or p is modular, stable, stably embedded, and has a unique non-forking extension to any set containing E. Also, $p \not\perp (\sigma(x) = x)$ if and only if $\deg_\sigma(p) = 1$ and there is an integer N such that $[E(a, \sigma^k(a)) : E(a)] \leq N$ for every $k \in \mathbb{Z}$.*

Stably embedded means (n the arity of p, P the set of realisations of p): if $S \subseteq K^{nm}$ is definable, then $S \cap P^m = S' \cap P^m$ for some S' definable with parameters from P.

Note that a type can be stably embedded even if it is unstable. Indeed, one can show that if P is the set of realisations of a type p containing the formula $\sigma(x) = x$, then the field generated by P is all of $\mathrm{Fix}(\sigma)$. Thus, by 2.9(2), p is stably embedded.

4.2. The result of 4.1 extends to formulas: if $\varphi(x) \perp (\sigma(x) = x)$, then the set of elements satisfying φ, with the structure inherited from K, is stable and modular. In the case of groups, this has the following striking consequence, by a theorem of Hrushovski and Pillay [1987]:

PROPOSITION. *Assume characteristic* 0. *Let G be a group of finite SU-rank definable in a model K of ACFA, and assume that the formula defining G is orthogonal to $(\sigma(x) = x)$, and has its parameters in $E = \mathrm{acl}_\sigma(E)$. Let $S \subseteq G^m$ be definable. Then S is a Boolean combination of cosets of E-definable subgroups of G^m.*

4.3. THEOREM (THE DICHOTOMY IN CHARACTERISTIC $p > 0$). *Let q be a type of SU-rank* 1. *Then either q is modular, or q is non-orthogonal to the formula $\sigma^m(x) = x^{p^n}$ for some $m > 0$ and $n \in \mathbb{Z}$.*

REMARKS. (1) The Frobenius automorphism $x \mapsto x^p$ is definable. Hence, for $m > 0$ and $n \in \mathbb{Z}$, the formula $\sigma^m(x) = x^{p^n}$ defines a pseudofinite subfield of K. We will refer to these fields as *fixed fields*.

(2) The result obtained in characteristic 0 does not generalise to characteristic $p > 0$. For instance, one can show that the set of realisations of $\sigma(x) = x^p - x$ is unstable, and not stably embedded either. However, any complete type containing this formula is modular. We will see below that this is enough for some applications.

(3) There is a criterion analogous to the one given in characteristic 0 for types non-orthogonal to $(\sigma(x) = x^{p^n})$: one replaces algebraic degree by separable degree. If the field is defined by the equation $\sigma^m(x) = x^{p^n}$ with $m > 1$, then the criterion has to be suitably modified.

4.4. The proof of the dichotomy in characteristic $p > 0$ is quite different from the one in zero characteristic. An essential ingredient of the proof is the central role played by certain reducts of the structure. If $\mathcal{M} = (K, \sigma)$ is a model of ACFA, we let $\mathcal{M}[n]$ be the structure (K, σ^n), which is also a model of ACFA by Proposition 1.6(7). While $\mathcal{M}[n]$ is a reduct of $\mathcal{M}$, certain definable sets appear to attain more structure. It turns out that $\mathcal{M}[n]$ behaves more and more smoothly as n approaches infinity (a phenomenon which already showed up in the proof of the characteristic 0 case). In the characteristic $p > 0$, the proof begins by defining a certain limit structure $\mathcal{M}[\infty]$ of the sequence $\mathcal{M}[n]$ (the "virtual structure"). This limit structure is shown to be very well-behaved, and some of its properties are translated back to the reducts $\mathcal{M}[n]$ and to $\mathcal{M}$. This role for reducts and the type of limit taken, appear to be new in model theory.

We put a topology on some definable subsets of $\mathcal{M}[\infty]$, and show that it satisfies an adapted version of the axioms of Zariski geometries. Then, given a non-modular definable subset X of $\mathcal{M}[\infty]$, we reproduce the proof of [Hrushovski and Zilber 1996] to obtain a field F of rank 1 interpretable in $\mathcal{M}[\infty]$ and non-orthogonal to X, and show that this field F is algebraically closed. The proof that this gives the theorem uses the following result, of independent interest:

4.5. PROPOSITION. *Let H be a simple algebraic group, and let G be a Zariski dense subgroup of $H(K)$ definable in $K \models$ ACFA. If $SU(G)$ is infinite, then $G = H(K)$. If $SU(G)$ is finite, then the generics of G are non-orthogonal to some fixed field F. Moreover, some subgroup of finite index of G is conjugate to a subgroup of $H(F)$.*

4.6. Semi-minimal analysis. Let $E = \mathrm{acl}_\sigma(E)$, and a a tuple with $SU(a/E) < \omega$. There are $a_1, \ldots, a_n \in \mathrm{acl}_\sigma(Ea)$, such that $a \in \mathrm{acl}_\sigma(Ea_1, \ldots, a_n)$, and for every i, either $\mathrm{tp}(a_{i+1}/E(a_i)_\sigma)$ is modular of SU-rank 1, or there is some finite set B, such that the set of realisations of $\mathrm{tp}(a_{i+1}/E(a_i)_\sigma)$ is contained in the perfect closure of the difference field generated by $E(a_i) \cup B \cup F$, where F is some fixed field.

4.7. One can show easily that if a formula $\varphi(x)$ is orthogonal to all fixed fields then $\varphi(x)$ is modular. While the full theory of the set of realisations of $\varphi(x)$ may be unstable, we have what is called quantifier-free ω-stability. Thus, in the case of groups, we get the analogue of 4.2 but only for subsets of G^m defined by quantifier-free formulas (within G^m).

I thought it worthwhile to give a proof of this result, for two reasons. The first is the reaction of the audience during my talk at MSRI: they weren't surprised by the dichotomy results but by their corollaries. The second is that in the particular context of a quantifier-free definable set X of an algebraic group H, the classical proof of [Hrushovski and Pillay 1987] becomes very short, and still retains many of the ingredients which demonstrate the strength of modularity.

PROPOSITION. *Let $K \models$ ACFA, H an algebraic group defined over K, and let G be a definable subgroup of finite SU-rank of $H(K)$. Assume that the formula defining G is orthogonal to all fixed fields, and has its parameters in $E = \mathrm{acl}_\sigma(E)$. Let $X \subseteq H(K)^m$ be a quantifier-free definable set. Then $X \cap G^m$ is a Boolean combination of cosets of subgroups of G^m which are defined in G^m by a quantifier-free formula with parameters in E.*

PROOF. The group G has finite index in the smallest quantifier-free definable group $\bar{G}$ containing it. By 4.3, the group G is modular, which implies that $\bar{G}$ is also modular. We may therefore assume that G is quantifier-free definable. By an easy reduction, we may also assume that $m = 1$ (work in G^m), and that X is an irreducible σ-closed set contained in G. We then want to show that X is the coset of a σ-closed subgroup S of G, and that S is defined over E.

We will assume that the difference field (K, σ) has sufficiently many automorphisms. If Z is a σ-closed set defined over some difference field F, we define $\deg_\sigma(Z) = \max\{\deg_\sigma(a/F) \mid a \in Z\}$. Note that $\deg_\sigma$ is invariant under translation, that is, $\deg_\sigma(Z) = \deg_\sigma(aZ)$ for any $a \in G$. In analogy with algebraic varieties, if Z is an irreducible σ-closed set defined over some field F, we will

say that a *is a generic of Z over F* if $a \in Z$ and every difference polynomial over F which vanishes at a, vanishes on Z. Equivalently, if $a \in Z$ and
$\deg_\sigma(a/F) = \deg_\sigma(Z)$. One can show that the generics of the group G in the
stability theoretic sense (i.e., of maximal SU-rank) are precisely the generics of
G in this sense.

Let F be the smallest algebraically closed difference field containing E and
over which X is defined. Let $S = \{h \in G \mid hX = X\}$, fix a generic a of X over
F, and a generic g of G over $F(a)_\sigma$. Then S is a σ-closed subgroup of G defined
over F, and $b = ga$ is a generic of G over $\mathrm{acl}_\sigma(Fa)$. Consider the set $Y = gX$;
then b is a generic of Y over $\mathrm{acl}_\sigma(Fg)$.

CLAIM 1. *Let τ be an automorphism of the difference field (K, σ), which is the
identity on F. Then $\tau(Y) = Y \iff \tau(gS) = gS$.*

PROOF. Indeed, using the fact that X and S are defined over F, we get: $\tau(Y) =
Y \iff \tau(gX) = gX \iff \tau(g)X = gX \iff g^{-1}\tau(g) \in S \iff \tau(gS) = gS$,
which gives the result. $\square$

CLAIM 2. *The fields of definition of Y and of gS are equi-algebraic over F.*

PROOF. This follows from Claim 1 and the following two observations: (1)
If τ does not fix the field of definition of an irreducible σ-closed set Z, then
$\tau(Z) \neq Z$. (2) Let $k_0 \subseteq k_1 \subseteq K$ be fields, with $k_1 \not\subseteq \mathrm{acl}_\sigma(k_0)$. Then there is
some automorphism τ of (K, σ) which fixes $\mathrm{acl}_\sigma(k_0)$ and does not fix k_1. $\square$

By modularity and because b is a generic of Y over $\mathrm{acl}_\sigma(Fg)$, the field of definition
of Y is contained in $\mathrm{acl}_\sigma(Fb) \cap \mathrm{acl}_\sigma(Fg)$. Choose a $c \in \mathrm{acl}_\sigma(Fb) \cap \mathrm{acl}_\sigma(Fg)$ that
generates the field of definition of Y (and the one of gS by Claim 2). Using $b = ga$
and the fact that g and b are independent from a over F, and hence over $F(c)_\sigma$,
we obtain that $\deg_\sigma(Y) = \deg_\sigma(b/F(c, a)_\sigma) = \deg_\sigma(g/F(c, a)_\sigma) = \deg_\sigma(gS)$.
Hence $\deg_\sigma(S) = \deg_\sigma(X)$, and from $Sa \subseteq X$ and the irreducibility of the
σ-closed set X, we deduce that $S = Xa$.

Because g is a generic of the coset gS and by modularity, gS is defined over
$\mathrm{acl}_\sigma(Eg)$, which implies that $S = g^{-1}(gS)$ is also defined over $\mathrm{acl}_\sigma(Eg)$. So, S
is defined over $\mathrm{acl}_\sigma(Eg) \cap F = E$. $\square$

5. Application: the Manin–Mumford Conjecture
over a Number Field

The result of Raynaud [1983] (which implies the Manin–Mumford conjecture)
states that if A is an abelian variety and X a subvariety of A, then $\mathrm{Tor}(A) \cap X$
is a finite union of sets of the form $a_i + \mathrm{Tor}(A_i)$, with A_i a group subvariety of
A. (Here, $\mathrm{Tor}(A)$ denotes the set of torsion points of A.) This result was later
extended by Hindry and McQuillan. It is a particular case of a conjecture of
Lang; for details see [Lang 1991, p. 37].

Hrushovski saw that the results on difference fields could be used to obtain a new proof of this theorem, for A a commutative algebraic group defined over a number field K. His proof gives an explicit bound on the number of cosets, of the form $M = c \deg(X)^e$, where c and e depend on A but not on X, and $\deg(X)$ is the degree of the variety X with respect to a fixed embedding of A into projective space. His result appears in [Hrushovski 1995]; see also [Pillay 1997]. His bound is explicit modulo the choice of two primes of good reduction for A; see Section 5.10 for the definition of good reduction. (If A is semi-abelian, let $h(A)$ denote the height of A in the sense of Faltings; according to one specialist, the order of magnitude of a bound for a prime of good reduction is likely to be $h(A)$.)

The strategy is very simple. Suppose we are given a commutative algebraic group A, a subvariety X of A, and some subgroup Γ of A. Then we would like to find an automorphism σ of some large model L of ACFA containing Γ, and a modular definable subgroup G of $A(L)$ containing Γ. The result then would follow by 4.2. This is however too simple to work. There are two problems:

- In characteristic 0, every proper definable subgroup of $\mathbb{G}_a(L)$ is non-orthogonal to the fixed field (see Section 5.9), and is therefore never modular. One gets around this difficulty by reducing to the case of a semi-abelian variety, using model theory.

- In order to get explicit bounds, one needs an explicit description of G (and not only its mere existence). When Γ is the subgroup $\mathrm{Tor}_{p'}(A)$ of prime-to-p-torsion, for p a prime with good properties, then Weil's result on abelian varieties defined over finite fields gives an equation of bounded complexity for σ a lifting of the Frobenius. However, this doesn't say anything on the points of order a power of p. A further trick is needed, involving some model theory and ugly computations.

We indicate below some of the ingredients involved in the proof of Hrushovski. This section is organised as follows. We first introduce some tools and definitions from algebraic geometry, and state the main results used in Hrushovski's proof. Of particular interest in my opinion is his description of definable subgroups of abelian varieties and of their definable endomorphisms. And of course, his criterion for modularity is absolutely fundamental in the proof; see Theorem 5.6. We then show in Section 5.13 how to obtain the qualitative result, and reduce the problem of finding an explicit bound for the number of cosets to the case where the group variety A is a semi-abelian variety.

We give a fairly detailed exposition on how to get the explicit bound. This part is essentially self-contained if one accepts the results stated earlier together with those of Section 5.12. We start with the "easy case" of the p'-torsion subgroup in Theorem 5.14. We then proceed slowly towards a proof of the full result, given in Theorem 5.17.

5.1. Degrees of varieties. We embed our algebraic group A in some projective space $\mathbb{P}_n$. By the degree of a subset of $\mathbb{P}_n$ we mean the degree of its Zariski closure. It is convenient to define a degree on algebraic subsets of cartesian powers of $\mathbb{P}_n$, in such a way that it satisfies the following conditions:

(1) Let $V_1, \ldots, V_r$ be algebraic subsets of $(\mathbb{P}_n)^l$, and let $Z_1, \ldots, Z_s$ be the irreducible components of $V_1 \cap \cdots \cap V_r$. Then

$$\sum_{i=1}^{s} \deg(Z_i) \leq \prod_{j=1}^{r} \deg(V_j).$$

(2) Let V be an algebraic subset of $(\mathbb{P}_n)^l \times (\mathbb{P}_n)^k$, and consider the projection $\rho : (\mathbb{P}_n)^l \times (\mathbb{P}_n)^k \to (\mathbb{P}_n)^l$. Then $\deg(\rho(V)) \leq \deg(V)$.

(3) Let V be an algebraic subset of $(\mathbb{P}_n)^l \times (\mathbb{P}_n)^k$, and let ρ be defined as above. For $a \in (\mathbb{P}_n)^l$ define $V(a) = V \cap \rho^{-1}(a)$. Suppose that $\dim(V(a)) = r$ for generic $a \in \rho(V)$. Then the Zariski closure V^* of the set $\{a \in (\mathbb{P}_n)^l \mid \dim(V(a)) > r\}$ has degree $\leq \deg(V)$.

For the definition of this degree and further properties, see [Fulton 1984, Example 8.4.4] or [Hrushovski 1995].

5.2. Let A be an algebraic group. Then A has a unique maximal connected linear subgroup H, and A/H is an *abelian variety*, i.e., a commutative projective group variety. If H is commutative then the simple factors of H are isomorphic to either the multiplicative group $\mathbb{G}_m$ or the additive group $\mathbb{G}_a$.

A *semi-abelian variety* is a commutative algebraic group A with no simple factor isomorphic to $\mathbb{G}_a$. If A is an abelian variety, then there is an *isogeny* (epimorphism with finite central kernel) from some product $A_1 \times \cdots \times A_n$ onto A, with the A_i's simple abelian subvarieties of A. A good reference for facts on abelian varieties is [Lang 1959].

5.3. We first show how to get from an effective description of G to an effective bound M. The group G will be described as $\{g \in A(K) \mid (g, \sigma(g), \ldots, \sigma^l(g)) \in S\}$ for some algebraic subgroup S of $A \times \sigma(A) \times \cdots \times \sigma^l(A)$. We view A as embedded in $\mathbb{P}_n$, and define $\deg(S)$ and $\deg(X)$ with respect to this embedding. Let $V = S \cap (X \times \sigma(X) \times \cdots \times \sigma^l(X))$. Then $\dim(V) \leq e = \min\{\dim(S), (l+1)\dim(X)\}$ and $\deg(V) \leq \deg(X)^{l+1} \deg(S)$. Thus an effective bound for the number of components of the Zariski closure of $G \cap X$ is given by the following result:

LEMMA. *Let $V \subseteq A^{l+1}$ be an algebraic set, and set*

$$\tilde{V} = \{g \in A(K) \mid (g, \sigma(g), \ldots, \sigma^l(g)) \in V\}.$$

Then the Zariski closure of $\tilde{V}$ has degree at most $\deg(V)^{2^{\dim(V)}}$. If V is defined over $L(c)$, where $\sigma(L) = L$, then Z is defined over

$$L\big(\sigma^{-\dim(V)}(c), \ldots, c, \sigma(c), \ldots, \sigma^{\dim(V)}(c)\big).$$

The proof of this result uses the properties of degrees of varieties stated in Section 5.1. We'll make a simple observation on how the irreducible components of Z are obtained. Let π_0 denote the projection on the first copy of $\mathbb{P}_n$, π_1 the projection on the first l copies of $\mathbb{P}_n$, and π_2 the projection on the last l. The irreducible components of Z are images by π_0 of algebraic subsets W of $\mathbb{P}_n^{l+1}$ satisfying $\sigma\pi_1(W) = \pi_2(W)$. Thus the procedure for getting the irreducible components of Z is as follows: start with some irreducible component W of V. If $\sigma\pi_1(W) = \pi_2(W)$, $\pi_0(W)$ will be an irreducible component of Z. If not, then consider $W \cap \pi_2^{-1}\sigma\pi_1(W)$, look at its irreducible components and repeat the procedure. This procedure stops after $\dim(V) + 1$ steps.

This result needs to be refined to give more information when X varies in a family of varieties (we are now thinking of $X + a$ for various a's in the p-torsion subgroup).

5.4. NOTATION. Let $\mathbb{P} = (\mathbb{P}_n)^k \times (\mathbb{P}_n)^m$, and $\rho : \mathbb{P} \to (\mathbb{P}_n)^k$ the projection. If Z is a subvariety of $\mathbb{P}$ and $a \in (\mathbb{P}_n)^k$ we define $Z(a) = \{b \in (\mathbb{P}_n)^m \mid (a,b) \in Z\}$, and if $r = \dim(Z(a))$ for generic $a \in \rho(Z)$, we set $Z^* = \{a \in \rho(Z) \mid \dim(Z(a)) > r\}$.

PROPOSITION. *Let $(K,\sigma) \models$ ACFA, and let V be a closed subset of $\mathbb{P}^l$ defined over K. There exist irreducible subvarieties Z_i of $\mathbb{P}$ satisfying these properties:*

(1) *If $a \in \tilde{V} =_{\mathrm{def}} \{x \in \mathbb{P}(K) \mid (x, \sigma(x), \ldots, \sigma^{l-1}(x)) \in V\}$, then there is an i such that $a \in Z_i$, $a \notin \rho^{-1}(Z_i^*)$.*
(2) *$Z_i(K) \cap \tilde{V}$ is dense in Z_i for every i.*
(3) *$\sum_i \deg(Z_i) \le \sum_{j=0}^{\dim(V)} \deg(V) 2^j \le 2 \deg(V)^{2^{\dim(V)}}$.*
(4) *If $i \ne j$ and Z_i is a proper closed subset of Z_j, then $\rho(Z_i) \subseteq Z_j^*$.*

5.5. Definable endomorphisms and definable subgroups of an abelian variety. Let A be an abelian variety defined over a model (K,σ) of ACFA, and let $\mathrm{End}(A)$ denote the ring of algebraic endomorphisms of A, $\mathrm{End}_\sigma(A)$ the ring of definable endomorphisms of A. Denote by $E(A)$ and $E_\sigma(A)$ the rings $\mathbb{Q} \otimes_\mathbb{Z} \mathrm{End}(A)$ and $\mathbb{Q} \otimes_\mathbb{Z} E_\sigma(A)$ respectively. Then $E(A)$ and $E_\sigma(A)$ have a description in terms of matrix rings over $E(A_i)$ and $E_\sigma(A_i)$ for some simple abelian subvarieties of A. The result is well-known for $E(A)$ because of Poincaré's reducibility theorem, and we will describe what happens for $E_\sigma(A)$:

PROPOSITION. *Let A be an abelian variety defined over K.*

(1) *Let $A_1, \ldots, A_n$ be abelian subvarieties of A such that A and $A_1 \times \cdots \times A_n$ are isogenous. Let $I \subseteq \{1, \ldots, n\}$ be maximal such that if $i \ne j$ are in I and $k \in \mathbb{N}$, then A_i and $\sigma^k(A_j)$ are not isogenous. For each $i \in I$ let $m(i)$ be the number of indices $j \le n$ such that A_j and $\sigma^k(A_i)$ are isogenous for some $k \in \mathbb{N}$. Then $E_\sigma(A) \simeq \prod_{i \in I} M_{m(i)}(E_\sigma(A_i))$.*
(2) *Let $k \ge 1$ and let B be a definable subgroup of $A^k(K)$. Then B is commensurable with a finite intersection H of kernels of definable homomorphisms*

$A^k(K) \to A(K)$ (*commensurable* means that $B \cap H$ has finite index in both B and H). *If $k = 1$, a single homomorphism suffices.*

Thus the study of definable subgroups of $A(K)$ reduces to the study of the rings $E_\sigma(A_i)$, $i \in I$. Of particular interest are the *c-minimal subgroups* of $A(K)$, i.e., definable subgroups which are minimal up to commensurability, because of the following result:

LEMMA. *Let B be a definable subgroup of $A(K)$ A. Then B is modular if and only if, for every $f \in \mathrm{End}_\sigma(A)$, every c-minimal subgroup of $f(B)$ is modular. If $B = \mathrm{Ker}(fg)$ for $f, g \in \mathrm{End}_\sigma(A)$ with infinite kernels, then B is modular if and only if $\mathrm{Ker}(f)$ and $\mathrm{Ker}(g)$ are modular.*

The next result gives a complete description.

5.6. THEOREM. *Let A be a simple abelian variety defined over $K \models \mathrm{ACFA}$.*

(1) *Assume that for every $n \in \mathbb{N}$, A and $\sigma^n(A)$ are not isogenous. Then $E_\sigma(A) = E(A)$, and every definable proper subgroup of $A(K)$ is finite.*

(2) *Assume that $n \geq 0$ is minimal such that there is an isogeny $h : A \to \sigma^n(A)$, and let $h' : \sigma^n(A) \to A$ and $m \in \mathbb{N}^{>0}$ be such that $h'h = [m]$. Define $\psi = \sigma^{-n}h$ and $\psi' = h'\sigma^n$. Then $E_\sigma(A)$ is isomorphic to the twisted Laurent polynomial ring $E(A)[\psi, \psi']$. Note that $\psi'\psi = [m]$.*

From now on we assume that the hypotheses in (2) hold.

(3) *Let $f \neq 0$ be an element of $\mathrm{End}_\sigma(A)$. Then f is onto and $\mathrm{Ker}(f)$ has finite rank.*

(4) *A definable subgroup B of $A(K)$ is c-minimal if and only if it is commensurable with $\mathrm{Ker}(f)$, for some $f \in \mathrm{End}_\sigma(A)$ which is irreducible in $E_\sigma(A)$. Thus, if $f \in \mathrm{End}_\sigma(A)$ is non-zero, then $\mathrm{Ker}(f)$ is modular if and only if $\mathrm{Ker}(g)$ is modular for every irreducible divisor g of f.*

(5) *Let B be a c-minimal subgroup of $A(K)$. If B is not modular, then there is an abelian variety A' defined over $\mathrm{Fix}(\tau)$ for some $\tau = \sigma^m \phi_p^{-n}$, and an algebraic isomorphism $\varphi : A \to A'$ such that $\varphi(B) \subseteq A'(\mathrm{Fix}(\tau^l))$ for some l.*

Similar results hold for the multiplicative group $\mathbb{G}_m$ (with $E_\sigma(\mathbb{G}_m) \simeq \mathbb{Q}[\sigma, \sigma^{-1}]$), and putting everything together, one obtains:

THEOREM. *Let A be a semi-abelian variety defined over $\mathrm{Fix}(\sigma)$, and let $f(T) \in \mathbb{Z}[T]$. Assume that $f(T)$ is relatively prime to all cyclotomic polynomials. Then $\mathrm{Ker}(f(\sigma))$ is orthogonal to the formula $\sigma(x) = x$, and therefore modular if the characteristic is 0.*

5.7. REMARK. In characteristic p, one obtains a similar criterion for semi-abelian varieties defined over $\mathrm{Fix}(\sigma)$ or over $\mathrm{Fix}(\tau)$ for some $\tau = \sigma^{-m}\phi_p^n$.

Also, observe that if A is a simple abelian variety defined over $\mathrm{Fix}(\sigma)$ and if $A(K)$ has a definable subgroup B of finite rank non-orthogonal to the formula

$\sigma^m(x) = x^{p^n}$ for some $m > 0$, $n \in \mathbb{Z}$ with $n \neq 0$, then A must be isomorphic to a variety A' defined over $\mathrm{Fix}(\tau)$, where $\tau = \sigma^{-m}\phi_p^n$. This implies that the field of definition of A is contained in a finite algebraic extension of $\mathrm{Fix}(\tau)$, and therefore is finite, since $\mathrm{Fix}(\sigma) \cap \mathrm{Fix}(\tau^l) = \mathrm{Fix}(\sigma, \phi_p^{nl}) \subseteq \mathbb{F}_{p^{nl}}$.

Assume that A is defined over the finite field $\mathbb{F}_q$ fixed by σ. Let $f(T) \in \mathbb{Z}[T]$, and consider the subgroup B of $A(K)$ defined by the equation $f(\sigma)(x) = 0$. Let $\alpha_1, \ldots, \alpha_r$ be the roots of $f(T) = 0$ (in $\mathbb{C}$). Going to some power of τ, we may assume that σ^m and ϕ_p^n commute with all elements of $\mathrm{End}(A)$ (and fix the field of definition of A). To finish the discussion we need the following result of Weil (see [Weil 1971] or [Mumford 1974, pp. 203 and 205]), which will also be used in the proof:

THEOREM. *Let A be an abelian variety defined over a finite field $\mathbb{F}_q$, and consider the endomorphism $\phi_q : x \mapsto x^q$ of A. If $\omega_1, \ldots, \omega_{2d}$ are the roots in $\mathbb{C}$ of the characteristic polynomial of ϕ_q on $A(\mathbb{F}_q^{\mathrm{alg}})$, then $d = \dim(A)$, the ω_i's are algebraic integers of modulus $|\omega_i| = q^{1/2}$, and q/ω_i is among the ω_j's.*

Hence, the endomorphism τ satisfies a functional equation $g(T) = 0$ on B, where the roots of $g(T)$ are of the form $\alpha_i^{-m}\omega_j^l$, where l is such that $q^l = p^n$. Thus, B is orthogonal to $\mathrm{Fix}(\tau)$ if and only if no $\alpha_i^{-m}\omega_j^l$ is a root of unity.

Thus we obtain: B is modular if and only if $\alpha_i^{-m}\omega_j^l \neq 1$ for every i and j, $m \neq 0$ and l.

5.8. Before going on with Hrushovski's proof, we mention an easy corollary of his characterisation of modular subgroups.

PROPOSITION. *Let A be a semi-abelian variety, and X a subvariety of A. Assume that m is an integer > 1 and prime to the characteristic of the field of definition of A, such that $[m]X = X$. Then $X = a + C$ for some group subvariety C of A and element $a \in A[m-1]$.*

PROOF. Let k be an algebraically closed field over which X and A are defined, and embed k in a model (L, σ) of ACFA, with σ being the identity on k. By assumption, if u is a generic of X, then so is $[m]u$, and they have the same type (in the language of fields) over k. Hence, in L there is a generic u of X such that $\sigma(u) = [m]u$. Consider the subgroup B of A defined by the equation $\sigma(x) = [m]x$. Since $m > 1$ is prime to the characteristic of k, B is modular. Hence, $B \cap X$ is a finite union of cosets of definable subgroups of B. On the other hand, $B \cap X$ contains a generic point of X, which implies that one of these cosets is Zariski dense in X. This shows that $X = a + C$ for some algebraic subgroup C of A. We also have: $[m]X = [m]a + C = X = a + C$, which implies that $[m-1]a \in C$. Since C is divisible, we may choose $a \in A[m-1]$. $\square$

5.9. Definable subgroups of $\mathbb{G}_a(K)$. The ring of endomorphisms of $\mathbb{G}_a$ definable in the model (K, σ) of ACFA contains the twisted ring $\mathrm{End}_K(\mathbb{G}_a)[\sigma, \sigma^{-1}]$, with the appropriate action of σ on $\mathrm{End}_K(\mathbb{G}_a)$. If the characteristic is 0, then

$\mathrm{End}_K(\mathbb{G}_a)$ is canonically isomorphic to K. If the characteristic is $p > 0$, then $\mathrm{End}_K(\mathbb{G}_a)$ is generated over K by the Frobenius $\phi_p : x \mapsto x^p$.

In characteristic 0, a definable subgroup of $\mathbb{G}_a(K)$ will be commensurable to a subgroup defined by an equation $\sum_{i=0}^{n} a_i \sigma^i(x) = 0$ for some n and $a_0, \ldots, a_n \in K$. One checks easily that a polynomial $\sum_{i=0}^{n} a_i \sigma^i$ with $a_n = 1$ can be written as a product of linear terms of the form $\sigma - a$. Furthermore, if $a \neq 0$ then the solution set of $\sigma(x) = ax$ is non-orthogonal to the fixed field: if b_1 and b_2 are two solutions then $\sigma(b_1/b_2) = b_1/b_2$.

5.10. Hypotheses and some notations. Let A be a commutative algebraic group defined over a number field K. Choose a sequence $(0) = D_0 \subset D_1 \subset \cdots \subset D_s = A$ of algebraic subgroups of A such that each factor D_{i+1}/D_i is a K-irreducible abelian variety or torus for $i \geq 1$. Let $\{E_j\}$ be a set of representatives of the K-isogeny classes of the factors D_{i+1}/D_i and define $d = \sum \dim(E_j)$. Note that this number d does not change when we take powers of A.

We fix a prime p of good reduction, by which we mean: if $\mathbb{F}_q$ is the residue field of K modulo p and $\bar{D}_i$ denotes the algebraic set obtained by reducing modulo p, then each $\bar{D}_i$ is a reduced connected algebraic group. Moreover, for each $i \geq 1$, $\bar{D}_{i+1}/\bar{D}_i$ and D_{i+1}/D_i have the same dimension and are of the same type (i.e., an abelian variety or a torus). We also request that $\bar{D}_1$ be a vector group.

We denote by $\mathrm{Tor}_{p'}(A)$ the subgroup of torsion elements of A of order prime to p, and by $\mathrm{Tor}_p(A)$ the subgroup of torsion elements of A of order a power of p. Then $\mathrm{Tor}(A) = \mathrm{Tor}_{p'}(A) \oplus \mathrm{Tor}_p(A)$.

5.11. PROPOSITION. *With notation as above, there is $\sigma \in \mathrm{Aut}(\mathbb{Q}^{\mathrm{alg}})$ and an integral polynomial $F(T)$ with no roots of unity among its roots, such that $F(\sigma)$ vanishes on $\mathrm{Tor}_{p'}(A)$. Furthermore, the degree of F is at most $2d$ and the sum of the absolute values of its coefficients is bounded by $(1 + q^{1/2})^{2d}$.*

PROOF. Consider the Frobenius map $\phi_q : x \mapsto x^q$ defined on $\mathbb{F}_p^{\mathrm{alg}}$, and let $\sigma \in \mathrm{Gal}(\mathbb{Q}^{\mathrm{alg}}/K)$ be a lifting of ϕ_q. Note first that our assumptions on p imply that reduction modulo p induces an isomorphism $\mathrm{Tor}_{p'}(A) \to \mathrm{Tor}_{p'}(\bar{A})$. Thus, for $F(T) \in \mathbb{Z}[T]$, if $F(\phi_q)$ vanishes on $\mathrm{Tor}_{p'}(\bar{A})$, then $F(\sigma)$ will automatically vanish on $\mathrm{Tor}_{p'}(A)$.

Note also that if $f(T), g(T) \in \mathbb{C}[T]$, then the sum of the absolute values of the coefficients of $fg(T)$ is no greater than the product of the sums of the absolute values of the coefficients of $f(T)$ and of $g(T)$. It therefore suffices to show the assertion for each of the factors $\bar{D}_{i+1}/\bar{D}_i$ and for $\bar{D}_1$. Since $\bar{D}_1$ has no points of order prime to p, we may take the constant polynomial 1. If $\bar{D}_{i+1}/\bar{D}_i$ is a K-simple abelian variety, then Weil's result 5.7 gives us a monic polynomial of degree $2 \dim(\bar{D}_{i+1}/\bar{D}_i)$, with roots of modulus $q^{1/2}$. Hence the sum of the absolute values of the coefficients of this polynomial is $\leq (1 + q^{1/2})^{2\dim(\bar{D}_{i+1}/\bar{D}_i)}$. Assume that $\bar{D}_{i+1}/\bar{D}_i$ is a torus, isomorphic to $\mathbb{G}_m^n$ via an algebraic map φ defined over the finite field $\mathbb{F}_{q^l}$. The Frobenius map on $\bar{D}_{i+1}/\bar{D}_i$ induces an automorphism ψ of $\mathbb{G}_m^n$, which is the composition of $\theta = \varphi \circ \phi_q(\varphi)^{-1}$ with raising to the q-th power in $\mathbb{G}_m^n$. Since $\theta \in \mathrm{End}(\mathbb{G}_m^n) \simeq \mathrm{GL}_n(\mathbb{Z})$, it is left fixed by ϕ_p. Hence,

$$\theta^l = \theta \circ \phi_q(\theta) \circ \cdots \circ \phi_q^{l-1}(\theta) = \mathrm{id},$$

which implies that the roots of the characteristic polynomial of θ in $\mathrm{GL}_n(\mathbb{Z})$ are roots of unity. Thus, the characteristic polynomial of ψ has degree n and its roots have absolute value q. Going back to $\bar{D}_{i+1}/\bar{D}_i$ we get the result.

Since we may choose the same polynomial within a K-isogeny class, we get the correct bounds. $\square$

5.12. Now comes the time to take care of the vector subgroup of G (a *vector group* is an algebraic group isomorphic to a product of copies of the additive group $\mathbb{G}_a$). For that, we need two results, which we state below. The proof of the Proposition uses model theory and the full strength of Theorem 4.1. In positive characteristic the proof seems to work for quantifier-free definable subsets of G. The proof of the lemma is purely algebraic.

DEFINITION. Let A be a commutative algebraic group, and let V be the maximal vector subgroup of A. A definable set $X \subseteq A(K)$ is *special* if it is of the form $Y + C$ where Y is a definable subset of $V(K)$ and C is a coset of a definable subgroup of $A(K)$. Similarly, an algebraic subset of A is *special* if it is of the form $Y + C$ where Y is an algebraic subset of V and C is a coset of an algebraic subgroup of A.

PROPOSITION. *Assume characteristic* 0. *let* $(K, \sigma) \models$ ACFA *and let* A *be a commutative algebraic group defined over* $\mathrm{Fix}(\sigma)$. *Let* $F(T) \in \mathbb{Z}[T]$ *be a polynomial with no root of unity among its roots, and let* $G = \{g \in A(K) \mid F(\sigma)(g) = 0\}$. *Then every definable subset of* G *is a finite Boolean combination of special subsets of* G. *If* X *is a subvariety of* A, *then* $X \cap G$ *is a finite union of special subvarieties of* A.

LEMMA. *Let* A *be a commutative algebraic group defined over* $\mathbb{Q}^{\mathrm{alg}}$, *and* T *the group of torsion points of* A *(or of prime-to-p torsion points of* A). *Let* X *be a subvariety of* A *and assume that* $X \cap T \subseteq \bigcup_{i=1}^M D_i$, *where each* D_i *is a special subvariety of* X. *Then the Zariski closure of* $X \cap T$ *is the union of at most* M *cosets of group subvarieties of* A. *More precisely, for every* i, $D_i \cap T$ *is either empty or its Zariski closure is the coset of a group subvariety of* A.

5.13. The qualitative result and reduction to the semi-abelian case. Let A be a commutative algebraic group defined over the number field K, let V be the maximal vector subgroup of A and $B = A/V$. We want to find $\sigma' \in \mathrm{Gal}(\mathbb{Q}^{\mathrm{alg}}/K)$ and $G(T) \in \mathbb{Z}[T]$ such that $G(\sigma')$ vanishes on $\mathrm{Tor}(A)$. Since the reduction map $A \to B$ is injective on $\mathrm{Tor}(A)$, it suffices to find σ' and $G(T)$ such that $G(\sigma')$ vanishes on $\mathrm{Tor}(B)$.

By Proposition 5.11 applied to two primes p and l of good reduction for B, there are $\sigma \in \mathrm{Gal}(K(\mathrm{Tor}_{p'}(B))/K)$ and $\tau \in \mathrm{Gal}(K(\mathrm{Tor}_p(B))/K)$, and polynomials $F_p(T), F_l(T) \in \mathbb{Z}[T]$, with no roots of unity among their roots, and such that $F_p(\sigma)$ vanishes on $\mathrm{Tor}_{p'}(B)$ and $F_l(\tau)$ vanishes on $\mathrm{Tor}_p(B)$.

Using a result of Serre [1985/86], one can show that the field

$$L = K(\mathrm{Tor}_{p'}(B)) \cap K(\mathrm{Tor}_p(B))$$

is a finite Galois extension of K, over which $K(\mathrm{Tor}_{p'}(B))$ and $K(\mathrm{Tor}_p(B))$ are linearly disjoint. Hence, for $m = [L : K]$, there is $\sigma' \in \mathrm{Gal}(\mathbb{Q}^{\mathrm{alg}}/L)$ that extends σ^m on $K(\mathrm{Tor}_{p'}(B))$ and τ^m on $K(\mathrm{Tor}_p(B))$. Let $\alpha_1, \ldots, \alpha_{2d}$ and $\beta_1, \ldots, \beta_{2d}$ be the roots of $F_p(T)$ and $F_l(T)$ respectively, and define

$$G(T) = \prod_{i=1}^{2d}(T - \alpha_i^m)(T - \beta_i^m).$$

Then $G(\sigma')$ vanishes on $\mathrm{Tor}(B)$, and $\mathrm{Ker}(G(\sigma'))$ defines a modular subgroup of B in any model of ACFA extending $(\mathbb{Q}^{\mathrm{alg}}, \sigma')$.

This shows immediately, by the two results in Section 5.12, that the Zariski closure of $X \cap \mathrm{Tor}(A)$ is the union of finitely many cosets of group subvarieties of A. However, since we don't know $[L : K]$, we cannot expect to get an explicit bound on the number of cosets. To get the explicit bound we reduce to the semi-abelian case via the following observation:

Let Y be the image of X in B. Then the map $A \to B$, which is injective on $\mathrm{Tor}(A)$, establishes a bijection between the irreducible components of the Zariski closure of $X \cap \mathrm{Tor}(A)$ and the irreducible components of the Zariski closure Z of $Y \cap \mathrm{Tor}(B)$. Thus the Zariski closure of $X \cap \mathrm{Tor}(A)$ is the union of at most $\deg(Z)$ cosets of algebraic subgroups of A. So, we have

THEOREM. *Let A be a commutative algebraic group defined over the number field K, let X be a subvariety of A. Then $X \cap \mathrm{Tor}(A) = \bigcup_{i=1}^{M} a_i + \mathrm{Tor}(A_i)$ where each A_i is an algebraic subgroup of A. Let V be the maximal vector subgroup of A, and Y the image of X in $B = A/V$. The number M is bounded by the number of irreducible components of the Zariski closure of $Y \cap \mathrm{Tor}(B)$.*

5.14. THEOREM (THE BOUND ON M IN THE CASE OF THE p'-TORSION SUBGROUP). *Let A be a commutative algebraic group over a number field K, let X be a subvariety of A, and fix a prime p such that A has good reduction at p. Let q be the size of the residue field of K at p. Then*

$$X \cap \mathrm{Tor}_{p'}(A) = \bigcup_{i=1}^{M} a_i + \mathrm{Tor}_{p'}(A_i),$$

where each A_i is an algebraic subgroup of A. If $d \leq \dim(A)$ is defined as in Section 5.10, and if d_+ is the degree of the graph of addition in A^3, then

$$M \leq \left(\deg(X)^{2d+1} d_+^{2d^2(2d+1)(\log_2(1+q^{1/2})+1)^2}\right)^{2^d \dim(X)}.$$

PROOF. Choose σ and $F(T) = \sum_{i=0}^{2d} m_i T^i$ as in 5.11, and work in a model of ACFA extending $(\bar{\mathbb{Q}}, \sigma)$. Let $\tilde{S} = \text{Ker } F(\sigma)$ and consider the subgroup S of A^{2d+1} defined by $\{(a_0, \ldots, a_{2d}) \in A^{2d+1} \mid \sum_{i=0}^{2d} [m_i] a_i = 0\}$. Using the fact that $\sum_i |m_i| \leq (1 + q^{1/2})^{2d}$, and that multiplication by a number $M \geq 2$ can be achieved with $\log_2(M)(\log_2(M) + 1)/2$ additions, one obtains $\deg(S) \leq d_+^{2d^2(2d+1)(\log_2(1+q^{1/2})+1)^2}$. Let Z be the Zariski closure of $\tilde{S} \cap X$. Then 5.3 gives

$$\deg(Z) \leq (\deg(X)^{2d+1} \deg(S))^{2^{2d \dim(X)}}.$$

Furthermore, by modularity of $\tilde{S}$, Z consists of cosets of algebraic subgroups of A. Each of these cosets intersects $\text{Tor}_{p'}(A)$ in either the empty set or a Zariski dense set. This gives the result.   □

5.15. The whole torsion subgroup.

Finding the bound on M in the case of all torsion is rather involved. By the qualitative result in Section 5.13 we may assume that A is semi-abelian. Fix two primes p and l of good reduction for A, and let $F_p(T), F_l(T) \in \mathbb{Z}[T]$ and $\sigma, \tau \in \text{Aut}(\mathbb{Q}^{\text{alg}})$ be as in 5.13. Choose also some (K_p, σ) and (K_l, τ) models of ACFA and extending $(\mathbb{Q}^{\text{alg}}, \sigma)$ and $(\mathbb{Q}^{\text{alg}}, \tau)$ respectively. We may, and will, identify the fields K_p and K_l. That is, we are working in a large algebraically closed field $K_p = K_l$, with two distinguished automorphisms σ and τ. Write $F_p(T) = \sum_{i=0}^{2d} m_i T^i$ and $F_l(T) = \sum_{i=0}^{2d} n_i T^i$. Put

$$S_q = \left\{ (a_0, \ldots, a_{2d}) \mid \sum_i [m_i] a_i = 0 \right\}$$

and

$$\tilde{S}_q = \{ a \in A(K_p) \mid (a, \sigma(a), \ldots, \sigma^{2d}(a)) \in S_q \},$$

and define the sets S_l and $\tilde{S}_l$ similarly. The groups $\tilde{S}_q$ and $\tilde{S}_l$ are modular in the structures (K_p, σ) and (K_l, τ) respectively.

Set also

$$\omega_1 = 2d \dim(X) \quad (\leq 2d \dim(A)), \quad \omega_2 = 2\omega_1 + 1, \quad \omega_3 = 2d\omega_2 \dim(A).$$

We know that if b is any element of A, then $(X - b) \cap \tilde{S}_q$ is of the form $C_b(K_p) \cap \tilde{S}_q$, with $C_b(K_p) \cap \tilde{S}_q$ Zariski dense in C_b, and where C_b is a finite union of cosets of algebraic subgroups of A; moreover we know that C_b is defined over $L(\sigma^{-\omega_1}(b), \ldots, \sigma^{\omega_1}(b))$, where $L = \sigma(L)$ is a finite Galois extension of K over which X is defined.

We first define the various components of C_b uniformly in b. For that we need to look at the Zariski closure of the set $(\sigma^{-\omega_1}(b), \ldots, \sigma^{\omega_1}(b), a)$ when b ranges over $A(K_p)$, $a \in \tilde{S}_q$ and $a + b \in X$, and more precisely at the algebraic set which defines it, i.e., at the algebraic subset S of $(A^{\omega_2} \times A)^{2d+1}$ defined by:

(1) $(x_0, \ldots, x_{2d}) \in S_q$;

(2) $(x_i + y_{0,i}) \in \sigma^i(X)$ for $0 \leq i \leq 2d$;

(3) $y_{j,i+1} = y_{j+1,i}$ for $0 \leq i \leq 2d - 1$, $-\omega_1 \leq j \leq \omega_1$.

A word about the indices: x_i corresponds to $\sigma^i(a)$, and $y_{j,i}$ to $\sigma^i(\sigma^j(b))$. One verifies that

$$\dim(S) = 2d \dim(A) + (2d + 1) \dim(X) + 2\omega_1 \dim(A) \leq (4d + \omega_2) \dim(A),$$
$$\deg(S) \leq \deg(A)^{2\omega_1} \deg(S_q)(\deg(X)d_+)^{2d+1}.$$

We now apply Lemma 5.4 to S and obtain a set of irreducible subvarieties W_i of $A^{\omega_2} \times A$, such that, if $\rho : A^{\omega_2} \times A \to A^{\omega_2}$ is the projection, the following conditions hold:

(i) If $(b, a) \in \tilde{S} =_{\text{def}} \{(b, a) \mid ((b, a), \ldots, \sigma^{2d}(b, a)) \in S\}$, then for some i we have $(b, a) \in W_i$ and $b \notin W_i^*$.

(ii) The set $\tilde{W}_i = W_i(K_p) \cap \tilde{S}$ is Zariski dense in W_i.

(iii) $\sum_i \deg(W_i) \leq \sum_{i=0}^{\dim(S)} \deg(S)^{2^i} \leq 2 \deg(S)^{2^{\dim(S)}}$.

By (ii) we may choose $(b, a) \in \tilde{S}$ which is a generic of W_i (in the sense of the Zariski topology). Since W_i is irreducible, we know that the irreducible components of $\rho^{-1}(b) \cap W_i$ are conjugate over $L(b)$. Since $\tilde{S}_q$ is modular we also know that these components are cosets of some algebraic subgroups of A. Let A_i be the algebraic subgroup of A such that the component of $\rho^{-1}(b) \cap W_i$ containing (b, a) is a coset of A_i. If c is a generic of A_i, then $(b, a+c)$ is a generic of $\rho^{-1}(b) \cap W_i$ and therefore $(b, a + c)$ is a generic of W_i. Since W_i is closed this shows that $W_i = W_i + ((0) \times A_i)$, and therefore that for every $y \in \rho(W_i)$, $\rho^{-1}(y) \cap W_i$ is a union of cosets of A_i. Furthermore these cosets are finite in number if $y \notin W_i^*$.

5.16. Working on W_i. Fix i, let $W_i^0 = W_i \setminus \rho^{-1}(W_i^*)$, $B_i = A/A_i$ and let $\theta_i : A \to B_i$ be the natural projection. For $j \in \mathbb{Z}$ let $\tau^j(B_i) = A/\tau^j(A_i)$ and $\tau^j(\theta_i) = \tau^j \theta_i \tau^{-j} : A \to \tau^j(B_i)$. Define also $B_i' = \prod_{j=0}^{\omega_3} \tau_j(B_i)$, $C = A^{\omega_2}$.

We are interested in the set $\Theta_i = \{(b, \theta_i(a)) \mid (b, a) \in \tilde{W}_i,\ b \notin W_i^*, b \in \tilde{S}_l^{\omega_2}\}$. Note that if $(b, c) \in \Theta_i$ then $c \in L(b)^{\text{alg}}$. From $\deg_\tau(b) \leq 2d\omega_2 \dim(A) = \omega_3$ we deduce that $\deg_\tau(c) \leq \omega_3$.

Let $R \subseteq C^{\omega_3+1}$ be defined by

$$R = \left\{ (y_0, \ldots, y_{\omega_3}) \in C^{\omega_3+1} \;\middle|\; \sum_{j=0}^{2d} [n_j] y_{i+j} = 0 \text{ for } 0 \leq i \leq \omega_3 - 2d \right\}.$$

Then $\dim(R) = \omega_3$ and $\deg(R) \leq \deg(S_l)^{\omega_2(\omega_3-2d+1)}$. Consider now the closed set $U_i \subseteq (C \times B_i) \times (C \times \tau(B_i)) \times \cdots \times (C \times \tau^{\omega_3}(B_i))$ which is the Zariski closure of the set of tuples $((y_0, z_0), \ldots, (y_{\omega_3}, z_{\omega_3}))$ satisfying:

- $(y_0, \ldots, y_{\omega_3}) \in R$.
- For every $0 \leq j \leq \omega_3$, and x_j such that $\tau^j(\theta)(x_j) = z_j$, $(y_j, x_j) \in \tau^j(W_i^0)$.

We also let V_i be the image of U_i in B_i' under the natural projection

$$(C \times A)^{\omega_3+1} \to A^{\omega_3+1} \to \prod_{j=0}^{\omega_3} \tau^j(B_i).$$

Then $\deg(V_i) \leq \deg(U_i) \leq \deg(R)\deg(W_i)^{\omega_3+1}$, and $\dim(U_i) = \dim(V_i) \leq \dim(R) \leq \omega_3$.

By Theorem 5.14 the Zariski closure Z_i of $V_i \cap T_{p'}(B_i')$ is a finite union of cosets of definable subgroups of B_i', and $\dim(Z_i) \leq \omega_3$. Moreover,

$$\deg(Z_i) \leq (\deg(V_i)^{2d+1} \deg(S_q)^{\omega_3+1})^{2^{2d\dim(R)}}.$$

CLAIM. $\tilde{Z}_{i,\tau} = \{a \in B_i \mid (a, \tau(a), \ldots, \tau^{\omega_3}(a)) \in Z_i\}$ is a finite union of cosets of τ-definable subgroups of B_i of finite SU-rank.

PROOF. Being a coset of a subgroup is a property preserved under homomorphisms and intersections. The first assertion follows since $\tilde{Z}_{i,\tau}$ is obtained from Z_i using projections, intersections, and the maps τ, τ^{-1}.

Let a be a generic of $\tilde{Z}_{i,\tau}$. Since $\dim(Z_i) \leq \omega_3$, we have

$$\operatorname{tr} \deg(a, \tau(a), \ldots, \tau^{\omega_3}(a)) \leq \omega_3,$$

which implies that $\deg_\tau(a) < \infty$ and that the SU-rank of $\tilde{Z}_{i,\tau}$ in (K_l, τ) is finite. □

Now consider the set

$$\tilde{U}_{i,\tau} = \{(b,a) \in \tilde{S}_l^{\omega_2} \times B_i \mid ((b,a), \ldots, \tau^{\omega_3}(b,a)) \in U_i, \ (a, \ldots, \tau^{\omega_3}(a)) \in Z_i\}.$$

Since $Z_i \subseteq B_i'$ and $((b,a), \ldots, \tau^{\omega_3}(b,a)) \in U_i$ implies $b \in \tilde{S}_l^{\omega_2}$, Lemma 5.3 implies that the Zariski closure of $\tilde{U}_{i,\tau}$ has degree $\leq (\deg(U_i)\deg(Z_i))^{2^{\dim(R)}}$, and $\tilde{U}_{i,\tau} \subseteq \tilde{S}_l^{\omega_2} \times \tilde{Z}_{i,\tau}$. By the claim, $\tilde{Z}_{i,\tau}$ is a union of cosets of definable subgroups of B_i of finite SU-rank. We also know that $\tilde{Z}_{i,\tau}$ is modular (since every element in it is algebraic over a tuple from $\tilde{S}_l$). Hence, every definable subset of $\tilde{S}_l^{\omega_2} \times \tilde{Z}_{i,\tau}$ is a Boolean combination of cosets of definable subgroups of $A^{\omega_2} \times B_i$.

This implies that $\{(\theta_i^{-1}(a) + b_0) \mid (b_{-\omega_1}, \ldots, b_{\omega_1}, a) \in \tilde{U}_{i,\tau}\}$ is the union of at most $(\deg(U_i)\deg(Z_i))^{2^{\dim(R)}}$ cosets of definable subgroups of A.

5.17. THEOREM. *Let A be a commutative algebraic group defined over a number field K and X a subvariety of A. Then the Zariski closure of $X \cap \operatorname{Tor}(A)$ consists of finitely many cosets of algebraic subgroups of A, and a bound on the number M of these cosets can be effectively computed (modulo the choice of the primes p and l).*

PROOF. The first assertion is proved in Section 5.13. It remains to show that the results of the previous paragraph give us the bound. For that we need to show:

CLAIM. If $c \in \mathrm{Tor}(A) \cap X$, if $c = a + b$ with $a \in \mathrm{Tor}_{p'}(A)$ and $b \in \mathrm{Tor}_p(A)$ and $b^* = (\sigma^{-\omega_1}(b), \ldots, \sigma^{\omega_1}(b))$, then $(b^*, \theta_i(a)) \in \tilde{U}_{i,\tau}$ for some i.

By definition, $a \in \tilde{S}_q$ and $a + b \in X$ so that $(b^*, a) \in \tilde{S}$ (see Section 5.15). Choose i such that $(b^*, a) \in W_i^0$. Then $\tau^j(b^*, a) \in \tau^j(W_i^0)$ and $\tau^j(b^*) \in \tilde{S}_l^{\omega_2}$ for every j. Hence, $((b^*, \theta_i(a)), \ldots, \tau^{\omega_3}(b^*, \theta_i(a)) \in U_i$, so that $(b^*, \theta_i(a)) \in \tilde{U}_{i,\tau}$. □

Note also that if $(c_{-\omega_1}, \ldots, c_{\omega_1}, d) \in \tilde{U}_{i,\tau}$, then $(c_0 + d) \in X$, so that the Zariski closure of the coset containing $a + b$ is contained in X.

To conclude, we obtain the following bound on the number of cosets: M is bounded by the sum over i of the degrees of the Zariski closures of $\tilde{U}_{i,\tau}$. Unwinding, we get

$$M \leq \sum_i \deg \tilde{U}_{i,\tau} \leq \sum_i (\deg(U_i) \deg(Z_i))^{2^{\omega_3}},$$

$$\sum_i \deg(U_i) \deg(Z_i) \leq \deg(S_q)^{(\omega_3+1)2^{2d\omega_3}} \sum_i \deg(V_i)^{(2d+1)2^{2d\omega_3}+1},$$

$$\sum_i \deg(V_i) \leq \deg(S_l)^{\omega_2(\omega_3-2d+1)} \sum_i \deg(W_i)^{\omega_3+1},$$

$$\sum_i \deg(W_i) \leq 2\deg(S)^{2^{\dim(S)}}$$

$$\leq 2(\deg(A)^{2\omega_1} \deg(S_q) \deg(X)^{2d+1} d_+^{2d+1})^{2^{(4d+\omega_2)\dim(A)}},$$

so that

$$M \leq 2^{M_1} \deg(S_q)^{M_2} \deg(S_l)^{M_3} \deg(A)^{M_4} (d_+ \deg(X))^{M_5},$$

where

$$M_1 = (\omega_3 + 1)((2d+1)2^{2d\omega_3} + 1)2^{\omega_3},$$
$$M_2 = 2^{(4d+\omega_2)\dim(A)} M_1 + (\omega_3 + 1)2^{(2d+1)\omega_3},$$
$$M_3 = \omega_2(\omega_3 - 2d + 1)((2d+1)2^{2d\omega_3} + 1)2^{\omega_3}$$
$$M_4 = 2\omega_1 2^{(4d+\omega_2)\dim(A)} M_1,$$
$$M_5 = (2d+1)2^{(4d+\omega_2)\dim(A)} M_1.$$

The order of magnitude of ω_3 is $8d^2 \dim(A) \dim(X) \leq 8d^2 \dim(A)^2$.

6. Some Other Applications

In this section we state without proofs some other applications of the results on difference fields. We start with a result of Hrushovski, and conclude with two results by T. Scanlon.

6.1. Reduction of a conjecture of Lang. Let A be a commutative algebraic group defined over a number field K, let Γ be the division group of $A(K)$, i.e., the set of elements $a \in A(K^{\mathrm{alg}})$ such that $[m]a \in A(K)$ for some non-zero integer m. A conjecture of Lang states that if X is a subvariety of A containing no cosets of infinite algebraic subgroups of A, then $X \cap \Gamma$ is finite.

The techniques used in the previous paragraph give the following reduction of the conjecture (also proved by Raynaud, Hindry, McQuillan), with effective bounds:

THEOREM [Hrushovski 1995]. *Let A be a commutative algebraic group defined over a number field, and let Γ be the division group of $A(K)$. Suppose that X is a subvariety of A, containing no cosets of infinite algebraic subgroups of A. One can effectively find an integer M such that $\Gamma \cap X(K^{\mathrm{alg}}) \subseteq \frac{1}{M} A(K)$. Moreover one can effectively find coset representatives r_i of $A(K)/M A(K)$ such that $\Gamma \cap X(K^{\mathrm{alg}}) \subseteq \bigcup_i \frac{r_i}{M} + A(K)$.*

IDEA OF THE PROOF. Fix a prime p of good reduction for A and let $\Gamma_{p'}$ denote the p'-division subgroup of $A(K)$, i.e. we require that the integer m in the definition of division group be prime to p. Let $\mathbb{F}_q$ be the residue field of K at p, σ a lifting of the Frobenius $\phi_q : x \mapsto x^q$ to K, and $F_p(T) \in \mathbb{Z}[T]$ the Weil polynomial.

One first shows that $(\sigma - 1)F_p(\sigma)$ vanishes on $\Gamma_{p'}$. Using the fact that $\mathrm{Ker}(F_p(\sigma))$ is orthogonal to the fixed field and the assumption on X, one then shows that that $X(K^{\mathrm{alg}}) \cap \mathrm{Ker}((\sigma - 1)F_p(\sigma))$ is contained in finitely many cosets of $\mathrm{Ker}(\sigma - 1) = A(\mathrm{Fix}(\sigma))$. From this one deduces a number M_p such that $M_p(X(K^{\mathrm{alg}}) \cap \Gamma_{p'}) \subseteq A(K)$.

Choosing another prime l of good reduction for A one obtains a number M_l such that $M_l(X(K^{\mathrm{alg}}) \cap \Gamma_{l'}) \subseteq A(K)$. Then $M_p M_l(X(K^{\mathrm{alg}}) \cap \Gamma) \subseteq A(K)$. The bound $M_p M_l$ is effective, modulo the choice of the two primes p and l. $\square$

6.2. CONJECTURE (TATE AND VOLOCH). *Let G be a semi-abelian variety defined over $\mathbb{C}_p$, and let X be a subvariety of G. There is a constant N such that for any $P \in \mathrm{Tor}(G)$, either $P \in X$ or $d(P, X) > N$.*

Here $\mathbb{C}_p$ is the completion of the algebraic closure of $\mathbb{Q}_p$ (with respect to the p-adic valuation on $\mathbb{Q}_p^{\mathrm{alg}}$), and $d(P, X)$ is a p-adic distance associated to the valuation. If X is a subvariety of an affine space, one defines $d(P, X) = \max\{p^{-v(f(P))} \mid f \in I\}$, where I is the ideal of polynomials defining X. In the general case, one extends the definition by using a cover by affine sets.

When G is a torus, this conjecture is a theorem [Tate and Voloch 1996]. Hrushovski [1996a] proved the conjecture when G is over $\mathbb{Q}_p^{\mathrm{alg}}$, has good reduction, and for prime-to-p torsion points. Scanlon [1998; 1999a] proved the conjecture when G is defined over $\mathbb{Q}_p^{\mathrm{alg}}$. He considers liftings σ of the Frobenius, a Weil polynomial $F_q(T)$, and uses the fact that $\mathrm{Tor}(G) \subseteq \mathrm{Ker}((\sigma - 1)F_q(\sigma))$.

6.3. Drinfeld modules. Let K be an algebraically closed field of positive characteristic p and of positive transcendence degree. Consider the ring $\mathrm{End}_K(\mathbb{G}_a)$ of endomorphisms of $\mathbb{G}_a$ defined over K. Then $\mathrm{End}_K(\mathbb{G}_a)$ is isomorphic to the twisted polynomial ring $K[\phi_p]$. Let $A = \mathbb{F}_p[T]$ and view it as a subring of K, by identifying T with some transcendental $t \in K$.

A *Drinfeld module* (*over* A) is given by a ring homomorphism $\varphi : A \to$ $\mathrm{End}_K(A)$ so that if $\varphi(T) = \sum_{i=0}^{n} a_i \phi_p^i$, then $a_0 = t$ and $a_n = 1$.

THEOREM [Scanlon 1999b]. *Let φ be a Drinfeld module. Consider K^N as an A-module via φ. If X is a subvariety of K^N then the intersection of X, the A-torsion subgroup of K^N (that is, $\{x \in K^N \mid \varphi(a)(x) = 0$ for some non-zero $a \in A\}$) is a finite union of translates of A-torsion subgroups of algebraic subgroups of K^N.*

References

[Ax 1968] J. Ax, "The elementary theory of finite fields", *Ann. of Math.* (2) **88** (1968), 239–271.

[Chatzidakis and Hrushovski 1999] Z. Chatzidakis and E. Hrushovski, "Model theory of difference fields", *Trans. Amer. Math. Soc.* **351**:8 (1999), 2997–3071.

[Chatzidakis et al. 1999] Z. Chatzidakis, E. Hrushovski, and Y. Peterzil, "The model theory of difference fields II: periodic ideals and the trichotomy in all characteristics", preprint, 1999. Available at http://www.logique.jussieu.fr/www.zoe/.

[Cherlin and Hrushovski 1998] G. Cherlin and E. Hrushovski, "Large finite structures", preprint, 1998. Available at http://www.math.rutgers.edu/~cherlin/Paper. Earlier version: "Smoothly approximable structures", 1994.

[Cohn 1965] R. M. Cohn, *Difference algebra*, Tracts in Math. **17**, Interscience, New York, 1965.

[Delon 1998] F. Delon, "Separably closed fields", pp. 143–176 in *Model theory and algebraic geometry*, edited by E. Bouscaren, Lecture Notes in Math. **1696**, Springer, Berlin, 1998.

[Duret 1980] J.-L. Duret, "Les corps faiblement algébriquement clos non séparablement clos ont la propriété d'indépendance", pp. 136–162 in *Model theory of algebra and arithmetic* (Karpacz, 1979), edited by L. Pacholski et al., Lecture Notes in Math. **834**, Springer, Berlin, 1980.

[Fulton 1984] W. Fulton, *Intersection theory*, Ergebnisse der Mathematik (3) **2**, Springer, Berlin, 1984. Second edition, 1998.

[Hrushovski 1991] E. Hrushovski, "Pseudo-finite fields and related structures", manuscript, 1991.

[Hrushovski 1995] E. Hrushovski, "The Manin–Mumford conjecture and the model theory of difference fields", preprint, 1995.

[Hrushovski 1996a] E. Hrushovski, 1996. E-mail to José Felipe Voloch.

[Hrushovski 1996b] E. Hrushovski, "The first-order theory of the Frobenius", preprint, 1996.

[Hrushovski 1996c] E. Hrushovski, "The Mordell–Lang conjecture for function fields", *J. Amer. Math. Soc.* **9**:3 (1996), 667–690.

[Hrushovski and Pillay 1987] U. Hrushovski and A. Pillay, "Weakly normal groups", pp. 233–244 in *Logic colloquium '85* (Orsay, 1985), edited by the Paris Logic Group, Stud. Logic Found. Math. **122**, North-Holland, Amsterdam, 1987.

[Hrushovski and Pillay 1994] E. Hrushovski and A. Pillay, "Groups definable in local fields and pseudo-finite fields", *Israel J. Math.* **85**:1-3 (1994), 203–262.

[Hrushovski and Pillay 1995] E. Hrushovski and A. Pillay, "Definable subgroups of algebraic groups over finite fields", *J. Reine Angew. Math.* **462** (1995), 69–91.

[Hrushovski and Ž. Sokolović 1994] E. Hrushovski and Ž. Sokolović, "Minimal subsets of differentially closed fields", 1994. To appear in *Trans. Amer. Math. Soc.*

[Hrushovski and Zilber 1996] E. Hrushovski and B. Zilber, "Zariski geometries", *J. Amer. Math. Soc.* **9**:1 (1996), 1–56.

[Kim and Pillay 1997] B. Kim and A. Pillay, "Simple theories", *Ann. Pure Appl. Logic* **88**:2-3 (1997), 149–164.

[Lang 1959] S. Lang, *Abelian varieties*, Interscience, New York, 1959. Reprinted by Springer, New York, 1983.

[Lang 1991] S. Lang, *Number theory III: Diophantine geometry*, Encyclopaedia of Math. Sciences **60**, Springer, Berlin, 1991.

[Macintyre 1997] A. Macintyre, "Generic automorphisms of fields", *Ann. Pure Appl. Logic* **88**:2-3 (1997), 165–180.

[Macintyre ≥ 2000] A. Macintyre, "Nonstandard Frobenius". In preparation.

[Mumford 1974] D. Mumford, *Abelian varieties*, 2nd ed., Tata Institute Studies in Mathematics **5**, Oxford U. Press, London, 1974.

[Pillay 1997] A. Pillay, "ACFA and the Manin–Mumford conjecture", pp. 195–205 in *Algebraic model theory* (Toronto, 1996), edited by B. Hart et al., NATO Adv. Studies Inst. Series, C **496**, Kluwer, Dordrecht, 1997.

[Pillay 1998] A. Pillay, "Definability and definable groups in simple theories", *J. Symbolic Logic* **63**:3 (1998), 788–796.

[Raynaud 1983] M. Raynaud, "Around the Mordell conjecture for function fields and a conjecture of Serge Lang", pp. 1–19 in *Algebraic geometry* (Tokyo/Kyoto, 1982), Lecture Notes in Math. **1016**, Springer, Berlin, 1983.

[Scanlon 1998] T. Scanlon, "p-adic distance from torsion points of semi-abelian varieties", *J. Reine Angew. Math.* **499** (1998), 225–236.

[Scanlon 1999a] T. Scanlon, "The conjecture of Tate and Voloch on p-adic proximity to torsion", *Internat. Math. Res. Notices* **1999**:17 (1999), 909–914.

[Scanlon 1999b] T. Scanlon, "Diophantine geometry of the torsion of a Drinfeld module", Preprint, 1999.

[Serre 1985/86] J.-P. Serre, "Résumés des cours au Collège de France", *Annuaire du Collège de France* (1985/86), 95–99.

[Shelah 1980] S. Shelah, "Simple unstable theories", *Ann. Math. Logic* **19**:3 (1980), 177–203.

[Suzuki 1982] M. Suzuki, *Group theory, I*, Grundlehren der mat. Wiss. **247**, Springer, Berlin, 1982.

[Tate and Voloch 1996] J. Tate and J. F. Voloch, "Linear forms in p-adic roots of unity", *Internat. Math. Res. Notices* **1996**:12 (1996), 589–601.

[Wagner 1997] F. Wagner, "Groups in simple theories", preprint, 1997. Available at
http://www.desargues.univ-lyon1.fr/home/wagner/publ.html.

[Weil 1971] A. Weil, *Courbes algébriques et variétés abéliennes*, Hermann, Paris, 1971.

ZOÉ CHATZIDAKIS
ÉQUIPE DE LOGIQUE MATHÉMATIQUE (CNRS - UPRESA 7056)
UFR DE MATHÉMATIQUES, CASE 7012
UNIVERSITÉ PARIS 7
2, PLACE JUSSIEU
75251 PARIS CÉDEX 05
FRANCE
    zoe@logique.jussieu.fr

Model Theory, Algebra, and Geometry
MSRI Publications
Volume **39**, 2000

Notes on o-Minimality and Variations

DUGALD MACPHERSON

ABSTRACT. The article surveys some topics related to o-minimality, and
is based on three lectures. The emphasis is on o-minimality as an analogue of strong minimality, rather than as a setting for the model theory
of expansions of the reals. Section 2 gives some basics (the Monotonicity
and Cell Decomposition Theorems) together with a discussion of dimension. Section 3 concerns the Peterzil–Starchenko Trichotomy Theorem (an
o-minimal analogue of Zil'ber Trichotomy). There follows some material
on definable groups, with powerful applications of the Trichotomy Theorem in work by Peterzil, Pillay and Starchenko. The final section introduces
weak o-minimality, P-minimality, and C-minimality. These are analogues
of o-minimality intended as settings for certain henselian valued fields with
extra structure.

1. Introduction

This paper is a survey of selected topics in and around o-minimality. The
emphasis is on analogies with stability theory, and there is little here on analytic
expansions of the reals. On the other hand, there is quite a lot on dimension in
o-minimal and related structures. I have concentrated on algebraic examples.

Section 2 is introductory, and covers definable functions, cell decomposition,
dimension for definable sets, prime models, and definable types. Section 3 is on
groups definable in o-minimal theories: there I describe a trichotomy theorem
due to Peterzil and Starchenko, and consequences, due to them and Pillay, for
definable groups. In Section 4, I leave o-minimality and turn to other settings
(weak o-minimality, C-minimality, P-minimality) which are superficially similar,
and survey some of the main examples and results.

Generally I have omitted proofs, but where possible try to give the idea of a
proof. As a general source for o-minimality, I recommend [van den Dries 1998;
1996]. Much of the material from Section 2 comes from [Knight et al. 1986] and
[Pillay and Steinhorn 1986], and the latter paper gives an excellent introduction
to the subject.

2. Basics of o-minimality

We consider first-order structures $\mathcal{M} = (M, <, \ldots)$, where M is the domain, $<$ is a binary relation symbol interpreted by a dense total order on M, and there may be other symbols for relations, functions or constants in the language. The assumption that $<$ is dense is not necessary for all the theory, but holds in the examples of interest to us. Indeed, by results from [Pillay and Steinhorn 1987; 1988], any discrete o-minimal structure is essentially trivial, in the sense that definable functions are given piecewise by translations.

DEFINITION 2.0.1. The above structure $\mathcal{M}$ is *o-minimal* if every definable subset of M is a finite union of singletons and open intervals (with endpoints in $M \cup \{\infty, -\infty\}$).

REMARKS. 1. Here, as throughout these notes, 'definable' means 'definable with parameters'.
2. It is crucial that the intervals are not just convex sets, but have endpoints in $M \cup \{+\infty, -\infty\}$. Without this we have weak o-minimality, with a much weaker structure theory (see Section 4).
3. There is an obvious question whether, if $\mathcal{M}$ is o-minimal and $\mathcal{N}$ is elementarily equivalent to $\mathcal{M}$, then $\mathcal{N}$ must also be o-minimal. The answer is positive (see Remark 2 after Theorem 2.1.3 below).
4. The class of o-minimal structures is closed under reducts (so long as $<$ stays in the language). Frequently a structure in a rich language is proved to be o-minimal by quantifier-elimination, and it follows that all reducts (with the ordering still in the language) are o-minimal. Also, o-minimality is closed under expansions by constants.
5. The definition says that every definable subset of M is quantifier-free definable just using the symbols $=$ and $<$. This suggests an analogy with strong minimality, which says that *in all models of the theory*, every definable set is quantifier-free definable just from $=$.
6. The order topology on M has a uniformly definable basis (of intervals). Likewise, the induced topology on M^n has a uniformly definable basis. Hence, given a definable function $f : U \to M^n$, say, where $U \subset M^m$, the condition 'f is continuous at $\bar{a}$' is first-order expressible, uniformly in $\bar{a}$.
7. Definable continuous partial functions $M \to M$ satisfy the intermediate value property.

EXAMPLES. The following structures are o-minimal. I emphasise that there is a large literature now on the rich supply of o-minimal expansions of the reals, not touched on here.

1. $(\mathbb{Q}, <)$.
2. $(\mathbb{Q}, <, +)$.
3. $\mathcal{R} = (\mathbb{R}, <, +, -, \cdot, 0, 1)$. By Tarski's quantifier elimination, we need only check that atomic formulas with parameters define finite unions of intervals.

This is clear, for one need only consider the formulas $\sum_{i=0}^{n} a_i x^i < 0$ and $\sum_{i=0}^{n} a_i x^i = 0$.

4. $(\mathfrak{R}, \exp)$. Here, o-minimality follows from Wilkie's model-completeness result [1996].

The following result is at the root of most of the theory of o-minimality.

THEOREM 2.0.2 [Pillay and Steinhorn 1986]. *Let* $\mathfrak{M}$ *be o-minimal, and* $f :$ $(a, b) \rightarrow M$ *be a definable function with domain* (a, b) *(possibly* $a = -\infty$, *or* $b = \infty$). *Then there are points* $a = a_0 < \cdots < a_{k+1} = b$ *such that for each* $j = 0, \ldots, k$, *the restriction* $f|_{(a_j, a_{j+1})}$ *is either constant, or a strictly monotonic and continuous bijection to an interval.*

SKETCH PROOF. It suffices to prove, for any definable function $f : I \rightarrow M$, where I is an interval, that

(i) there is an infinite subinterval of I on which f is constant or injective;
(ii) if f is injective, then it is strictly monotonic on a subinterval,
(iii) if f is strictly monotonic, then f is continuous on a subinterval.

For given (i)–(iii), let X be the set of $x \in (a, b)$ such that on some open interval containing x, f is constant, or strictly monotonic and continuous. By (i)-(iii) above, $(a, b) \setminus X$ is finite, so we may assume (by throwing away finitely many points and replacing (a, b) by subintervals) that $(a, b) \setminus X = \varnothing$, so f is continuous on (a, b). There are finitely many possible kinds of local behaviour, so, after partitioning (a, b) further we may suppose that f has the same local behaviour throughout (a, b). If, for example, f is locally strictly increasing everywhere, then it follows easily by o-minimality that f is strictly increasing everywhere.

I sketch a proof of (i) above. We may assume that all sets $f^{-1}(x)$ are finite (for otherwise by o-minimality some set $f^{-1}(x)$ contains an interval, and f is constant on this interval). Hence $f(I)$ is infinite, so contains an interval J. Define $g : J \rightarrow I$ by $g(y) := \text{Min}\{x \in I : f(x) = y\}$, find an infinite interval $K \subset g(J)$, and observe that $f|_K$ is injective. $\qquad\square$

REMARKS. 1. If $\mathfrak{M}$ is an expansion of an ordered field, then the notion of differentiability makes sense, and we can sharpen the above theorem to arrange that f is continuously differentiable on each (a_j, a_{j+1}) (see Chapter VII of [van den Dries 1998]). In fact, for any n, we can arrange that f is $C^{(n)}$ (that is, n times continuously differentiable) on each (a_j, a_{j+1}). However, as n increases we may need more and more intervals, so we cannot expect to arrange that f is $C^{(\infty)}$ on each interval.

2. Here, as elsewhere in the theory, we have good control over parameters. In particular, we may choose the a_i so that they are definable over the parameters used to define f.

2.1. Cell decomposition. The notion of o-minimality tells us about definable sets in one variable. The cell decomposition theorem (and its variants) carry

such information to definable sets in several variables. I follow the treatment
from [van den Dries 1998].

Given a definable $X \subseteq M^n$, let

$$C(X) := \{f : X \to M, f \text{ is definable and continuous}\},$$

and let $C_\infty(X) := C(X) \cup \{-\infty, +\infty\}$ (here, ∞ denotes the 'function' on X
taking value ∞ everywhere, and $-\infty$ is defined similarly.) Suppose that $f, g \in
C_\infty(X)$ and that $(\forall \bar{x} \in X)(f(\bar{x}) < g(\bar{x}))$. Then

$$(f, g)_X := \{(\bar{x}, y) \in X \times M : f(\bar{x}) < y < g(\bar{x})\}.$$

DEFINITION 2.1.1. Let $(i_1, \ldots, i_m)$ be a sequence of zeros and ones. Then an
$(i_1, \ldots, i_m)$-*cell* is a definable subset of M^m, defined as follows by induction on
m.

(i) A (0)-cell is a singleton of M, and a (1)-cell is a non-empty open interval,
    possibly unbounded.
(ii) Suppose $(i_1, \ldots, i_m)$-cells have been defined. Then an $(i_1, \ldots, i_m, 0)$-cell
    is the graph of a function $f \in C(X)$, where X is an $(i_1, \ldots, i_m)$-cell. An
    $(i_1, \ldots, i_m, 1)$-cell is a set $(f, g)_X$, with X some $(i_1, \ldots, i_m)$-cell and $f, g \in
    C_\infty(X)$ with $f(\bar{x}) < g(\bar{x})$ for all $\bar{x} \in X$.

A *cell* is an $(i_1, \ldots, i_m)$-cell for some $i_1, \ldots, i_m \in \{0, 1\}$.

REMARKS. 1. The numbers $i_1, \ldots, i_m$ are uniquely determined by the cell.
2. A cell in M^m is open if and only if it is a $(1, \ldots, 1)$-cell.
3. More generally, let X be an $(i_1, \ldots, i_m)$-cell, let $k := i_1 + \cdots + i_m$, and suppose
   we have $\lambda(1) < \cdots < \lambda(k)$ and $i_{\lambda(1)} = \ldots = i_{\lambda(k)} = 1$. Let $\pi : M^m \to M^k$
   project to the $\lambda(1), \ldots, \lambda(k)$ coordinates. Then π is a homeomorphism onto
   an open cell in M^k.
4. Cells are *definably connected*; that is, a cell X cannot be expressed as the
   disjoint union of two non-empty definable sets which are open in X. If $M = \mathbb{R}$,
   then cells are even connected.

DEFINITION 2.1.2. A *decomposition* of M^m is a partition of M^m into finitely
many cells, defined as follows by induction.

(i) Any partition of M into finitely many disjoint cells is a decomposition.
(ii) A decomposition of M^{m+1} is a finite partition of M^{m+1} into cells, such that
    if $\pi : M^{m+1} \to M^m$ is the projection onto the first m coordinates, then the
    set of π-projections of the cells forms a decomposition of M^m.

THEOREM 2.1.3 [Knight et al. 1986]. *For each $n > 0$, the following statements
hold.*

(I)$_n$ *Given definable $A_1, \ldots, A_k \subseteq M^n$, there is a decomposition of M^n which
    partitions each of the A_i.*

$(II)_n$ *Given definable $A \subseteq M^n$ and a definable function $f : A \to M$, there is a decomposition $\mathcal{D}$ of M^n partitioning A, such that for each $B \in \mathcal{D}$ with $B \subseteq A$, $f|_B : B \to M$ is continuous.*

$(III)_n$ *Suppose that $Y \subseteq M^{n+1}$ is definable. For any $\bar{a} \in M^n$, let*

$$Y_{\bar{a}} := \{x \in M : (\bar{a}, x) \in Y\}.$$

Then there is a number N (depending on Y) such that any finite set of the form $Y_{\bar{a}}$ has size at most N.

REMARKS. 1. If condition $(III)_n$ holds for all n for a structure $\mathcal{M}$ (not necessarily o-minimal), then we say that $\mathcal{M}$ is *uniformly bounded*. This is a property of $\mathrm{Th}(\mathcal{M})$, that is, it is preserved by elementary equivalence.

2. Uniform boundedness ensures that if $\mathcal{M}$ is o-minimal, and $\mathcal{N}$ is elementarily equivalent to $\mathcal{M}$, then $\mathcal{N}$ is o-minimal. For suppose that $Y \subset N$ is definable by a formula $\phi(x, \bar{a})$, and let $\psi(x, \bar{a})$ define (uniformly in $\bar{a}$) the boundary of Y. By uniform boundedness, there is a natural number K such that for any $\bar{b}$ from M, $\psi(x, \bar{b})$ has at most K realisations. Now $\mathrm{Th}(\mathcal{M})$ says this, so it holds in $\mathcal{N}$, so $\psi(x, \bar{a})$ has at most K realisations, and hence (as $\mathrm{Th}(\mathcal{M})$ says that any maximal convex definable set has a supremum and infimum in $M \cup \{\infty, -\infty\}$) $\phi(x, \bar{a})$ is a union of finitely many singletons and intervals.

3. If $\mathcal{M}$ is an expansion of an ordered field, then for any $p > 0$ one can require that the definable functions in the cell decomposition are all $C^{(p)}$.

4. Because of the inductive definition of cells, the cell decomposition theorem makes possible many proofs by induction on the dimension of a definable set.

5. In $(I)_n$ the cells in the decomposition can be chosen definable over the parameters used to define the A_i (and a similar statement holds for $(II)_n$).

SKETCH PROOF. The proof of Theorem 2.1.3 is by simultaneous induction on n. $(I)_1$ holds by definition of o-minimality, and $(II)_1$ follows by the Monotonicity Theorem, whilst $(III)_1$ requires a direct argument which is really the crux of the whole proof, but which we omit. At the inductive step, we prove:

$(I)_m$, $(II)_m$, $(III)_m$ (for $m < n$) $\implies (I)_n$.

$(I)_m$ (for $m \leq n$), $(II)_m$ (for $m < n$) $\implies (II)_n$.

$(I)_m$, $(II)_m$ (for $m \leq n$) and $(III)_m$ (for $m < n$) $\implies (III)_n$.

I sketch the proof of $(I)_n$. For simplicity we suppose that $k = 1$ and $A := A_1$ (as the general case is similar). Let $\pi : M^n \to M^{n-1}$ drop the last coordinate, and for each $\bar{a} \in \pi(A)$ consider the fibre $A_{\bar{a}} := \{y \in M : (\bar{a}, y) \in A\}$, a definable subset of M. By o-minimality, $A_{\bar{a}}$ is a finite union of singletons and intervals, and by $(III)_{n-1}$ there is a bounded number of these. Inductively, we use $(I)_{n-1}$ to decompose the base, partitioning $\pi(A)$, to ensure that, for each cell of the base, all fibres $A_{\bar{a}}$ look the same as $\bar{a}$ ranges through the cell. (Essentially, this means that all these fibres must have the same number of singletons and intervals, occurring in the same order.) Hence there are finitely many definable partial functions $M^{n-1} \to M$ picking out these singletons and the endpoints of the intervals, and by $(II)_{n-1}$ we can ensure that these functions are continuous, partitioning the base further if necessary. Now piece this information together.

The proof of $(II)_n$ is also relatively straightforward, but requires the Monotonicity Theorem (not just piecewise continuity of unary functions). The idea is to reduce to the situation where for any $\bar{a} \in M^{n-1}$, the partial function $f(\bar{a}, y) : M \to M$ is continuous and monotonic (where defined), and for any $b \in M$, the function $f(\bar{x}, b) : M^{n-1} \to M$ is continuous where defined.

The proof of $(III)_n$ is intricate.   □

2.2. Definable closure and dimension. The next task is to describe dimension in o-minimal structures, in a way relevant also to Section 4. For an alternative treatment of model-theoretic dimension, see [van den Dries 1989].

Recall that if $A \subset M$ then the *algebraic closure* acl(A) of A is the union of the finite A-definable sets, and the *definable closure* dcl(A) is the union of the finite A-definable singletons. In general, dcl(A) $\subseteq$ acl(A), but in an o-minimal structure $\mathcal{M}$, they are equal (because in a finite set, we can define the least element, the next least element, and so on).

NOTATION. If $A \subseteq M$ and $\bar{a} \in M^n$ with $\bar{a} = (a_1, \ldots, a_n)$, I abuse notation by writing $A\bar{a}$ for $A \cup \{a_1, \ldots, a_n\}$.

DEFINITION 2.2.1. A *pregeometry* on a set X is a function cl : $\mathcal{P}(X) \to \mathcal{P}(X)$ (where $\mathcal{P}(X)$ denotes the power set of X) which satisfies the following conditions:

(i) for all $A \subseteq X$, $A \subseteq$ cl(A);
(ii) for all $A \subseteq X$, cl(cl(A)) = cl(A);
(iii) for all $A \subseteq X$, cl(A) = $\bigcup\{$cl(F) : $F \subseteq A, F$ finite$\}$;
(iv) (exchange) if $A \subseteq X$ and $b, c \in X$ with $b \in$ cl(Ac) \ cl(A), then $c \in$ cl(Ab).

In any structure, not necessarily o-minimal, algebraic closure satisfies (i)–(iii), and we show next that in the o-minimal case (iv) holds also. The exchange property holds in many other nice model-theoretic classes (for example strongly minimal sets, and some of the structures discussed in Section 4 below).

THEOREM 2.2.2 [Pillay and Steinhorn 1986]. *Let $\mathcal{M}$ be o-minimal, $A \subseteq M$, and $b, c \in M$. If $b \in$ dcl(Ac) \ dcl(A), then $c \in$ dcl(Ab).*

PROOF. We may suppose the base set $A = \varnothing$ (by adding constants for elements of A to the language), so that $b \in$ dcl(c) \ dcl($\varnothing$). There is a 0-definable (partial) function $f : M \to M$ with $b = f(c)$. We apply the Monotonicity Theorem to f. Since $b \notin$ dcl($\varnothing$), c lies in the interior of an open 0-definable interval I on which f is strictly monotonic. Now $J := f(I)$ is 0-definable, and since $c := f|_J^{-1}(b)$, $c \in$ dcl(b).   □

Theorem 2.2.2 gives us an important notion of dimension in o-minimal structures. Much of what follows is folklore, and possible sources are [Pillay 1988; Hrushovski and Pillay 1994]. We do not restrict to o-minimal structures in 2.2.3 and 2.2.5–2.2.8 below.

DEFINITION 2.2.3. A first-order structure $\mathcal{M}$ is *geometric* if algebraic closure has the exchange property (so defines a pregeometry) in all models of $\text{Th}(\mathcal{M})$, and $\mathcal{M}$ is uniformly bounded.

From Theorems 2.1.3 (III) and 2.2.2, we now have

COROLLARY 2.2.4. *Every o-minimal structure is geometric.*

There is a general dimension theory for geometric structures which I now sketch (some of it does not require uniform boundedness). First, observe that in a geometric structure $\mathcal{M}$, there is a general notion of independence: $I \subseteq M$ is *independent* if, for all $x \in I$, $x \notin \text{acl}(I \setminus \{x\})$. If $A \subseteq M$, then we can also talk of I being 'independent over A' (regard the elements of A as being interpreted by new constants). If $A \subseteq M$ is algebraically closed, then any two maximal independent subsets of A have the same size (by the proof that any two bases of a vector space have the same size), and we may call this size the *rank* of A (but we will not use this).

DEFINITION 2.2.5. Let $\mathcal{M}$ be geometric, $A \subseteq M$, and $\bar{a} \in M^n$. Then $\dim(\bar{a}/A)$ is the least cardinality of a subtuple $\bar{a}'$ of $\bar{a}$ such that $\bar{a} \subseteq \text{acl}(A\bar{a}')$. If $p(\bar{x}) \in S_n(A)$ (the set of complete n-types over A), then $\dim(p) = \dim(\bar{a}/A)$, for any $\bar{a}$ realising p in an elementary extension of $\mathcal{M}$.

LEMMA 2.2.6 [Pillay 1988]. *Let $\mathcal{M}$ be a geometric structure.*

(i) $\dim(\bar{a}/A)$ *is the cardinality of any maximal independent (over A) subtuple of $\bar{a}$.*

(ii) *If $A \subseteq B$ then $\dim(\bar{a}/A) \geq \dim(\bar{a}/B)$;*

(iii) $\dim(\bar{a}\bar{b}/A) = \dim(\bar{a}/A\bar{b}) + \dim(\bar{b}/A)$;

(iv) *If $p(\bar{x}) \in S_n(A)$ and $A \subseteq B$ then there is $p'(\bar{x}) \in S_n(B)$ such that $p \subseteq p'$ and $\dim(p) = \dim(p')$.*

REMARK. In the above, we can think of p' as a kind of 'non-forking extension of p', but we cannot control the number of non-forking extensions. For example, in the o-minimal structure $(\mathbb{Q}, <)$, if p is the unique 1-type over $\varnothing$, then any of the $2^{\aleph_0}$ non-algebraic extensions of p over $\mathbb{Q}$ is non-forking in this sense. Furthermore, the 'type amalgamation condition' or 'independence theorem' of simple theories (see Theorems 3.5 and 5.8 of [Kim and Pillay 1997]) cannot hold in any o-minimal structure.

We can also use algebraic closure to obtain a notion of dimension for definable sets, mimicking Zariski dimension for constructible sets in algebraically closed fields. In 2.2.7–2.2.10, we shall assume that $\mathcal{M}$ is *sufficiently saturated*; that is, $|A|^+$-saturated for any parameter set A which we might care to mention. This will ensure that certain tuples in a definable set exist *in our model*. Without the saturation assumption, we can still define dimension for definable sets, but have to quantify over elementary extensions of the model, as a realisation in the definable set of the appropriate dimension may not exist *in the model*. It is simplest, whenever talking of generics and dimension in definable sets, to assume enough saturation.

DEFINITION 2.2.7. Let $\mathcal{M}$ be geometric and sufficiently saturated, and $X \subseteq M^n$ be A-definable. Then $\dim(X) := \text{Max}\{\dim(\bar{a}/A) : \bar{a} \in X\}$ (so $\dim(X) = \text{Max}\{\dim(p) : p \in S_n(A), p \text{ realised in } X\}$). In particular, if $\bar{a} \in X$, then $\bar{a}$ is a *generic of X over A* (and $\text{tp}(\bar{a}/A)$ is a *generic type in X over A*), if $\dim(\bar{a}/A) = \dim(X)$.

There is a possible confusion here, since the notion of *rank* defined after Corollary 2.2.4 is sometimes called dimension. The ordered field of reals has rank $2^{\aleph_0}$ (its transcendence degree) but dimension 1 (as a set defined by the formula $x = x$).

EXAMPLE. In the ordered field $\mathbb{R}$, if A is a finite subset of $\mathbb{R}$, and $\bar{a} \in \mathbb{R}^n$, then $\dim(\bar{a}/A)$ is the transcendence degree of $\mathbb{Q}(A)(\bar{a})$ over $\mathbb{Q}(A)$. Hence, if X is a definable subset of $\mathbb{R}^n$, then $\dim(X)$ is the algebraic-geometric dimension of the Zariski closure of X in $\mathbb{R}^n$.

We sometimes call the above notion of dimension *geometric* dimension, to distinguish it from another (topological) notion defined after Lemma 2.2.8. It is easily checked that this definition is independent of the choice of the defining set A, provided $|A|$ is not too large. The following lemma lists some properties of this dimension. Uniform boundedness is used essentially in (iv).

LEMMA 2.2.8. *Let $\mathcal{M}$ be a sufficiently saturated geometric structure.*

(i) $\dim(\{a\}) = 0$ *(for $a \in M$) and* $\dim(M) = 1$.

(ii) *If $X, Y \subseteq M^n$ are definable, then* $\dim(X \cup Y) = \text{Max}\{\dim(X), \dim(Y)\}$.

(iii) $\dim$ *is invariant under permutation of coordinates.*

(iv) *(Definability of dimension.) Let $X \subseteq M^{m+n}$ be definable, and for each $\bar{a} \in M^m$ let $X_{\bar{a}} := \{\bar{y} \in M^n : (\bar{a}, \bar{y}) \in X\}$. For each $i = 0, \ldots, n$ let $X(i) := \{\bar{x} \in M^m : \dim(X_{\bar{x}}) = i\}$. Then each set $X(i)$ is definable, and for each i,*

$$\dim(\{(\bar{x}, \bar{y}) \in X : \bar{x} \in X(i)\}) = \dim(X(i)) + i.$$

(v) *If $f : M^n \to M^m$ is a definable partial function, and $A \subseteq M^n$ is definable, then $\dim(f(A)) \leq \dim(A)$, with equality if A is injective. In particular, definable bijections preserve dimension.*

In a strongly minimal structure $\mathcal{M}$ (which is geometric), a definable subset of X^n has (geometric) dimension equal to its Morley rank.

In a structure $\mathcal{M}$ which carries a topology with a uniformly definable basis, there is also a notion of *topological dimension* for definable sets. If $X \subseteq M^n$ is definable, then the *topological dimension* $\text{tdim}(X)$ is the greatest $k \leq n$ such that for some projection $\pi : M^n \to M^k$, $\pi(X)$ has non-empty interior in M^k. The following is quite easy to prove.

THEOREM 2.2.9. *Let $\mathcal{M}$ be o-minimal, and $X \subseteq M^n$ a definable set. Then* $\dim(X) = \text{tdim}(X)$.

If X is A-definable, and $\bar{b} \in X$, then $\bar{b}$ is a generic of X over A if and only if $\bar{b}$ does not lie in any A-definable set of dimension less than $\dim(X)$. The last theorem gives the following useful topological characterisation of genericity. It says, very roughly, that if an A-definable property holds of a generic over A, then it holds throughout a neighbourhood of it in the definable set.

LEMMA 2.2.10. *Let $\mathcal{M}$ be o-minimal, $X \subset M^n$ be A-definable, and $\bar{b}$ be a generic of X over A. Then $\dim(X) = k$ if and only if there is an open rectangular neighbourhood $Y \subset M^n$ of $\bar{b}$ and a projection $\pi : M^n \to M^k$ inducing a homeomorphism from $X \cap Y$ onto an open subset of M^k.*

In any o-minimal expansion of the ordered set of reals (in a countable language), any definable set X has a generic *in the model* over any countable set of parameters, even though $(\mathbb{R}, <)$ is not ω_1-saturated. This is easy to prove using the Baire Category Theorem and the fact that $\dim(X) = \text{tdim}(X)$ (see Lemma 2.17 of [Hrushovski and Pillay 1994]).

Finally, I give a rapid consequence of the cell decomposition theorem. If $Y \subseteq X \subseteq M^n$, we say that Y is *large in* X if $\dim(X \setminus Y) < \dim(X)$.

LEMMA 2.2.11. *Let $\mathcal{M}$ be o-minimal, D be a subset of M^k with $\dim(D) = k$, and $f : D \to M^n$ be a definable function. Assume that both D and f are definable over a set A. Then there is an A-definable large subset S of D, open in M^k, such that $f|_S$ is continuous. In addition, if $\mathcal{M}$ expands a real closed field, then for any $k > 0$, we can choose S so that f is $C^{(k)}$ on S.*

2.3. Prime models.

DEFINITION 2.3.1. A model M is *prime* over a subset A if for every $N \models \text{Th}(M, a)_{a \in A}$, there is an elementary embedding $f : M \to N$ over A.

By a result of Shelah, prime models exist (and are unique up to isomorphism over A) over arbitrary sets A in ω-stable theories. In particular, in an algebraically closed field, the prime model over a set A is just its field-theoretic algebraic closure. Prime models are a tool for classification of the models in certain classes of ω-stable theories (for example, uncountably categorical theories). The following result draws out the analogy between o-minimality and stability.

THEOREM 2.3.2 [Pillay and Steinhorn 1986]. *If $\mathcal{M}$ is o-minimal, and $A \subseteq M$, then $\text{Th}(\mathcal{M})$ has a prime model over A, unique up to A-isomorphism.*

We shall denote the prime model over A by $\mathcal{M}(A)$, or $\mathcal{M}(\bar{a})$ if A is a tuple $\bar{a}$.

SKETCH PROOF. For existence, it suffices, by general model theory, to show that isolated types are dense in the Stone space $S_1(A)$; that is, for any formula $\phi(x)$ over A there is a formula $\psi(x)$ over A such that

(i) $\mathcal{M} \models \forall x \, (\psi(x) \rightarrow \phi(x))$, and

(ii) $\psi(x)$ isolates a complete type over A.

But this is straightforward: we may suppose that $\phi(x)$ defines an interval I; either this interval is already a complete type over A, or some A-formula defines a proper subinterval, in which case an endpoint in I of that subinterval will be A-definable, so realise a complete type over A.

The proof of uniqueness of prime models is much harder, and is omitted. An easy back-and-forth argument ensures that any two *countable* prime models over A are A-isomorphic.   □

The o-minimal structures most commonly considered are expansions of ordered groups. If $\mathcal{M}$ is an expansion of an ordered group with at least two 0-definable elements (for example, if $\mathcal{M}$ expands an ordered field), then we may uniformly pick the midpoint of a bounded interval. Likewise, since there is a positive 0-definable element a, in any unbounded interval (x, ∞) we may uniformly pick out $x + a$, and in $(-\infty, x)$ we may pick out $x - a$. This means that $\mathrm{Th}(\mathcal{M})$ has definable Skolem functions; for in an $\bar{a}$-definable set we may pick out a cell, then a midpoint b_1 of its projection onto the first coordinate, then a midpoint b_2 of the first coordinate of the fibre above b_1, and so on (only using the parameters $\bar{a}$). The existence of definable Skolem functions ensures that the prime model over A is precisely $\mathrm{dcl}(A)$. Thus for example, in the ordered field $\mathbb{R}$, the prime model over a set A is precisely its real closure. Incidentally, the above hypotheses on $\mathcal{M}$ also ensure that $\mathrm{Th}(\mathcal{M})$ has elimination of imaginaries. This is because, given a 0-definable equivalence relation, by the above we can uniformly pick out an element of each equivalence class.

There are also good prime model theorems in some other classes of unstable algebraic structures. For example, in $\mathrm{Th}(\mathbb{Q}_p)$, the prime model over a set is just its p-adic closure.

2.4. Definable types

DEFINITION 2.4.1. Let $\mathcal{M}$ be an ordered structure. Then a *cut* of M is a maximal consistent set of formulas each of the form $a < x$ or $x < a$ (where $a \in M$).

LEMMA 2.4.2 [Pillay and Steinhorn 1986]. *Suppose $\mathcal{M}$ is o-minimal. Then for each cut of M there is a unique 1-type over M extending it.*

PROOF. Let C be a cut of M, and let $\phi(x)$ be a formula over M such that ϕ is consistent with C. Now ϕ partitions M into intervals, and by o-minimality, only one of these intervals is consistent with C. Hence, just one of ϕ, $\neg\phi$ is consistent with C.   □

The following definition is usually associated with stability theory.

DEFINITION 2.4.3. A type $p(\bar{x}) \in S_n(A)$ (where A is a subset of an ambient structure $\mathcal{M}$) is *definable* if, for every formula $\phi(\bar{x}, \bar{y})$ over $\varnothing$, there is a formula $\psi(\bar{y})$ over A, such that for any tuple $\bar{a}$ from A, $\phi(\bar{x}, \bar{a}) \in p$ if and only if $\mathcal{M} \models \psi(\bar{a})$.

Observe that if A is $\bar{b}$-definable, and all types over A realised in $\mathcal{M}$ are definable, then for each n every $\mathcal{M}$-definable subset of A^n is $A\bar{b}$-definable.

If C is a cut of M, then $a \in M$ is a *standard part of C* (written $a = \mathrm{st}(C)$, or $a = \mathrm{st}(b)$, if b realises C) if, for any b realising C (in an elementary extension), there is no element of M between a and b. Now we easily obtain the following.

LEMMA 2.4.4. *If $\mathcal{R}$ is an o-minimal expansion of $\mathbb{R}$, then every 1-type over $\mathbb{R}$ is definable.*

PROOF. If the 1-type p over $\mathbb{R}$ is algebraic (i.e. realised in $\mathbb{R}$) this is obvious, so suppose that p is non-algebraic. Then either p is the unique type consisting of infinitely large or infinitely small elements, or determines a cut bounded above and below by elements of $\mathbb{R}$. If say p is the type consisting of infinitely large elements, then for any formula $\phi(x, \bar{y})$, and any $\bar{a}$ from $\mathbb{R}$, $\phi(x, \bar{a}) \in p$ if and only if $\phi(x, \bar{a})$ holds on a cofinal subset of $\mathbb{R}$, so the set of such $\bar{a}$ is $\mathbb{R}$-definable. In p determines a bounded cut C, then, as $\mathbb{R}$ is Dedekind complete, C has a standard part, and we may use this to define p. $\qquad\square$

We can extend this slightly. Suppose $\mathcal{M} \preceq \mathcal{N}$, both o-minimal, and $b \in N \setminus M$ with $\mathrm{tp}(b/M)$ definable. Then $\{x \in M : x < b\}$ and $\{x \in M : b < x\}$ are M-definable. It follows by o-minimality that b has a standard part in M. The most general statement of this form is the following, proved in [Marker and Steinhorn 1994], and given a different treatment in [Pillay 1994].

THEOREM 2.4.5 [Marker and Steinhorn 1994]. *Let $\mathcal{M}$ be o-minimal, and let $p(\bar{x}) \in S_n(M)$. Then p is definable if and only if, for every $\bar{a}$ realising p over M, M is Dedekind-complete in $M(\bar{a})$, that is, every non-algebraic 1-type over M which is realised in the prime model $M(\bar{a})$ over $M\bar{a}$ has a standard part in M.*

3. Definable groups and fields, and a trichotomy theorem

3.1. Ordered groups and fields, and a Trichotomy Theorem.

The main goal of this section is to describe a Trichotomy Theorem of Peterzil and Starchenko, and some results of them and Pillay on definably simple groups in o-minimal structures. I begin with two elementary results.

PROPOSITION 3.1.1 [Pillay and Steinhorn 1986]. *Let G be an o-minimal ordered group (so the group operation is part of the language). Then G is divisible abelian.*

PROOF. We first claim that G has no proper non-trivial definable subgroups. For let H be such a subgroup. As G is torsion-free, H is infinite, so contains an infinite interval J, and hence has a maximal non-trivial interval about 0, of the form $[-h, h]$ or $(-h, h)$. Both cases are easily eliminated. For example, if $J = (-h, h)$, pick h' such that $0 < h' < h$, and observe that both h' and $h - h' \in H$, so $h \in H$, a contradiction.

Given the claim, G is abelian (for the centraliser of any element is a non-trivial and definable subgroup, so equal to G). For any $n > 0$, nG is a definable subgroup of G, which cannot be $\{0\}$ (as G is torsion-free), so equals G. Hence G is divisible. $\qquad\qquad\qquad\qquad\qquad\qquad\qquad\qquad\qquad\qquad\qquad\quad$ □

PROPOSITION 3.1.2 [Pillay and Steinhorn 1986]. *Let R be an o-minimal ordered field. Then R is real closed, that is, elementarily equivalent to the reals.*

PROOF. It suffices to show that polynomials satisfy the intermediate value theorem. But this follows immediately from o-minimality. $\qquad\qquad\qquad\qquad\quad$ □

QUESTION 1. Is there a sense in which an o-minimal structure is either 'trivial' (like $(\mathbb{Q}, <)$), or grouplike, or fieldlike?

QUESTION 2. What can we say about definable groups, or fields, in an o-minimal structure? (Note that by a remark after Theorem 2.3.2, many o-minimal structures, such as expansions of ordered fields, admit elimination of imaginaries, and in such structures definability is equivalent to interpretability.)

Question 1 suggests the Zil'ber Conjectures for strongly minimal sets, another class of geometric structures — see the discussion before Theorem 3.1.6 for more on these conjectures. However, there is a major difference. For among other things these conjectures asserted that a (strongly minimal) algebraically closed field has no proper expansions other than those obtained by naming constants. Under extra hypotheses, positive results in this direction are obtained in work on Zariski structures [Hrushovski and Zilber 1996], and the only known counterexamples are artificial [Hrushovski 1992]. On the other hand, the field of reals has many natural o-minimal proper expansions, such as the real exponential field.

Any answer to Question 1 has to be *local*. For example, one could form an o-minimal structure with three parts, L (the leftmost part), M (the middle part), and R (the right part), with L carrying the structure of a pure dense linear order, M that of a divisible ordered abelian group, and R that of a real closed field (it is necessary to put a point between L and M, and one between M and R, to ensure o-minimality). And indeed, M might just be an *interval* of an ordered group, with the induced structure.

I will describe a beautiful answer to Question 1. It is the main theorem of [Peterzil and Starchenko 1998].

DEFINITION 3.1.3. Let $\mathfrak{M}$ be o-minimal, and $a \in M$.

(i) The point a is *non-trivial* if there is an open interval $I \subset M$ such that $a \in I$, and a definable continuous function $I \times I \to M$ which is strictly monotonic in each variable.

(ii) A *convex $\bigwedge$-definable group* in M is a group $(G, *)$, where $G \subseteq M$ is convex, and the group operation $*$ (regarded as a ternary relation) is the intersection of a definable set with G^3.

(iii) If $(G, <, +, 0)$ is a convex $\bigwedge$-definable group, and $p \in G$ with $p > 0$, then a *group interval* is a structure $([-p, p], <, +, 0)$, where $+$ is the induced *partial* function $[-p, p] \times [-p, p] \to [-p, p]$.

(iv) If I is an interval of M, then $\mathcal{M}|I$ is the structure with domain I, whose 0-definable sets are those of the form $I^k \cap U$ for definable $U \subseteq M^k$ (in fact, by Lemma 2.5 of [Peterzil and Starchenko 1998], if I is closed then such sets are I-definable).

Note that the domain of a convex infinitely definable group is an intersection of intervals, but may not be definable. The model theory of group intervals in o-minimal structures was described in [Loveys and Peterzil 1993]. They are all elementarily equivalent, essentially, because in $(\mathbb{Q}, <, +)$, all positive elements have the same type.

It is easy to see that if $(G, *)$ is a convex $\bigwedge$-definable ordered group in M, and $a \in G$, then $*$ witnesses that a is non-trivial. The following converse is much deeper.

THEOREM 3.1.4 [Peterzil and Starchenko 1998]. *Let $\mathcal{M}$ be ω^+-saturated, and $a \in M$ be non-trivial. Then there is a convex $\bigwedge$-definable infinite group $G \subseteq M$, such that $a \in G$ and G is a divisible ordered abelian group.*

In particular, even without the saturation assumption, there is a closed interval I with $a \in I \subset M$, and a definable group interval structure induced on I. Saturation enables us to find the whole domain of a group, on say an infinitesimal neighbourhood (with respect to an elementary substructure) about a, but this infinitesimal neighbourhood may not be definable.

THEOREM 3.1.5 [Peterzil and Starchenko 1998]. *Suppose that $(I, <, +, 0)$ is a 0-definable group interval in a sufficiently saturated o-minimal structure $\mathcal{M}$. Then precisely one of the following statements holds.*

(i) *There is an ordered vector space $\mathcal{V} = (V, +, c, d(x))_{c \in C, d \in D}$ (with C a set of constants) over an ordered division ring D, an interval $[-p, p]$ in V, and an order-preserving isomorphism of group-intervals $\sigma : I \to [-p, p]$, such that $\sigma(S)$ is 0-definable in $\mathcal{V}$ for every 0-definable $S \subseteq I^n$.*

(ii) *There is a real closed field $\mathcal{R}$ definable in $\mathcal{M}$ with its domain a subinterval of I and its order compatible with $<$.*

The point is that in (i), the division ring D is not a definable structure, so $\mathcal{V}$ is essentially a 'pure' linear structure. I emphasise that in (ii) we get the whole of a field, not just a field interval. Incidentally, if *every* point of M is trivial, then by [Mekler et al. 1992], every definable set is a boolean combination of binary relations.

I describe next an alternative, more algebraic-geometric, treatment of this theorem, given in the introduction to [Peterzil and Starchenko 1998].

Let $\mathcal{M}$ be a geometric structure (see Definition 2.2.3). A *curve* is a 1-dimensional subset of M^2. A set $\mathcal{F}$ of curves is said to be *definable* if there are definable $U \subseteq M^k$ and $F \subseteq U \times M^2$ such that $\mathcal{F} = \{C_{\bar{u}} : \bar{u} \in U\}$, where $C_{\bar{u}} := \{(x, y) : (\bar{u}, x, y) \in F\}$. We say $C_{\bar{u}}$ is generic in $\mathcal{F}$ if $\bar{u}$ is generic in U over any relevant parameters. Also, $\mathcal{F}$ is *normal of dimension n* if $\dim(U) = n$ and $C_{\bar{u}} \cap C_{\bar{v}}$ is finite for any distinct $\bar{u}, \bar{v}$ in U. There is a similar notion of *interpretable* normal family of curves.

In a geometric structure $\mathcal{M}$, one of the following must hold.

(Z1) If $\mathcal{F}$ is an infinite interpretable normal family of curves, and $C \in \mathcal{F}$ is generic, and (a, b) is generic in C, then either $\dim(C \cap (\{a\} \times M)) = 1$ or $\dim(C \cap (M \times \{b\})) = 1$.

(Z2) (Z1) fails, but every interpretable normal family of curves has dimension at most 1.

(Z3) There is an interpretable normal family of curves of dimension greater than 1.

Zil'ber's Conjecture (which is false in general, though parts are true) was that in the strongly minimal case, (Z1) should correspond to the case when there are no interpretable groups, (Z2) to the case when definable sets arise from a module, and the structures satisfying (Z3) should be bi-interpretable with (or at least interpret) an algebraically closed field. If $\mathcal{M}$ is a group, then (Z1) is clearly false (consider for each $g \in G$ the curve $C_g := \{(x, y) : xg = y\}$). Likewise, if $\mathcal{M}$ is a field, then for each $a, b \in M$ we have a curve $\{(x, y) : y = ax + b\}$, and this family of curves has dimension 2.

Suppose now that $\mathcal{M}$ is o-minimal. It is quite easy to see that $a \in M$ is trivial if and only if some induced structure $\mathcal{M}|I$ on a neighbourhood of a has type (Z1). If $a \in M$ is non-trivial, we say a has type (Z2) if there is an open interval I containing a such that the induced structure $\mathcal{M}|I$ satisfies (Z2), and that a has type (Z3) otherwise. Now we can state the following version of the trichotomy theorem (which gives more of a guide to its proof).

THEOREM 3.1.6 [Peterzil and Starchenko 1998]. *Let $\mathcal{M}$ be sufficiently saturated and o-minimal, and $a \in M$ be non-trivial. Then either*

(a) *a has type (Z2) and the structure induced on some closed interval I (whose interior contains a) satisfies Theorem 3.1.5(i), or*

(b) *a has type (Z3), and some open interval containing a satisfies Theorem 3.1.5(ii).*

REMARKS ON THE PROOF OF THEOREM 3.1.4 AND 3.1.5. First of all, a general mechanism for defining groups is given. Given a definable normal family of dimension greater than one of functions on an interval I, one can obtain, for some open interval $J \subset I$, a very well-behaved 'nice' family of functions of dimension 2, parametrised by an open subset of I^2. It turns out that this is a powerful abstract notion: in the final section of the paper, it is shown that without any algebraic assumptions one can use a nice family of functions to define tangency, and develop elementary differential calculus - the nice family replaces the family of functions of form $f(x) = ax + b$ in the usual definition of differentiation. For such a nice family, there is a technical device for defining a certain quaternary relation (a 'q-relation') on a convex subset of I, and from this a convex $\bigwedge$-definable ordered group (assuming enough saturation). The arguments here are similar to familiar group constructions in stable theories. One starts with a family $\mathcal{F}$ of functions from a set A to a set B, and by taking compositions fg^{-1} obtains a family of partial definable functions $B \to B$. The defining parameters for these compositions (taken up to some equivalence relation corresponding to tangency at a 'fixed' generic point) are projected to an interval to obtain the convex $\bigwedge$-definable ordered group, and the group operation is essentially composition of functions. To prove Theorem 3.1.4, one takes a definable function

$$F : I \times I \to M$$

which witnesses non-triviality of a (here, $a \in I$). Massaging this function using composition, a function $G(x, y, z)$ is obtained, defined on an open subset of I^3, continuous, and strictly increasing in each variable. This gives either a situation where methods from [Peterzil 1994] apply, or a normal family of functions of dimension 2. Either way, a group interval is obtained.

Assume now that a is non-trivial but the conclusion of Theorem 3.1.5(i) does not hold around a. By the last paragraph, there is an interval I containing a such that the structure induced on I is an o-minimal expansion of a group interval. Furthermore, by our case assumption and Proposition 4.2 of [Loveys and Peterzil 1993], there is a definable function on a subinterval J of I which is not 'linear' on any subinterval of J. One can use this to construct a new nice family of curves. From a nice family living on a group interval, one can construct two different q-relations, one using the operation of the group interval, the other using composition. From this, it is possible to define on a convex subset of I two new groups G, H, corresponding to field addition and multiplication. It is also possible to define a continuous faithful action H on G. (Care is needed with the notion of definability, as G and H are convex $\bigwedge$-definable ordered groups, but are not in general definable.) This gives a certain convex $\bigwedge$-definable ring of definable endomorphisms. The fraction field is a real closed field, which turns out to be *definable*, essentially, because any element of the fraction field is represented by infinitely many pairs from the ring, and some of these pairs lie in arbitrary small boxes $J \times J$, where J is an interval about a. $\qquad\qquad$ $\square$

I conclude this subsection by stating a global dichotomy theorem of Miller and Starchenko which looks similar to Theorem 3.1.5, but is global rather than local. If $\mathcal{G} = (G, <, +, \ldots)$ is an expansion of an ordered group, we say that $\mathcal{G}$ is *linearly bounded* if for any definably function $f : G \to G$ there is a definable endomorphism λ of G such that $|f(x)| \leq \lambda(x)$ for sufficiently large x.

THEOREM 3.1.7 [Miller and Starchenko 1998]. *Let* $\mathcal{R} = (R, <, +, \ldots)$ *be an o-minimal expansion of an ordered group. Then if* $\mathcal{R}$ *is not linearly bounded, there is a definable binary operation* $\cdot$ *such that* $(R, <, +, \cdot)$ *is an ordered field.*

3.2. Definable groups and fields. I first describe some results from [Pillay 1988], based on the dimension theory of Section 2. I begin with Proposition 1.8 of that reference, which is based on Proposition 2.1 of [Pillay 1986].

LEMMA 3.2.1. *Let* $\mathcal{M}$ *be o-minimal, and* $X \subseteq M^n$. *Then* $\dim(X) \geq k + 1$ *if and only if there is a definable equivalence relation* E *on* X *with infinitely many classes of dimension at least* k.

Here, the left-to-right direction is trivial and holds for topological dimension in any reasonable class of structures. The right-to-left direction, however, is specific to o-minimality, and fails for the classes considered in Section 4.

COROLLARY 3.2.2 [Pillay 1988]. *Let* $G \subseteq M^n$ *be a group definable in an o-minimal structure, that is, the group and its operation are definable. Then if* $H \leq G$ *is definable, then* $\dim(H) = \dim(G)$ *if and only if* $|G : H| < \infty$.

PROOF. If $|G : H| = \infty$, then as the cosets of H form the classes of a definable equivalence relation on G, $\dim(H) < \dim(G)$ by Lemma 3.2.1. The other direction follows from Corollary 2.2.8, parts (ii) and (v). $\qquad\square$

I now describe some results from [Pillay 1988], extended slightly in [Otero et al. 1996] and [Peterzil et al. 2000] (whose terminology I follow). Fix an o-minimal structure $\mathcal{M}$, and $p \geq 0$ (if $p > 0$, then we assume that $\mathcal{M}$ expands a real closed field). Let X be a definable set. We wish to endow X with a kind of manifold structure.

A *definable chart* on X is a triple $\mathbf{c} = \langle U, \phi, n \rangle$, where U is a definable subset of X, $n \geq 0$, and ϕ is a definable bijection from U to an open subset of M^n. Two charts $\mathbf{c} = \langle U, \phi, n \rangle$ and $\mathbf{c}' = \langle U', \phi', n' \rangle$ are $C^{(p)}$-*compatible* if either $U \cap U' = \varnothing$, or $\phi(U \cap U'), \phi'(U \cap U')$ are open, and the transition mappings $\phi \circ \phi'^{-1}, \phi' \circ \phi^{-1}$ are $C^{(p)}$ on their domains. A *definable* $C^{(p)}$-*atlas* on X is a finite pairwise $C^{(p)}$-compatible set of definable charts covering X. A *definable* $C^{(p)}$-*manifold* is a pair consisting of X, and a definable $C^{(p)}$-atlas on X. Note that given a definable manifold on X, we can talk about the dimension at any point of X.

Using the charts as coordinate systems, one can talk about a definable $C^{(p)}$-function between manifolds, and (if $p > 0$) its *differential* at a point (a linear map given by a matrix of partial derivatives). The following result is essentially Proposition 2.5 of [Pillay 1988] (for the case $p > 0$, it was stated in [Otero et al.

1996], but the proof is essentially that of [Pillay 1988]). Behind the proof lies Lemma 2.2.11.

THEOREM 3.2.3. *If G is a group definable in an o-minimal structure $\mathfrak{M}$, then there is an atlas on G making G into a definable $C^{(p)}$-group, i.e., the group operation $G \times G \to G$, and inversion, are $C^{(p)}$.*

As a corollary in the case $p = 0$, it follows (see [Pillay 1988]) that if G is as in the theorem, then G has a unique smallest definable subgroup G^o of finite index, and that this is also its connected component in the above manifold topology. Furthermore, by this and Lemma 3.2.1 again, G has the descending chain condition on definable subgroups; in particular, if $X \subseteq G$ then there is finite $X_0 \subseteq X$ such that $C_G(X) = C_G(X_0)$, so centralisers are definable. Thus, definable groups in o-minimal structures share many properties of groups of finite Morley rank.

Analogous results are shown in [Pillay 1988] to hold for definable fields, and yield (quite rapidly) that any definable infinite field in an o-minimal structure is real closed of dimension 1 or algebraically closed. In the algebraically closed case, by an easy Euler characteristic argument the characteristic is 0, and in fact the field has dimension 2 [Peterzil and Steinhorn 1999]. (The Euler characteristic argument is suggested by the beautiful paper [Strzebonski 1994], which develops a Sylow theory based on Euler characteristic for groups definable in o-minimal structures.) More precisely, it is shown in Theorem 4.1 of [Peterzil and Steinhorn 1999] that if K is any infinite definable ring without zero-divisors in an o-minimal structure $\mathfrak{M}$, then K is a division ring and there is a one-dimensional $\mathfrak{M}$-definable subring R of K which is a real closed field such that K is either R, or $R(i)$ (where i denotes $\sqrt{-1}$)), or the ring of quaternions over R. Earlier, under the additional assumption that $\mathfrak{M}$ is an o-minimal expansion of a real closed field R_0, it was shown in [Otero et al. 1996] that such a ring K is *definably* isomorphic to R_0, or $R_0(i)$, or the quaternions over R_0.

All this suggests that an o-minimal analogue of Cherlin's Conjecture might hold. This conjecture states that any infinite simple group of finite Morley rank is definably isomorphic to a simple algebraic group over an algebraically closed field. Such a result has now been proved, by Peterzil, Pillay and Starchenko [Peterzil et al. 2000], and for the rest of this section I sketch it and the finer structure theory from [Peterzil et al. 1997]. These notes also use the summary [Peterzil et al. 1998].

Before stating the main theorem of [Peterzil et al. 2000], we need some definitions. Let $\mathfrak{R}$ be an o-minimal expansion of a real closed field. In the context of $\mathfrak{R}$, a *semialgebraic* set or group is just a set or group definable in the *pure* field $\mathfrak{R}_0 := (R, <, +, \cdot)$. A semialgebraic linear group over R is a subgroup of $\mathrm{GL}(n, R)$, for some n, which is definable in $\mathfrak{R}_0$. It is semialgebraically connected if it has no proper semialgebraic subgroup of finite index. If $H \leq \mathrm{GL}(n, R)$ is a semialgebraic linear group over R, then H has Zariski closure in $\mathrm{GL}(n, R(i))$ defined over R by the vanishing of polynomial equations. We let $\bar{H}$ denote the subgroup of this Zariski closure consisting of matrices with entries in R. Clearly we have $\bar{H} \leq \mathrm{GL}(n, R)$, and H and $\bar{H}$ both have a dimension in the sense of Definition 2.2.7. These dimensions are equal to the algebraic-geometric dimension of $\bar{H}$, and so equal to each other. Hence, by Corollary 3.2.2, $|\bar{H} : H| < \infty$.

If $\mathbf{G} = (G, \cdot)$ is a group definable in a structure $\mathcal{M}$, then we say that G is *definably simple* if G has no proper non-trivial $\mathcal{M}$-definable normal subgroups, and that G is $\mathbf{G}$-definably simple if it has no proper non-trivial normal subgroups which are definable just in the structure $\mathbf{G}$ (so $\mathbf{G}$-definable simplicity is the weaker condition). Likewise, G is *definably connected* if it has no proper definable subgroups of finite index, and $\mathbf{G}$-*definably connected* if it has no such subgroups definable in the structure $\mathbf{G}$. In general, 'definable' means 'definable in the sense of $\mathcal{M}$'.

Now I state the main theorems of [Peterzil et al. 2000].

THEOREM 3.2.4. *Assume that G is an infinite $\mathbf{G}$-definably connected group definable in an o-minimal structure $\mathcal{M}$, with no non-trivial abelian normal subgroup. Then there is $k > 0$, and for each $i = 1, \ldots, k$ a definable real closed field R_i, a $\mathbf{G}$-definable subgroup $H_i \leq G$, and a definable isomorphism between H_i and a semialgebraic subgroup of $\mathrm{GL}(n, R_i)$, such that $G = H_1 \times \cdots \times H_k$. Each H_i is $\mathbf{H}_1$-definably simple, and its definably connected component in the sense of $\mathcal{M}$ is definably simple.*

THEOREM 3.2.5. *Let G be a non-abelian infinite $\mathbf{G}$-definably simple group definable in an o-minimal structure $\mathcal{M}$. Then there is a definable real closed field $\mathcal{R}$ such that G is definably isomorphic to a semialgebraic linear group over R.*

REMARK. In the finite Morley rank context, any non-abelian definably simple group is simple, by Zil'ber Indecomposability. In the o-minimal context, there is only an infinitesimal version of Zil'ber Indecomposability, given in Section 2 of [Peterzil et al. 1997]. There are examples of non-abelian groups which are definably simple but not simple. For example, if R is a non-archimedean real closed field, then $SO(3, R)$ is definably simple (in the sense of R) but not simple: it has a normal subgroup consisting of matrices $A + I$, where A has infinitesimal entries.

SKETCH OF THE PROOF OF THEOREM 3.2.5. The first step, given an o-minimal expansion $\mathcal{R}$ of a real closed field, is to develop a general Lie theory. If X is a definable $C^{(1)}$-manifold, with a chart $\langle U, \phi, n \rangle$, and $m \in U$, then the *tangent space at m* is the set of definable $C^{(1)}$-functions $f : R \to X$ with $f(0) = m$, modulo an equivalence relation $f_1 \sim f_2$, which holds if $\phi \circ f_1$ and $\phi \circ f_2$ have the same differential at 0; it is in canonical bijection with R^n, so has a vector space structure. Formally, $T_m(X)$ is not a definable object, but the vector space R^n is, and we often identify $T_m(X)$ with R^n. Given a definable $C^{(1)}$-mapping between definable $C^{(p)}$-manifolds $f : X \to Y$, and $m \in X$, there is a (linear) differential

$d_m(f) : T_m(X) \to T_{f(m)}(Y)$. The rank of $d_m(f)$ equals the dimension of $f(X)$ at $f(m)$.

Suppose that G is a group definable in $\mathcal{R}$, so G carries a definable manifold structure. We consider the tangent space $T_e(G)$ at the identity e of G. Let $n := \dim(G)$. Then $\dim(T_e(G)) = n$. For any $g \in G$, conjugation by g induces an inner automorphism of G, whose differential on $T_e(G)$ is non-singular, so (identifying $T_e(G)$ with R^n), induces an element of $\mathrm{GL}(n, R)$. The induced map $G \to \mathrm{GL}(n, R)$ is a homomorphism. Thus, if G is centreless, we have a definable embedding

$$f : G \to \mathrm{GL}(n, R)$$

(the *adjoint representation*: see [Otero et al. 1996]), but we do not yet know that its image is semialgebraic in the sense of $\mathcal{R}$, that is, that the image is definable in the *pure* field R.

There is a Lie algebra structure on the tangent space $T_e(G)$, given in a standard way. Since the Lie operation is bilinear, this gives an *R-definable* Lie algebra structure $L := L(G)$ on R^n, and $\mathrm{Aut}(L)$ is an *algebraic* subgroup of $\mathrm{GL}(n, R)$. If G is assumed to be semisimple, that is, it has no infinite abelian normal subgroup, then L is semisimple, that is, its only abelian ideal is $\{0\}$. In this case an easy argument shows that $\dim(\mathrm{Aut}(L)) = \dim(L)$ (which equals n). Also the adjoint representation gives a definable embedding $f : G \to \mathrm{Aut}(L)$, so, as the dimensions are equal, by Corollary 3.2.2 $f(G)$ has finite index in $\mathrm{Aut}(L)$.

A useful fact (given in a more general context in Claim 1.3 of [Peterzil et al. 2000]) is that the connected component $\mathrm{Aut}(L)^o$ of $\mathrm{Aut}(L)$ in the sense of the *expansion* $\mathcal{R}$ is the same as the semialgebraic connected component in the sense of the *pure field* R, that is, $\mathrm{Aut}(L)^o$ is semialgebraic. Hence, since $f(G)$ is definable in $\mathcal{R}$, it is a union of finitely many cosets of $\mathrm{Aut}(L)^o$, so $f(G)$ is semialgebraic. This argument shows that if G is centreless and definable, with semisimple connected component, then G is definably isomorphic to a linear semialgebraic group over $\mathcal{R}$.

In the proof of Theorem 3.2.5, one first uses the Trichotomy Theorem 3.1.5 (and the existence of the group G) to find a real closed field $\mathcal{R}$ on a definable interval I of $\mathcal{M}$. This can be used to define a chart on G at e with image in I^n, such that the group multiplication and inversion are $C^{(1)}$ near e. This enables us to use the adjoint representation to embed G definably into $\mathrm{GL}(n, R)$. Thus, the image $h(G)$ of this embedding in $\mathrm{GL}(n, R)$ is a definable group in an o-minimal expansion of $\mathcal{R}$. Since $h(G)$ is definably simple, we can apply the previous paragraph to it to obtain Theorem 3.2.5. For Theorem 3.2.4, it is also necessary to develop an orthogonality theory between intervals, and a notion of a *unidimensional* group.   □

I now turn to the finer structure theory for definable groups, from [Peterzil et al. 1997]. So far, we have an o-minimal structure $\mathcal{M}$, and a **G**-definably simple group G, definable in $\mathcal{M}$. We know about the abstract group structure of G, but not much about the model theory of **G**. In [Peterzil et al. 1997], the structure **G** is identified up to bi-interpretability.

A structure $\mathcal{M}$ is *interpretable* in $\mathcal{N}$ if there is an isomorphic copy $f(\mathcal{M})$ definable in $\mathcal{N}^{\mathrm{eq}}$. If $\mathcal{M}$ is interpretable in $\mathcal{N}$ by f, and $\mathcal{N}$ is interpretable in $\mathcal{M}$ by g, then in $\mathcal{N}^{\mathrm{eq}}$ there is an isomorphic copy $f(g(\mathcal{N}))$ of $\mathcal{N}$, and in $\mathcal{M}^{\mathrm{eq}}$ there is an isomorphic copy $g(f(\mathcal{M}))$ of $\mathcal{M}$. In general, the isomorphism $f \circ g$ will not be definable in $\mathcal{N}$. To illustrate this, consider a group G, with a definable subgroup H, which itself has a definable subgroup G^* isomorphic to G; then the domain of G^* is G-definable, but there is no reason why there should be a G-*definable* isomorphism $G \to G^*$. We say that $\mathcal{M}$ and $\mathcal{N}$ are *bi-interpretable* if the isomorphism $f \circ g$ is definable in $\mathcal{N}$ and $g \circ f$ is definable in $\mathcal{M}$. We can now state the main theorem of [Peterzil et al. 1997].

THEOREM 3.2.6. *Let G be a non-abelian infinite group which is **G**-definably simple, and is definable in an o-minimal structure. Then there is a real closed field $\mathcal{R} = (R, +, \cdot)$ such that **G** is bi-interpretable either with $\mathcal{R}$ or with its degree 2 algebraically closed extension $(R(i), +, \cdot)$.*

As a curious corollary, one can bypass Cherlin's Conjecture for groups of finite Morley rank to obtain a model-theoretic characterisation (among infinite simple groups) of algebraic groups over algebraically closed fields of characteristic 0: these are precisely the stable groups definable in some o-minimal structure.

The following further corollary is one of many results on the relationship between abstract homomorphisms and algebraic (or analytic) homomorphisms between algebraic groups. It was already known, by a combination of results in [Borel and Tits 1973] and [Weisfeiler 1979].

COROLLARY 3.2.7 [Peterzil et al. 1997]. *Let G_1 be an unstable semialgebraic G_1-definably simple group over an real closed field R_1, and G_2 another semialgebraic group over a real closed field R_2. Then any abstract group isomorphism $f : G_1 \to G_2$ has the form $f = g \circ h$, where h is induced by an (abstract) field isomorphism $R_1 \to R_2$, and g is an R_2-semialgebraic group isomorphism.*

REMARKS ON THE PROOF OF THEOREM 3.2.6.. The first step is to find an infinite field interpretable in **G**. There is a real closed field R_1 provided by Theorem 3.2.5, such that G is definably isomorphic to a semialgebraic linear group over R_1. Let $K_1 := R_1(i)$, let $\bar{G}$ be the Zariski closure of G in K_1, and let H be the minimal *algebraic* subgroup of $\bar{G}$ of finite index (so H is a connected linear algebraic group defined over R_1). We say that H is R_1-*isotropic* if it has an R_1-defined algebraic subgroup T which is rationally R_1-isomorphic to a direct product of at least one copy of the multiplicative group of the field K_1, and R_1-*anisotropic* otherwise.

The argument splits into two cases, according to whether H is $\mathcal{R}_1$-isotropic or $\mathcal{R}_1$-anisotropic. If H is $\mathcal{R}_1$-anisotropic, then $H(R_1)$ is closed and bounded in $\mathrm{GL}(n, R_1)$, and this makes possible a model-theoretic transfer of results from

[Nesin and Pillay 1991] on compact Lie groups, which give an interpretable real closed field K.

In the other case, H is R_1-isotropic. Now, arguments familiar from the finite Morley rank case are applicable. Inside an 'R_1-parabolic subgroup of H', using the Levi decomposition one can define in H a connected soluble non-nilpotent group. From this it is is possible to find an infinite definable abelian group M acting faithfully and definably on a definable abelian group A, such that A has no infinite definable M-invariant proper subgroups. Using a local version of Zil'ber Indecomposability proved in [Peterzil et al. 1997], one can interpret an infinite field K in $\mathbf{G}$ (which may be real closed or algebraically closed). The field K is definable in the original field R_1, so, by results of [Otero et al. 1996], is semialgebraically (in the sense of $\mathcal{R}_1$) isomorphic to R_1 or $R_1(i)$. A short argument shows that every $\mathbf{G}$-definable subset of K^n, for any n, is definable in $(K, +, \cdot)$. Here, one may have first to replace $R_1(i)$ by R_1 and apply [Marker 1990].

By the last two paragraphs, there is a real closed field R and an interpretable field K which is equal to R or $R(i)$. By a general model-theoretic argument, to show that $\mathbf{G}$ and $(K, +, \cdot)$ are bi-interpretable it is now necessary to show that G is K-internal, that is, there is a G-definable surjection $K^r \to G$ for some r. By an application of the infinitesimal Zil'ber Indecomposability Theorem and a Lie algebra argument, there is definable and K-internal $U \subseteq G$ with $\dim(U) = \dim(G)$. By elimination of imaginaries in $(K, +, \cdot)$, U is in definable bijection with a subset of K^r for some r. We may suppose that U contains an open neighbourhood of the identity e of G. If K is real closed, we obtain a chart with differential structure on U, and the adjoint representation gives a definable embedding $G \to \mathrm{GL}(r, K)$, so G is K-internal. If K is algebraically closed, then an argument with Morley rank and degree shows that $G = U \cdot U$, so again G is K-internal. $\square$

4. Variants of o-minimality

There are several structures related to topological fields whose model theory is similar to that of $\mathbb{R}$, but which are not o-minimal. Examples include the p-adic field $\mathbb{Q}_p$, any algebraically closed valued field, and any real closed field with a definable convex valuation ring. In each case, the model theory is made manageable by a quantifier elimination theorem. All these structures are geometric in the sense of Section 2, and there is also a good dimension theory for expansions of them by subanalytic sets. (In the real case, one expands by *finitely* subanalytic sets: see [van den Dries 1986], [van den Dries and Miller 1996], or [van den Dries 1996].) I describe here attempts to set their model theory in a general context.

We consider here some model theoretic notions, akin to o-minimality, for some classes of structures with a topology. These are *weak o-minimality*, *C-*

minimality, and *P-minimality*. In each case, the general theory is not nearly as well-developed as o-minimality, and the results obtained are still rather haphazard. Unfortunately, the definable connectedness properties of the topology are not nearly as good as in the o-minimal case. This leads to a number of problems: for example, we cannot expect analogues of the theorem that if a structure is o-minimal then so is any structure elementarily equivalent to it.

4.1. Weak o-minimality

DEFINITION 4.1.1 (Dickmann). A totally ordered structure $\mathcal{M} = (M, <, \dots)$ is *weakly o-minimal* (weakly o-minimal) if every parameter-definable subset of M is a finite union of convex sets. We say that a complete theory T is *w.o.m* if every model of T is weakly o-minimal.

There is an example in [Macpherson et al. 1999] of a weakly o-minimal structure whose theory is not weakly o-minimal. The following is the main motivating example for weak o-minimality.

EXAMPLE 4.1.2. Let R be a real closed field with a proper convex subring V which is a valuation ring, and induced valuation map v to the value group. Let L_{rcvf} denote the language $(<, +, -, \cdot, 0, 1, D)$ of ordered rings with an additional binary relation symbol D. We interpret D by putting Dxy whenever $v(x) \leq v(y)$. Then, by results from [Cherlin and Dickmann 1983], the theory of all such structures in the language L_{rcvf} is complete and has quantifier elimination, and (by [Dickmann 1985]), is weakly o-minimal.

As a concrete example of such a structure, let R be any non-archimedean real closed field, and V the set of its finite elements, that is, elements bounded in absolute value by a natural number. The unique maximal ideal of V consists of the infinitesimals.

In Example 4.1.2, a weakly o-minimal structure is obtained from an o-minimal one essentially by adding a unary predicate interpreted by a convex set, namely, the valuation ring. In [Macpherson et al. 1999], another example is given of this phenomenon. Let R be the field of real algebraic numbers, and let $\mathcal{R}$ be the structure $(R, <, +, -, \cdot, 0, 1, P)$, where P is a unary predicate interpreted by the convex set $(-\pi, \pi) \cap R$. Then $\mathcal{R}$ has weakly o-minimal theory (in fact, here, π could be any real transcendental).

Cherlin asked whether this phenomenon holds generally, that is, whether *any* expansion of any o-minimal structure by a predicate for a convex set is weakly o-minimal. The following positive answer is given in [Baizhanov 1999] and in [Baisalov and Poizat 1998] (in which the proof uses Theorem 2.4.5).

THEOREM 4.1.3. *Let $\mathcal{M}$ be an o-minimal structure, let $\{C_i : i \in I\}$ be a family of convex subsets of M, and let $\mathcal{M}^*$ be the expansion of $\mathcal{M}$ obtained by adding unary predicates interpreted by the C_i. Then $\text{Th}(\mathcal{M}^*)$ is weakly o-minimal.*

By this theorem, if $\mathcal{R}$ is an o-minimal expansion of $\mathbb{R}$ and $\mathcal{R}'$ is a non-archimedean elementary extension, then the structure $(\mathcal{R}', V)$ has weakly o-minimal theory, where V denotes the convex valuation ring of finite elements of R'. Structures of this sort have been investigated further in [van den Dries and Lewenberg 1995; van den Dries 1997]. For example, the latter paper shows that if $\mathcal{R}$ is an o-minimal expansion of a real closed field and V is a proper non-empty convex subring closed under 0-definable continuous functions, then $(\mathcal{R}, V)$ is weakly o-minimal (as follows also from Theorem 4.1.3), with weakly o-minimal value group and residue field (as is fairly clear). Furthermore the theory T_{convex} of such expansions of models of T is complete, and has a relative quantifier-elimination (relative to quantifier-elimination for T, and assuming T is universally axiomatised). In addition, the structure induced on the value group is o-minimal precisely if T is *power-bounded* (a generalisation of 'polynomially bounded' for arbitrary o-minimal expansions of real closed fields), precisely if $\mathcal{R}$ has no definable order-preserving isomorphism from $(R, +)$ to the multiplicative group $R^{>0}$.

The structure theory for o-minimal structures begins with the Monotonicity Theorem. In the weakly o-minimal case, we can only hope for a local version of this. For example, let $\mathcal{M} = (M, <, f)$, where $(M, <)$ is naturally identified with $\mathbf{Z} \times \mathbb{Q}$, ordered lexicographically, and, for all $(z, q) \in M$, $f((z, q)) = (-z, q)$. Then $\text{Th}(\mathcal{M})$ is weakly o-minimal, and the function f is locally strictly monotonic, but not piecewise monotonic in the sense of Theorem 2.0.2.

In order to obtain a reasonable cell decomposition theorem, we need to consider not merely definable functions $M \to M$, but definable functions $M \to \bar{M}$, where $\bar{M}$ denotes the Dedekind completion of M. There is a natural notion of *definable sort* in $\bar{M}$. Let $Y \subset M^{n+1}$ be 0-definable, let $\pi : M^{n+1} \to M^n$ be the projection dropping the last coordinate, let $Z := \pi(Y)$, and for each $\bar{a} \in Z$ let $Y_{\bar{a}} := \{y : (\bar{a}, y) \in Y\}$. Suppose that each set $Y_{\bar{a}}$ is bounded above but does not have a supremum in M. Define an equivalence relation $\sim$ on Z, putting $\bar{a} \sim \bar{b}$ if $Y_{\bar{a}}, Y_{\bar{b}}$, have a common final segment. Then $Z/\sim$ (which is a sort in $\mathcal{M}^{\text{eq}}$) is naturally identified with a subset of $\bar{M}$.

If I, K are sets each endowed with a dense total order, then we say that a function $f : I \to K$ is *tidy* if each element of I lies in the interior of an open interval on which f is strictly monotonic or constant, with the same possibility (i.e., increasing, decreasing, or constant) holding for each $x \in I$. The following result was proved under extra hypotheses in [Macpherson et al. 1999], and in general by Arefiev [1997].

THEOREM 4.1.4. *Let $\mathcal{M}$ be weakly o-minimal, and $f : M \to \bar{M}$ a definable partial function (to a definable sort). Then there is a partition $\text{dom}(f) = X \cup I_1 \cup \cdots \cup I_m$, where X is finite, and for each j the set I_j is definable and convex, and $f|_{I_j}$ is tidy.*

Recall from Section 2 the definition of the *topological dimension* $\text{tdim}(X)$ of a definable set X. We say that *topological dimension is well-behaved in a class* $\mathcal{K}$

of structures, if for all $\mathcal{M} \in \mathcal{K}$, $m, n > 0$, and definable $X_1, \ldots, X_m \in M^n$, $\mathrm{tdim}(X_1 \cup \ldots \cup X_m) = \max(\mathrm{tdim}(X_1), \ldots, \mathrm{tdim}(X_m))$. By Corollary 2.2.8 and Theorem 2.2.9, topological dimension is well-behaved in the class of o-minimal structures. By results from [Macpherson et al. 1999] together with Theorem 4.1.4, we can extend this to obtain

THEOREM 4.1.5. *Topological dimension is well-behaved in the class of weakly o-minimal structures.*

From this (and related results in Section 4 of [Macpherson et al. 1999]) there follows a rather weak cell decomposition theorem for weakly o-minimal theories. It is weak in the sense that the boundary functions defining the cells may be functions to definable sorts in $\bar{M}$, which are not assumed to be continuous. If the *theory* of $\mathcal{M}$ is weakly o-minimal, then we can arrange in addition that each cell has a homeomorphic projection to an open set. Furthermore, in this case, definable bijections preserve topological dimension. There are weakly o-minimal theories in which algebraic closure does not have the exchange property (for example, the *contraction groups* of F.-V. Kuhlmann [1995]). However, models of any weakly o-minimal *theory* are uniformly bounded, so if algebraic closure does have the exchange property in such a theory, then its models are geometric structures in the sense of Section 2. If algebraic closure has the exchange property in a weakly o-minimal theory, then by Theorem 4.12 of [Macpherson et al. 1999], geometric dimension for definable sets is equal to topological dimension as in the o-minimal case (see Theorem 2.2.9).

In the o-minimal case, it is easy to show that any o-minimal ordered group is divisible abelian, and any o-minimal ordered field is real closed (see Propositions 3.1.1 and 3.1.2). The same theorems hold in the weakly o-minimal case:

THEOREM 4.1.6. (i) *Any weakly o-minimal ordered group is divisible abelian.* (ii) *Any weakly o-minimal ordered field is real closed.*

These results are proved in [Macpherson et al. 1999]. In the group case it is easy, but the argument in the field case is substantial, and uses the fact that topological dimension is well-behaved, a kind of inverse function theorem, and some valuation theory. Observe that it is only assumed that the *structure* is weakly o-minimal, not the *theory*. A slight extension of Theorem 4.1.6, Corollary 5.13 of [Macpherson et al. 1999], states that any weakly o-minimal ordered commutative ring with a unit is a real closed ring, that is, a convex valuation ring of a real closed field (or possibly the whole field).

For a weakly o-minimal expansion $\mathcal{M}$ of an ordered field, quite a nice dichotomy emerges from results in Section 6 of [Macpherson et al. 1999]. Either there is a definable convex valuation ring in $\mathcal{M}$, or $\mathcal{M}$ shares many properties of o-minimality: the Monotonicity Theorem is piecewise, and not just local; there is a cell decomposition theorem in which the boundary functions of the cells are continuous (to a sort in $\bar{M}$); algebraic closure has the exchange property; and the *theory* of $\mathcal{M}$ is weakly o-minimal (and in particular, $\mathcal{M}$ is uniformly bounded). The case when there *is* a definable convex valuation ring also has a combinatorial characterisation: it occurs precisely when there is a definable equivalence relation on M with infinitely many infinite classes.

In another direction, there is a structure theory (with several interesting examples) for ω-categorical weakly o-minimal structures [Herwig et al. 2000]. In particular, it is shown that if such a structure is *3-indiscernible* (that is, there is a unique type of strictly increasing triple) then it is indiscernible. For a typical example of the structures which arise, consider a countable non-archimedean real closed field $\mathcal{R}$ with the archimedean valuation v corresponding to the valuation ring of finite elements, and define a ternary relation C on R, putting $C(x; y, z)$ whenever

$$v(y - x) < v(y - z).$$

Then $(R, <, C)$ is ω-categorical and weakly o-minimal (and C is a C-relation in the sense of the next subsection). In contrast, there are no interesting ω-categorical *o-minimal* structures [Pillay and Steinhorn 1986].

Finally, I comment that, if M is a weakly o-minimal theory which is a geometric structure, then by Chapter 8 of [Mosley 1996], the analogue of Lemma 3.2.3 holds (that is, any definable group is definably topologisable). Analogues of this also hold for C- and P-minimal structures discussed below (but the statement in the P-minimal case is weaker).

The theory of weak o-minimality has been developed further by several authors (Aref'ev, Baizhanov, Baisalov, Kulpeshov, Nurtazyn, Verbovsky) in Almaty.

4.2. *C* and *P*-minimality.

We now consider a different generalisation of o-minimality, from [Macpherson and Steinhorn 1996], and again obtain general settings for a model theory of certain valued fields.

Suppose $L \subset L^+$ are languages, and $\mathcal{K}$ is an elementary class of L-structures. We say that an L^+-structure $\mathcal{M}$ is $\mathcal{K}$-*minimal* if the reduct $\mathcal{M}|_L \in \mathcal{K}$ and every L^+-definable subset of M is definable by a quantifier-free L-formula. A complete L^+-theory is $\mathcal{K}$-minimal if all its models are. It is easily checked that if $\mathcal{M}$ is $\mathcal{K}$-minimal then its theory is $\mathcal{K}$-minimal if and only if the following condition holds: for any $m > 0$ and L^+-definable subset S of M^{m+1}, there is $m' > 0$ and a quantifier-free L-definable subset $S' \subseteq M^{m'+1}$ such that for each $\bar{a} \in M^m$ there is $\bar{a}' \in M^{m'}$ such that $S_{\bar{a}} = S'_{\bar{a}'}$ (here, as usual, $S_{\bar{x}} := \{y : (\bar{x}, y) \in S\}$).

This setting includes o-minimality as a special case (where L has just a single binary relation, and $\mathcal{K}$ is the class of all dense total orders). If L has no relation, function, or constant symbols (apart, of course, from $=$), and $\mathcal{K}$ is the class of all infinite L-structures, then a theory is $\mathcal{K}$-minimal if and only if it is strongly minimal. However, weak o-minimality does not quite fit into this setting. Like o-minimality, $\mathcal{K}$-minimality is closed under reducts to languages containing L, and under expansions by constants. It is a condition on definable sets in one variable, which makes it easy to verify if, for example, one has quantifier elimination. Like o-minimality, it often gives strong information for definable sets in several variables, such as a dimension theory. It is observed in Theorem 3.2 of [Macpherson and Steinhorn 1996] that if a theory in L^+ is $\mathcal{K}$-minimal, then various stability properties lift from the L-reducts.

Our task is to find classes $\mathcal{K}$ such that there is a reasonable model theory of $\mathcal{K}$-minimality and there are interesting $\mathcal{K}$-minimal structures. This was initiated in [Macpherson and Steinhorn 1996], and developed in [Haskell and Macpherson 1994; 1997; van den Dries et al. 1999; Lipshitz and Robinson 1998].

C-**minimality.** The symbol C here denotes a ternary relation $C(x; y, z)$ (the semicolon indicates that the first variable is distinguished). We let $L = \{C\}$, and let $\mathcal{K}_C$ be the class of L-structures which satisfy the following axioms, where the free variables are universally quantified. The axioms were isolated by Adeleke and Neumann in work on Jordan permutation groups, and much more information on them can be found in [Adeleke and Neumann 1998].

(C1) $C(x; y, z) \rightarrow C(x; z, y)$

(C2) $C(x; y, z) \rightarrow \neg C(y; x, z)$

(C3) $C(x; y, z) \rightarrow (C(w; y, z) \vee C(x; w, z))$

(C4) $x \neq y \rightarrow (\exists z \neq y) \, C(x; y, z)$

(C5) $\exists x \, \exists y \, (x \neq y)$.

As an example, let $(T, \leq)$ be a semilinearly ordered set, that is, a partial order such that any two elements have a common lower bound, but the set of all lower bounds of an element is totally ordered. Suppose that T is infinite there is branching arbitrarily far up every maximal chain of T. Let M be the set of maximal chains of $(T, \leq)$, and interpret $C(x; y, z)$ to hold if either $y = z \neq x$, or x, y, z are distinct and x branches below where y and z branch (that is, $y \cap x \subset y \cap z$, where we regard elements of M as subsets of T). Then (M, C) satisfies (C1)–(C5). A converse to this was shown in [Adeleke and Neumann 1998]: namely, if $(M, C) \in \mathcal{K}_C$, then there is a semilinear order $(T, \leq)$ interpretable in (M, C), living on a quotient of M^2, such that M consists of a set of maximal chains of $(T, \leq)$ with union T, and C is interpreted as above. Since we think of $\mathcal{K}_C$-structures in this way as sets of chains in a semilinear order, we often talk of *nodes* of (M, C), meaning internal nodes of the underlying semilinear order. If $(M, C) \in \mathcal{K}_C$ then there is a Hausdorff topology on M with a uniformly definable basis: each internal node a determines the basic open set consisting of maximal chains which pass through it. This gives an analogy between $\mathcal{K}_C$ and the class of infinite dense total orders, but observe that unlike in the totally ordered case, the above basis consists of clopen sets and so in particular the topology is totally disconnected.

With this class $\mathcal{K}_C$, we now say that a structure $\mathcal{M} = (M, C, \dots)$ is C-*minimal* if its theory is $\mathcal{K}_C$-minimal (unlike 0-minimality, we choose *by definition* to close the condition under elementary equivalence). This notion was introduced in [Macpherson and Steinhorn 1996], where a number of examples were given, and a reasonable structure theory found for C-minimal groups. (A *C-minimal group* is the C-analogue of an o-minimal ordered group; it is a C-minimal structure $\mathfrak{M} = (M, C, *)$, where $(M, *)$ is a group, $(M, C) \in \mathcal{K}_C$, and the C-relation is preserved by left and right multiplication.) For example, such a group must have a definable abelian normal subgroup with quotient of finite exponent.

There are nice connections between C-minimality and strong minimality and o-minimality. If $\mathfrak{M}$ is C-minimal with underlying semilinear order $(T, \leq)$, then any element of M corresponds to a subset of $S \subset T$, that is, the set of nodes on the chain. Such a set S is interpretable in $\mathfrak{M}$ and S together with the induced structure on it is o-minimal. Likewise, if $a \in T$, there is an equivalence relation E_a on the set $\{x \in M : a \in x\}$: put $E_a xy$ if $x \cap y$ contains a node strictly greater than a. The E_a-classes are called *cones at* a, and the set of cones at a is interpretable in $\mathfrak{M}$, and, if infinite, is strongly minimal.

More general results were obtained in [Haskell and Macpherson 1994]. We have a notion of topological dimension as in Section 2, and, as in Theorem 4.1.5, obtain:

THEOREM 4.2.1. *Topological dimension is well-behaved in any C-minimal structure.*

The proof of this is by induction, where simultaneously a cell decomposition theorem is proved (as in Theorem 2.1.3). However, the notion of 'cell' is very cumbersome. With this notion of 'cell' one also proves in the induction the following result.

THEOREM 4.2.2. *Let n be a positive integer, X a definable subset of M^n, and $f : M^n \to M$ a definable partial function. Then X can be expressed as the disjoint union of finitely many cells on each of which f is continuous.*

The proof also uses a local version of the o-minimal Monotonicity Theorem, where 'monotonic function' is replaced by 'isomorphism' (of neighbourhoods, endowed with the relation C).

It is shown in [Macpherson and Steinhorn 1996] that in a C-minimal structure $\mathfrak{M}$, algebraic closure need not have the exchange property. However, by Proposition 6.1 of [Haskell and Macpherson 1994], the exchange property can only fail in one way, namely if there is a certain kind of definable 'bad' function, between M and the set of internal nodes. Furthermore, any C-minimal structure $\mathfrak{M}$ is uniformly bounded, as otherwise, in an elementary extension, there would be an infinite definable set with empty interior, contrary to C-minimality. Hence, if algebraic closure in $\mathfrak{M}$ does have the exchange property, then $\mathfrak{M}$ is a geometric structure in the sense of Section 2, so algebraic closure in this case also provides a notion of dimension for definable sets. By Proposition 6.3 of [Haskell

and Macpherson 1994], in this situation the topological and geometric notions
of dimension coincide.

There is a natural class of C-minimal structures. Let

$$\mathcal{F} := (F, V, +, \cdot)$$

be a non-trivially valued field, where V is a valuation ring with corresponding
valuation map v to the value group. Define C on F by putting $C(x; y, z)$ if and
only if $v(y - x) < v(y - z)$. Then the relation C is invariant under addition, and
under multiplication by non-zero elements, and $(F, C) \in \mathcal{K}_C$. In fact, as noted
in [Macpherson and Steinhorn 1996], the converse holds: if $\mathcal{F} = (F, +, \cdot)$ is a
field, $(F, C) \in \mathcal{K}_C$, and C is preserved by the field operations in the above sense,
then C comes from a valuation as above, and the valuation is definable in (F, C).
In this situation, the C-relation and the semilinear order provide a natural way
of viewing the value group and residue field. For let $(T, \leq)$ be the semilinear
order underlying (F, C) (so members of F are maximal chains in T). Now the
value group Γ of F is identified with the set of nodes on the chain 0_F (the zero
of F), with the natural induced order. If $x \in F$, then $v(x) = \max(x \cap 0_F)$, that
is, the node at which the chains x and 0_F meet (there will be such a node). In
particular, the zero 0_Γ of the value group is the node at which the chains 1_F
and 0_F meet, the valuation ring V is the set of chains in F which pass through
this node, and the maximal ideal consists of those chains lying in the cone at
0_Γ which contains 0_F. The residue field consists of the set of cones at 0_Γ. The
picture is fairly clear in say the valued power series field $F_p[[t]]$, where we may
think of the internal nodes as given by polynomials in t, t^{-1}.

By a quantifier-elimination result of A. Robinson, if $\mathcal{F}$ is an algebraically
closed valued field, then $(\mathcal{F}, C)$ is C-minimal [Macpherson and Steinhorn 1996].
In [Haskell and Macpherson 1994], the converse was proved, namely:

THEOREM 4.2.3. *Every C-minimal field F is an algebraically closed non-trivially
valued field.*

In the proof it follows from the above identification of the value group Γ and
residue field $\bar{F}$ that Γ is an o-minimal ordered group, so is divisible abelian,
and $\bar{F}$ is finite or strongly minimal. By the divisibility of the value group, F
is closed under Kummer and Artin–Schreier extensions, and this forces $\bar{F}$ to be
infinite, so algebraically closed. The proof then uses valuation theory, together
with Theorem 4.2.1.

Given the rich supply of o-minimal expansions of the field of reals, it is natural
to ask for C-minimal *expansions* of algebraically closed valued fields. The model
theory of rigid analytic geometry was first developed by Lipshitz [1993], who
proved quantifier elimination for an expansion of an algebraically closed valued
field (complete with respect to a definable non-archimedean norm), by a rich and
rather complicated non-archimedean structure. Lipshitz and Robinson [1998]
have shown that this expansion is C-minimal. This means that subanalytic sets

in one variable are uniformly (in the parameters) definable in the pure valued field.

P-**minimality.** In this subsection I use 'semialgebraic' to mean 'definable in the pure field $\mathbb{Q}_p$.' (It is well-known that the natural valuation on $\mathbb{Q}_p$ is definable in the pure field.) As discussed in the survey [Macintyre 1986], there are many model-theoretic analogues between $\mathbb{Q}_p$ and $\mathbb{R}$, at both the semialgebraic and the subanalytic level (see [Denef and van den Dries 1988] for the latter). It is therefore natural to look for versions of o-minimality which support $\mathbb{Q}_p$, and one such, *P*-minimality, was proposed in [Haskell and Macpherson 1997].

DEFINITION 4.2.4. Let L be the language $(+, -, \cdot, 0, 1, P_n)_{n>1}$ (where the P_n are unary predicates). Regard $\mathbb{Q}_p$ as an L-structure, letting P_n pick out the nth powers in $\mathbf{Q_p}$. Let $\mathcal{K}_P$ be the class of L-structures elementarily equivalent to $\mathbb{Q}_p$. Then if $L^+ \supseteq L$, an L^+-structure is *P*-*minimal* if all models of its theory are $\mathcal{K}_P$-minimal.

By [Macintyre 1976], $\mathbb{Q}_p$ has quantifier elimination in the above language L. The version of *P*-minimality described above differs slightly from that in [Haskell and Macpherson 1997], where, for example, *p*-adically closed fields in the more general sense of [Prestel and Roquette 1984] are considered *P*-minimal. I emphasise that unlike the other model-theoretic classes considered in this paper, a *P*-minimal structure is *always* an expansion of a field.

The model theory of *P*-minimality has not been developed far, but I sketch some results. In any *P*-minimal structure F with value group Γ, the valuation topology has a uniformly definable basis of clopen sets, namely, sets of the form $B_\gamma(a) = \{x : v(x - a) \geq \gamma\}$ where $a \in F$ and $\gamma \in \Gamma$. We again obtain that topological dimension is well-behaved in *P*-minimal structures. As in the *C*-minimal case, *P*-minimal structures are uniformly bounded. Furthermore, unlike in the weakly o-minimal and *C*-minimal cases, algebraic closure has the exchange property in *any* *P*-minimal structure, so such structures are geometric. The resulting geometric notion of dimension coincides with topological dimension. There are also theorems about continuity of definable functions. No cell decomposition theorem for *P*-minimal structures has been proved (though it might not be difficult, and a cell decomposition and related results for $\mathbb{Q}_p$ are developed in [Denef 1986; Scowcroft and van den Dries 1988]).

Let L_{an}^D be the language introduced in [Denef and van den Dries 1988] to describe the subanalytic structure on the *p*-adic integers $\mathbf{Z}_p$. The following is an analogue of the theorem of Lipshitz and Robinson mentioned above, and is the main theorem of [van den Dries et al. 1999].

THEOREM 4.2.5. *The L_{an}^D-structure $\mathbb{Q}_p$ is P-minimal.*

It was shown earlier in [Denef and van den Dries 1988] (Corollary 3.32) that any L_{an}^D-definable subset of $\mathbb{Q}_p$ is semialgebraic. Theorem 4.2.5 shows that such sets are semialgebraic *uniformly* in the defining parameters. The point is that

P-minimality is a property of the *theory*; however, the uniformity provides new information also in the standard model $\mathbb{Q}_p$.

The proof of Theorem 4.2.5 uses the quantifier elimination of [Denef and van den Dries 1988]. One has a quantifier-free L_{an}^D-formula $\phi(x, \bar{y})$, and has to show that the set defined by $\phi(x, \bar{a})$ is semialgebraic, uniformly in $\bar{a}$. The formula $\phi(x, \bar{y})$ is assumed to be atomic, and the proof is based on induction on the complexity of a term $t(x)$ in the language. The main problems are posed by occurrences in a term of the binary function symbol D for division. One works in an elementary extension K, and, using a parametric Weierstrass Preparation Theorem, obtains a Preparation Theorem for a certain ring $K\{Y_1, \ldots, Y_n\}$ of definable functions on K^n. The idea is, given a term $t(x)$, to cover the valuation ring R of K with particularly nice sets known as 'connected affinoids'. Each connected affinoid F has a well-behaved associated ring $\mathcal{O}(F)$ of definable functions on it. The ring $\mathcal{O}(F)$ is a quotient of $K\{Y_1, \ldots, Y_n\}$, where $n-1$ is the number of 'holes' in F. It has nice divisibility properties — any non-zero element of $\mathcal{O}(F)$ has just finitely many zeros, and if it has no zeros then it is a unit of $\mathcal{O}(F)$. One uses this to show, by induction on the complexity, that any term is given piecewise by members of $\mathcal{O}(F)$ for various F. Theorem 4.2.5 follows immediately from this.

I conclude with an amalgam of results from [Macpherson et al. 1999] (Proposition 7.3), [Macpherson and Steinhorn 1996] (Proposition 3.4) and [Haskell and Macpherson 1997] (Proposition 7.1), which is suggested by Corollary 3.10 of [Pillay and Steinhorn 1986]. Recall that a theory T has the *independence property* [Shelah 1978] if there is $\mathcal{M} \models T$ and a formula $\phi(\bar{x}, \bar{y})$ with $l(\bar{x}) = m, l(\bar{y}) = n$, and $\bar{a}_i \in M^m$ (for all $i \in \omega$) such that: for any $S \subseteq \omega$, there is $\bar{b}_S \in M^n$ such that for all $i \in \omega$, $\mathcal{M} \models \phi(\bar{a}_i, \bar{b}_S)$ if and only if $i \in S$. No stable theory can have the independence property, but for example, pseudofinite fields do have it.

THEOREM 4.2.6. *No weakly o-minimal, C-minimal or P-minimal theory can have the independence property.*

References

[Adeleke and Neumann 1998] S. A. Adeleke and P. M. Neumann, *Relations related to betweenness: their structure and automorphisms*, Mem. Amer. Math. Soc. **623**, Amer. Math. Soc., Providence, RI, 1998.

[Aref'ev 1997] R. Aref'ev, "On the monotonicity property of weakly o-minimal structures", in *Algebra and model theory*, edited by A. G. Pinus and K. N. Ponomaryov, Novosibirsk, 1997. In Russian.

[Baisalov and Poizat 1998] Y. Baisalov and B. Poizat, "Paires de structures o-minimales", *J. Symbolic Logic* **63**:2 (1998), 570–578.

[Baizhanov 1999] B. Baizhanov, "Expansion of a model of o-minimal theory by a family of unary predicates", preprint, National Academy of Sciences, Almaty, 1999.

[Borel and Tits 1973] A. Borel and J. Tits, "Homomorphismes "abstraits" de groupes algébriques simples", *Ann. of Math.* (2) **97** (1973), 499–571.

[Cherlin and Dickmann 1983] G. Cherlin and M. A. Dickmann, "Real closed rings, II: Model theory", *Ann. Pure Appl. Logic* **25**:3 (1983), 213–231.

[Denef 1986] J. Denef, "p-adic semi-algebraic sets and cell decomposition", *J. Reine Angew. Math.* **369** (1986), 154–166.

[Denef and van den Dries 1988] J. Denef and L. van den Dries, "p-adic and real subanalytic sets", *Ann. of Math.* (2) **128**:1 (1988), 79–138.

[Dickmann 1985] M. A. Dickmann, "Elimination of quantifiers for ordered valuation rings", pp. 64–88 in *Proceedings of the third Easter conference on model theory* (Gross Köris, 1985), Fachbereich Mathematik, Humboldt Univ. Berlin, 1985.

[van den Dries 1986] L. van den Dries, "A generalization of the Tarski–Seidenberg theorem, and some nondefinability results", *Bull. Amer. Math. Soc. (N.S.)* **15**:2 (1986), 189–193.

[van den Dries 1989] L. van den Dries, "Dimension of definable sets, algebraic boundedness and Henselian fields", *Ann. Pure Appl. Logic* **45**:2 (1989), 189–209.

[van den Dries 1996] L. van den Dries, "O-minimal structures", pp. 137–185 in *Logic: from foundations to applications* (Staffordshire, 1993), edited by W. Hodges et al., Oxford Univ. Press, New York, 1996.

[van den Dries 1997] L. van den Dries, "T-convexity and tame extensions, II", *J. Symbolic Logic* **62**:1 (1997), 14–34. Correction in **63**:4 (1998), 1597.

[van den Dries 1998] L. van den Dries, *Tame topology and o-minimal structures*, London Math. Soc. Lecture Note Series **248**, Cambridge Univ. Press, Cambridge, 1998.

[van den Dries and Lewenberg 1995] L. van den Dries and A. H. Lewenberg, "T-convexity and tame extensions", *J. Symbolic Logic* **60**:1 (1995), 74–102.

[van den Dries and Miller 1996] L. van den Dries and C. Miller, "Geometric categories and o-minimal structures", *Duke Math. J.* **84**:2 (1996), 497–540.

[van den Dries et al. 1999] L. van den Dries, D. Haskell, and D. Macpherson, "One-dimensional p-adic subanalytic sets", *J. London Math. Soc.* (2) **59**:1 (1999), 1–20.

[Haskell and Macpherson 1994] D. Haskell and D. Macpherson, "Cell decompositions of C-minimal structures", *Ann. Pure Appl. Logic* **66**:2 (1994), 113–162.

[Haskell and Macpherson 1997] D. Haskell and D. Macpherson, "A version of o-minimality for the p-adics", *J. Symbolic Logic* **62**:4 (1997), 1075–1092.

[Herwig et al. 2000] B. Herwig, H. D. Macpherson, G. Martin, A. Nurtazyn, and J. K. Truss, "Omega-categorical weakly o-minimal structures", *Ann. Pure Appl. Logic* **101** (2000), 65–93.

[Hrushovski 1992] E. Hrushovski, "Strongly minimal expansions of algebraically closed fields", *Israel J. Math.* **79**:2-3 (1992), 129–151.

[Hrushovski and Pillay 1994] E. Hrushovski and A. Pillay, "Groups definable in local fields and pseudo-finite fields", *Israel J. Math.* **85**:1-3 (1994), 203–262.

[Hrushovski and Zilber 1996] E. Hrushovski and B. Zilber, "Zariski geometries", *J. Amer. Math. Soc.* **9**:1 (1996), 1–56.

[Kim and Pillay 1997] B. Kim and A. Pillay, "Simple theories", *Ann. Pure Appl. Logic* **88**:2-3 (1997), 149–164.

[Knight et al. 1986] J. F. Knight, A. Pillay, and C. Steinhorn, "Definable sets in ordered structures, II", *Trans. Amer. Math. Soc.* **295**:2 (1986), 593–605.

[Kuhlmann 1995] F. Kuhlmann, "Abelian groups with contractions, II: Weak o-minimality", pp. 323–342 in *Abelian groups and modules* (Padova, 1994), edited by A. Facchini and C. Menini, Mathematics and its applications **343**, Kluwer, Dordrecht, 1995.

[Lipshitz 1993] L. Lipshitz, "Rigid subanalytic sets", *Amer. J. Math.* **115**:1 (1993), 77–108.

[Lipshitz and Robinson 1998] L. Lipshitz and Z. Robinson, "One-dimensional fibers of rigid subanalytic sets", *J. Symbolic Logic* **63**:1 (1998), 83–88.

[Loveys and Peterzil 1993] J. Loveys and Y. Peterzil, "Linear o-minimal structures", *Israel J. Math.* **81**:1-2 (1993), 1–30.

[Macintyre 1976] A. Macintyre, "On definable subsets of p-adic fields", *J. Symbolic Logic* **41**:3 (1976), 605–610.

[Macintyre 1986] A. Macintyre, "Twenty years of p-adic model theory", pp. 121–153 in *Logic colloquium '84* (Manchester, 1984), edited by J. B. Paris et al., Stud. Logic Found. Math. **120**, North-Holland, Amsterdam, 1986.

[Macpherson and Steinhorn 1996] H. D. Macpherson and C. Steinhorn, "On variants of o-minimality", *Ann. Pure Appl. Logic* **79**:2 (1996), 165–209.

[Macpherson et al. 1999] H. D. Macpherson, D. Marker, and C. Steinhorn, "Weakly o-minimal structures and real closed fields", preprint, 1999. Available at http://amsta.leeds.ac.uk/pure/staff/macpherson/preprints.html. To appear in *Transactions Amer. Math. Soc.*

[Marker 1990] D. Marker, "Semialgebraic expansions of ℂ", *Trans. Amer. Math. Soc.* **320**:2 (1990), 581–592.

[Marker and Steinhorn 1994] D. Marker and C. I. Steinhorn, "Definable types in o-minimal theories", *J. Symbolic Logic* **59**:1 (1994), 185–198.

[Mekler et al. 1992] A. Mekler, M. Rubin, and C. Steinhorn, "Dedekind completeness and the algebraic complexity of o-minimal structures", *Canad. J. Math.* **44**:4 (1992), 843–855.

[Miller and Starchenko 1998] C. Miller and S. Starchenko, "A growth dichotomy for o-minimal expansions of ordered groups", *Trans. Amer. Math. Soc.* **350**:9 (1998), 3505–3521.

[Mosley 1996] A. Mosley, *Groups definable in topological structures*, Ph.D. thesis, Queen Mary and Westfield College, London, 1996.

[Nesin and Pillay 1991] A. Nesin and A. Pillay, "Some model theory of compact Lie groups", *Trans. Amer. Math. Soc.* **326**:1 (1991), 453–463.

[Otero et al. 1996] M. Otero, Y. Peterzil, and A. Pillay, "On groups and rings definable in o-minimal expansions of real closed fields", *Bull. London Math. Soc.* **28**:1 (1996), 7–14.

[Peterzil 1994] Y. Peterzil, "Constructing a group-interval in o-minimal structures", *J. Pure Appl. Algebra* **94**:1 (1994), 85–100.

[Peterzil and Starchenko 1998] Y. Peterzil and S. Starchenko, "A trichotomy theorem for o-minimal structures", *Proc. London Math. Soc.* (3) **77**:3 (1998), 481–523.

[Peterzil and Steinhorn 1999] Y. Peterzil and C. Steinhorn, "Definable compactness and definable subgroups of o-minimal groups", *J. London Math. Soc.* (2) **59**:3 (1999), 769–786.

[Peterzil et al. 1997] Y. Peterzil, A. Pillay, and S. Starchenko, "Simple algebraic and semialgebraic groups over real closed fields", Preprint, 1997. Available at http://www.math.uiuc.edu/Reports/pillay/97-028.html. To appear in *Trans. Amer. Math. Soc.*

[Peterzil et al. 1998] Y. Peterzil, A. Pillay, and S. Starchenko, "Simple groups definable in o-minimal structures", pp. 211–218 in *Logic Colloquium '96* (San Sebastián, 1996), Lecture Notes in Logic **12**, Springer, Berlin, 1998.

[Peterzil et al. 2000] Y. Peterzil, A. Pillay, and S. Starchenko, "Definably simple groups in o-minimal structures", *Trans. Amer. Math. Soc.* (2000).

[Pillay 1986] A. Pillay, "Some remarks on definable equivalence relations in o-minimal structures", *J. Symbolic Logic* **51**:3 (1986), 709–714.

[Pillay 1988] A. Pillay, "On groups and fields definable in o-minimal structures", *J. Pure Appl. Algebra* **53**:3 (1988), 239–255.

[Pillay 1994] A. Pillay, "Definability of types, and pairs of o-minimal structures", *J. Symbolic Logic* **59**:4 (1994), 1400–1409.

[Pillay and Steinhorn 1986] A. Pillay and C. Steinhorn, "Definable sets in ordered structures, I", *Trans. Amer. Math. Soc.* **295**:2 (1986), 565–592.

[Pillay and Steinhorn 1987] A. Pillay and C. Steinhorn, "Discrete o-minimal structures", *Ann. Pure Appl. Logic* **34**:3 (1987), 275–289.

[Pillay and Steinhorn 1988] A. Pillay and C. Steinhorn, "Definable sets in ordered structures, III", *Trans. Amer. Math. Soc.* **309**:2 (1988), 469–476.

[Prestel and Roquette 1984] A. Prestel and P. Roquette, *Formally p-adic fields*, Lecture Notes in Math. **1050**, Springer, Berlin, 1984.

[Scowcroft and van den Dries 1988] P. Scowcroft and L. van den Dries, "On the structure of semialgebraic sets over p-adic fields", *J. Symbolic Logic* **53**:4 (1988), 1138–1164.

[Shelah 1978] S. Shelah, *Classification theory and the number of non-isomorphic models*, Stud. Logic Found. Math. **92**, North-Holland, Amsterdam, 1978.

[Strzebonski 1994] A. W. Strzebonski, "Euler characteristic in semialgebraic and other o-minimal groups", *J. Pure Appl. Algebra* **96**:2 (1994), 173–201.

[Weisfeiler 1979] B. Weisfeiler, "On abstract homomorphisms of anisotropic algebraic groups over real-closed fields", *J. Algebra* **60**:2 (1979), 485–519.

[Wilkie 1996] A. Wilkie, "Model completeness results for expansions of the real field by restricted Pfaffian functions and the exponential function", *J. Amer. Math. Soc.* **9**:2 (1996), 1051–1094.

Dugald Macpherson
Department of Pure Mathematics
University of Leeds
Leeds LS2 9JT
England
pmthdm@amsta.leeds.ac.uk

Model Theory, Algebra, and Geometry
MSRI Publications
Volume **39**, 2000

Stability Theory and its Variants

BRADD HART

ABSTRACT. Dimension theory plays a crucial technical role in stability theory and its relatives. The abstract dependence relations defined, although combinatorial in nature, often have surprising geometric meaning in particular cases. This article discusses several aspects of dimension theory, such as categoricity, strongly minimal sets, modularity and the Zil'ber principle, forking, simple theories, orthogonality and regular types and in the third, stability, definability of types, stable groups and 1-based groups.

One of the achievements of the branch of model theory known as stability theory is the use of numerical invariants, dimensions, in a broad setting. In recent years, this dimension theory has been expanded to include the so-called simple theories. In this paper, I wish to give just a brief overview of the elements of this theory. In the first section, the special case of strongly minimal sets is considered. In the second section, the combinatorial definition of dividing is given and how it leads to a general independence relation is outlined. Only in the third section do stable theories appear and the theory surrounding them is developed there with an eye to other papers in this volume.

1. Strongly Minimal Sets

Categorical Theories. One of the simplest questions one can ask about a first order theory is how many models it has of a given cardinality. If T is a countable theory with an infinite model then, by the Lowenheim–Skolem Theorem, it will have at least one model of every infinite power. The situation we will look at first is when a theory has exactly one model of some fixed power.

DEFINITION 1.1. A theory T is λ-*categorical* if T has exactly one model up to isomorphism of cardinality λ. T is said to be *totally categorical* if T is λ-categorical for every infinite cardinal λ. We will say that T is *uncountably categorical* if T is λ-categorical for all uncountable λ.

EXAMPLE 1.2. 1. The theory of a set in a language which has only equality is totally categorical.

1. Suppose that D is a countable division ring. The theory of a vector space
   over D is uncountably categorical and is totally categorical if D is finite (that
   is, D is a finite field). Here the language contains a binary function symbol
   for addition and a unary function symbol for each scalar in D.

Two variants of this example are the following; fix an infinite dimensional vector
space V over a finite field F:

2. Infinite dimensional projective space over a finite field. $P(V)$ is the set of
   1-dimensional subspaces of V. Define an $n + 1$-ary relation R_n on $P(V)$
   as follows: $R_n(X_1, X_2, \ldots, X_n, Y)$ holds for $X_1, \ldots, X_n, Y \in P(V)$ if Y is
   contained in the subspace generated by $X_1, \ldots, X_n$. The theory of $P(V)$
   together with all the R_n's is totally categorical.

A more general version of this example involves the Galois group of F; we describe it via its automorphism group. By a projective geometry, we will mean
a structure whose underlying set is $P(V)$ and whose automorphism group is
$\mathrm{PGL}(V) \rtimes \mathrm{Gal}(F/L)$ where L is some subfield of F. All of these examples are
totally categorical.

3. Infinite dimensional affine space over a finite field. Let $\tau(x, y, z) = x - y + z$ and
   for every $f \in F$, let $\lambda_f(x, y) = \lambda x + (1 - \lambda)y$. The theory of $(V, \tau, \{\lambda_f : f \in F\})$
   is also totally categorical.

As with the projective space examples above, there are more general affine space
examples. By an affine geometry we will mean a structure whose underlying
set is V and whose automorphism group is $\mathrm{AGL}(V) \rtimes \mathrm{Gal}(F/L)$ where L is a
subfield of F. Again, all these examples are totally categorical.

   Here are a few other examples that will be commented on later:

5. The theory of $(Z/4Z)^\omega$ as an abelian group; this theory is totally categorical.
6. The theory of an algebraically closed field of a fixed characteristic; this theory
   is uncountably categorical.

It is possible for a theory to be ω-categorical without being totally categorical;
here are two examples:

7. The theory of the rationals as a linear order and the theory of an equivalence
   relation with two classes, each infinite.

Of course there are many theories which are not categorical in any power, for
example:

8. the theory of the real field, Peano arithmetic and the theory of the integers
   as an abelian group.

Los conjectured that there were four possibilities for the categoricity spectrum:
a theory would be either totally categorical, uncountably categorical but not ω-
categorical, ω-categorical but not uncountably categorical or not categorical in

any power. This was proved by Morley and is known as the Morley Categoricity Theorem:

THEOREM 1.3. *If T is a countable theory which is λ-categorical for some uncountable λ then T is λ-categorical for all uncountable λ.*

One intuition behind this conjecture is that for an uncountably categorical theory, the uncountable models are completely controlled by a single dimension and that by specifying that dimension, one specifies the isomorphism type of the model. I want to discuss the key ingredients of the proof of this theorem but first I want to indicate another theorem related to categoricity.

THEOREM 1.4 (LOS–VAUGHT TEST). *If T is categorical and has no finite models then T is complete.*

It follows then that examples 5, 6 and 7 above are complete. The first four examples above have finite models and are not complete. One can check that these four examples are all finitely axiomatized. The theory of the infinite models of these theories is then axiomatized by these finitely many axioms together with infinitely many sentences expressing the fact that the models are infinite. Various conjectures arose of which I will only state two:

1. Is there a finitely axiomatized, totally categorical theory?
2. Is every totally categorical theory finitely axiomatized modulo "axioms of infinity"?

We will say more about these conjectures later but to answer them, one needs a very detailed understanding of the models of totally categorical theories. Historically, this arose through an understanding of the proof of the Morley Categoricity Theorem.

   We will work throughout this paper with a countable, complete theory T with infinite models. For convenience, fix a structure $\mathcal{M}$ which is a λ-saturated model of T for some big cardinal λ. λ will be larger than any set of parameters or any submodel we will choose throughout our discussions. The following is a critical definition.

DEFINITION 1.5. An infinite definable set $X \subseteq M^n$ for some n is called *strongly minimal* if every definable subset of X is either finite or cofinite. Here the definable sets may be definable with parameters.

1. In the examples we saw early, the entire model (the set defined by $x = x$) is strongly minimal in the following cases: an infinite pure set, an infinite vector space over a countable division ring, an infinite projective space over a finite field, an infinite affine space over a finite field and any algebraically closed field of a fixed characteristic. The case of either projective or affine space over a finite field follows from the corresponding case of an infinite vector space. For an infinite set, an infinite vector space or an algebraically closed field, it

is enough to check atomic formulas by quantifier elimination in each case and the solution set of an atomic formula is finite in all cases.

2. In the case of the theory of $(Z/4Z)^\omega$ as an abelian group, the entire model is not strongly minimal but the set defined by $x + x = 0$ is. Again, this example has quantifier elimination which makes it easy to check this set is strongly minimal.

3. A theory does not have to be uncountably categorical to have a strongly minimal set: in the theory of an equivalence relation with two classes each infinite, either equivalence class is strongly minimal.

I wish to discuss dimension theory for strongly minimal sets so let me remind you of the definition of a pregeometry.

DEFINITION 1.6. A pregeometry (A, cl) is a set A together with a closure operator $\text{cl} : \mathcal{P}(A) \to \mathcal{P}(A)$ such that:

1. for any $B \subseteq A$, $B \subseteq \text{cl}(B) = \text{cl}(\text{cl}(B))$,
2. if $B \subseteq C \subseteq A$ then $\text{cl}(B) \subseteq \text{cl}(C)$,
3. if $B \subseteq A$ and $b \in \text{cl}(B)$ there is a finite $B_0 \subseteq B$ such that $b \in \text{cl}(B_0)$, and
4. (Steinitz exchange) if $B \subseteq A$, $b \in \text{cl}(B \cup \{c\}) \setminus \text{cl}(B)$ then $c \in \text{cl}(B \cup \{b\})$.

DEFINITION 1.7. If (A, cl) is a pregeometry, we say that $B \subseteq A$ is *independent* if $b \notin \text{cl}(B \setminus \{b\})$ for any $b \in B$. A basis for a set C is an independent set $B \subseteq C$ such that $C \subseteq \text{cl}(B)$.

PROPOSITION 1.8. *Suppose (A, cl) is a pregeometry. Then for any $B \subseteq A$, any maximal independent subset of B is a basis for B. Moreover, any two bases for B have the same cardinality.*

We now want to consider the closure operator acl_X on a definable set X. Suppose that X is defined over a set of parameters $\bar{d}$. Then $\text{acl}_X(Y) = \text{acl}(Y \cup \{\bar{d}\}) \cap X$ for $Y \subseteq X$. It is routine to check that acl_X is a closure operator on X which satisfies the first three properties from the definition of pregeometry.

PROPOSITION 1.9. *If X is a strongly minimal set then acl_X defines a pregeometry on X.*

PROOF. We need only to check the exchange property. Suppose that $a \in \text{acl}_X(C \cup \{b\}) \setminus \text{acl}_X(C)$. Suppose that $\varphi(x, y)$ is a formula with parameters from C which expresses that fact that a is algebraic over b. We can choose φ so that $\varphi(a, b)$ holds and there is an N so that for every b' there are at most N realizations of $\varphi(x, b')$. Now suppose that $\{a_i : i < N + 1\}$ is a set of distinct realizations of $\text{tp}(a/C)$ (remember that a is not algebraic over C).

For the sake of contradiction, suppose that $b \notin \text{acl}_X(C \cup \{a\})$. This means, in particular, that $\varphi(a, M) \cap X$ is infinite. Since X is strongly minimal, this set is cofinite, so $\varphi(a_i, M) \cap X$ is cofinite for each $i < N + 1$. Hence there is $b' \in X$ so that $\varphi(a_i, b')$ holds for all $i < N + 1$ which contradicts the choice of N. So in fact, $b \in \text{acl}_X(C \cup \{a\})$. $\qquad\square$

The following is easy by compactness.

PROPOSITION 1.10. *If X is strongly minimal and $\varphi(x,y)$ is any formula then there is a number N so that for any b, if $\left|\{a \in X : \varphi(a,b) \text{ holds}\}\right| > N$ then $\varphi(X,b)$ is infinite.*

How will strongly minimal sets be used in trying to understand the models? Suppose that X is a strongly minimal set defined over parameters $\bar{d}$ and further, that I and J are two sets of elements from X, independent with respect to acl_X. The claim is that any injection from I to J is a $\bar{d}$-elementary map. What this amounts to showing is that for any set $B \subseteq X$ there is a unique nonalgebraic type in X over $B \cup \{\bar{d}\}$. The latter is true because for any formula over $B \cup \{\bar{d}\}$, either it or its negation is an algebraic formula.

To summarize then, suppose X is a strongly minimal set defined over $\bar{d}$. Now if N is a model containing $\bar{d}$ then to understand the elementary type of $X \cap N$ over $\bar{d}$ one needs only to know the dimension of $X \cap N$ for then $X \cap N$ is algebraic over any basis.

But do uncountably categorical theories contain strongly minimal sets? Suppose that M itself is not strongly minimal. Then one can find an infinite, co-infinite definable set $X \subseteq M$. Let $X_0 = X$ and $X_1 = M \setminus X$. In general suppose that we have defined an infinite set X_η where η is a finite sequence of 0's and 1's. If X_η is not strongly minimal then it contains an infinite, co-infinite subset Y which we label $X_{\eta 0}$ and let $X_{\eta 1} = X_\eta \setminus Y$. In this way, we produce uncountably many consistent partial types over countably many parameters. Compare this with the following fact due to Ehrenfeucht.

FACT 1.11. *For any countable theory T with infinite models and any uncountable cardinal λ, there is a model of T, M of size λ with the property that for every countable set $A \subseteq M$, the number of types realized in M over A is countable.*

This naturally leads to the following definition.

DEFINITION 1.12. *A theory T is said to be ω-stable if for every countable $A \subseteq M$, $S(A)$ is countable.*

So above we have proved the following proposition.

PROPOSITION 1.13. *A countable theory T which is uncountably categorical is ω-stable.*

COROLLARY 1.14. *Any countable, uncountably categorical theory has strongly minimal sets.*

Now there are two questions I want to address in the rest of this section and in some sense they represent two different directions in what is known as stability theory:

1. In an uncountably categorical theory, how is the rest of the model related to the strongly minimal set?

2. What can the strongly minimal set look like/be?

Let's consider the first question. One of the means of measuring dimension in model theory is via ranks. The first and the one which most closely resembles dimension from algebraic geometry is Morley rank (written RM for Rang Morley). In the following definition, remember that we are working in a saturated model of the theory at hand and that definable means definable in that model.

DEFINITION 1.15. For a nonempty definable set X, we define $\text{RM}(X)$ inductively:

1. $\text{RM}(X) \geq 0$;
2. for a limit ordinal δ, $\text{RM}(X) \geq \delta$ if and only if $\text{RM}(X) \geq \alpha$ for all $\alpha < \delta$;
3. $\text{RM}(X) \geq \alpha + 1$ if and only if there are definable subsets Y_i of X for $i \in \omega$ which are pairwise disjoint and such that $\text{RM}(Y_i) \geq \alpha$.

$\text{RM}(X) = \alpha$ if $\text{RM}(X) \geq \alpha$ and $\text{RM}(X) \not\geq \alpha$. $\text{RM}(X) = \infty$ if $\text{RM}(X) \geq \alpha$ for all α. If $\varphi(x)$ is a consistent formula then by $\text{RM}(\varphi(x))$ we mean the Morley rank of the set this formula defines.

FACT 1.16. 1. *Morley rank is invariant; that is, if $\bar{a}$ and $\bar{b}$ have the same type over the empty set and $\varphi(x, \bar{a})$ is consistent then $\text{RM}(\varphi(x, \bar{a})) = \text{RM}(\varphi(x, \bar{b}))$.*
2. *If $X \subseteq Y$ then $\text{RM}(X) \leq \text{RM}(Y)$.*
3. $\text{RM}(\bigcup_{i=1}^n X_i) = \max\{\text{RM}(X_i) : i = 1, \ldots, n\}$.
4. *There is an ordinal α that depends only on the theory T such that if $\text{RM}(X) \geq \alpha$ then $\text{RM}(X) = \infty$.*

REMARKS 1.17. In the case of algebraically closed fields, Morley rank and algebro-geometric dimension coincide, that is, $\text{RM}(V) = \dim(V)$ for V an algebraic variety.

DEFINITION 1.18. If $\text{RM}(X) = \alpha < \infty$ then the Morley degree of X, $\text{dM}(X)$, is the largest number k so that there are pairwise disjoint definable subsets of X, Y_i, with $\text{RM}(Y_i) = \alpha$ for $i = 1, \ldots, k$.

FACT 1.19. *If $\text{RM}(X) < \infty$ then $\text{dM}(X)$ is well-defined.*

PROOF. Suppose that $\text{RM}(X) = \alpha$. For this proof say that a definable subset Y of X is α-irreducible if $\text{RM}(Y) = \alpha$ and there do not exist $Z_1, Z_2 \subseteq Y$ which are disjoint and have Morley rank α. It is easy to show that the fact that $\text{RM}(X) \not\geq \alpha$ implies both that there are α-irreducible subsets of X and that any maximal collection of such is finite. So suppose that $\{Y_1, \ldots, Y_m\}$ and $\{Z_1, \ldots, Z_n\}$ are two maximal collections of α-irreducible subsets of X. We may assume that $\bigcup_{i=1}^m Y_i = \bigcup_{j=1}^n Z_j = X$. So

$$Z_j = \left(\bigcup_{i=1}^m Y_i\right) \cap Z_j = \bigcup_{i=1}^m (Y_i \cap Z_j)$$

so there is at least one i so that $\mathrm{RM}(Y_i \cap Z_j) = \alpha$. Since Z_j is α-irreducible, there is at most one such i. So the map sending j to that i such that $\mathrm{RM}(Z_j \cap Y_i) = \alpha$ is well-defined and injective. Symmetry shows that it is a bijection and so $k = m$.

$\square$

Note that in the previous proof we don't show that the decomposition of X into Morley degree 1 pieces is unique but only that it is "α-unique": $\mathrm{RM}(Y_i \triangle Z_j) < \alpha$. It is also worth noting that a strongly minimal set has Morley rank and Morley degree 1.

PROPOSITION 1.20. *T is ω-stable if and only if $\mathrm{RM}(X) < \infty$ for every X.*

PROOF. Suppose there is a definable set X such that $\mathrm{RM}(X) = \infty$. By Fact 1.16 there is an α such that if $\mathrm{RM}(Y) \geq \alpha$ then $\mathrm{RM}(Y) = \infty$. So since $\mathrm{RM}(X) \geq \alpha + 1$, choose two disjoint subsets of X both of which have Morley rank greater than or equal to α. By the choice of α, both of these sets have Morley rank ∞. Repeating this argument one can build a binary tree of height ω of definable sets such that each branch is consistent and no two branches are mutually consistent. This contradicts ω-stability.

Now suppose that $\mathrm{RM}(X) < \infty$ for every X. Fix a countable model M of T. For every $p \in S(M)$, associate a formula $\varphi_p \in p$ so that φ_p has the least Morley rank of all formulas in p and among those has the least Morley degree. It is easy to see then that this formula uniquely determines p. But there are only countably many formulas over M so $S(M)$ is countable. $\square$

Now we will address how a structure is built or constructed over a strongly minimal set.

DEFINITION 1.21. 1. A type $p \in S_n(A)$ is said to be isolated if there is a formula $\varphi \in p$ such that if $\varphi \in q \in S_n(A)$ then $p = q$ (p is isolated in the Stone space topology on $S_n(A)$).
2. If $A \subseteq N \prec M$ then N is said to be a prime model over A if whenever $A \subseteq N' \prec M$ then there is an elementary map $f : N \to N'$ fixing A.
3. N is said to be constructible over A if $N = \{a_i : i < \alpha\}$ and, for every i,

$$\mathrm{tp}(a_i/A \cup \{a_j : j < i\})$$

is isolated.

FACT 1.22. 1. *If N is constructible over A then N is prime over A.*
2. *If T is ω-stable then there are constructible models over all sets.*
3. *If N_1 and N_2 are constructible over A then N_1 and N_2 are isomorphic over A.*

PROOF. The first is straightforward. For the second, the main point is that if you fix a set A and any consistent formula $\varphi(x)$ over A then the type of least Morley rank and degree containing $\varphi(x)$ is isolated. The third fact was proved by J. P. Ressayre. $\square$

The main structure theorem for uncountably categorical theories is:

THEOREM 1.23 [Baldwin and Lachlan 1971]. *If T is uncountably categorical then there is an isolated type $p(\bar{y})$ over the empty set and a formula $\varphi(x, \bar{y})$ such that whenever N is a model of T and $\bar{a} \in N$ realizes p then $\varphi(x, \bar{a})$ defines a strongly minimal set and N is constructible and minimal over $\varphi(N, \bar{a})$.*

That is, the isomorphism type of N is determined by the dimension of the set $\varphi(N, \bar{a})$.

EXAMPLE 1.24. An example of what the above theorem does not say is given by the theory of $(Z/4Z)^\omega$ as an abelian group. It would be nice if every model of an uncountably categorical theory was algebraic over a basis for a strongly minimal subset of the model (as in the case of algebraically closed fields) but $(Z/4Z)^\omega$ is a counterexample to this. For in this group, every element a is isolated over the strongly minimal set $2x = 0$ by the formula, $2x = 2a$. But this formula has infinitely many solutions.

Returning now to the question of what a strongly minimal set can be, we make a definition which makes sense for any pregeometry.

DEFINITION 1.25. A pregeometry (X, cl) is said to be modular if whenever $A, B \subseteq X$ are finite dimensional, closed subsets of X then

$$\dim(A) + \dim(B) = \dim(A \cap B) + \dim(A \cup B)$$

(X, cl) is said to be locally modular if the above equality holds whenever $A \cap B$ is nonempty.

A strongly minimal set is said to be modular or locally modular if the associated pregeometry is.

EXAMPLE 1.26. 1. An infinite set, an infinite vector space over a division ring or a projective space over a finite field are all examples of modular strongly minimal sets.

   If the pregeometry (X, cl) satisfies $\text{cl}(A) = \bigcup_{a \in A} \text{cl}(a)$ for all $A \subseteq X$ then the pregeometry is said to be trivial or degenerate. An infinite set is such a pregeometry.

2. An affine space over a finite field is locally modular but not modular; two distinct parallel lines are a counter-example to modularity.

3. An algebraically closed field is not locally modular (see Example 1.8 on page 77 of [Pillay 1996]).

One of the guiding principles of geometric model theory is the Zil'ber Principle (for a general discussion, see [Peterzil and Starchenko 1996]):

ZIL'BER PRINCIPLE. *Under suitable geometric or topological conditions, a non-locally modular strongly minimal set interprets an infinite field.*

A particular instance of this principle can be applied to simple, algebraic groups. Suppose that G is a simple, algebraic group over an algebraically closed field F.

Viewed as an abstract group G is not locally modular and in fact, G interprets a field F' which is isomorphic to F.

More information regarding the Zil'ber Principle appears in [Marker 2000]. In the remaining part of this section, I wish to point out a special case of the Zil'ber Principle in ω-categorical theories. The following theorem is due to Cherlin, Harrington and Lachlan [Cherlin et al. 1985] and Zil'ber [1984].

THEOREM 1.27. *An ω-categorical, strongly minimal set is locally modular.*

One can in fact say more but we need a definition first.

DEFINITION 1.28. A strongly minimal set X is said to be strictly minimal if $\mathrm{acl}_X(a) = \{a\}$ for all $a \in X$.

For instance, a projective or affine geometry over a finite field is strictly minimal but a vector space over a finite field is not. In general, to obtain a strictly minimal set from an ω-categorical strongly minimal set, one first removes the algebraic closure of the empty set (a finite set in the ω-categorical case) and then quotients by the definable equivalence relation of interalgebraicity. In this way, one can see that, except for finitely many points, an ω-categorical, strongly minimal set is a finite cover of a strictly minimal set.

The following theorem provides an enumeration of all ω-categorical, strictly minimal sets.

THEOREM 1.29. *An ω-categorical strictly minimal set is either a pure set, or a projective or affine geometry over a finite field.*

One can use this more precise information about ω-categorical, strictly minimal sets to understand the structure of the prime model discussed above in the case of a totally categorical theory. Fix a totally categorical theory T and any model, M, of T.

FACT 1.30. *For any finite, algebraically closed set $B \subseteq M$ and $a \notin B$, there is $c \in \mathrm{acl}(B \cup \{a\})$ such that $\mathrm{tp}(c/B)$ is strictly minimal.*

A more global version of this local fact is:

FACT 1.31. *For any $a \in M$, there is a number n and $a_0, \ldots, a_n \in \mathrm{acl}(a)$ such that $a = a_n$ and for every i, $\mathrm{tp}(a_i/a_0, \ldots, a_{i-1})$ is either algebraic or strictly minimal.*

EXAMPLE 1.32. To see how this works, let's return to the example of the abelian group, $(Z/4Z)^\omega$. Recall that it is not algebraic over the strongly minimal set $2x = 0$. However, if we fix any element a and let $2a = b$ then we see that the number n in the previous fact is 2: b is either 0 (and hence algebraic over the empty set) or $\mathrm{tp}(b/\varnothing)$ is strictly minimal. In either case, $\mathrm{tp}(a/b)$ is strictly minimal; in the latter case, the strictly minimal set is an affine geometry.

The picture then to have of this example is a base set represented by the strictly minimal set $2x = 0$ and then above each element of this set, a "fibre" which is itself a strictly minimal set. To completely understand the structure then one must know what the strictly minimal sets are (and this is given by the Theorem above) and how the fibres interact (often the hardest part).

Hrushovski used this type of analysis to prove the following:

THEOREM 1.33 [Hrushovski 1989]. *Any totally categorical theory is finitely ax-iomatized modulo "axioms of infinity" (which express the fact that the strongly minimal set is infinite).*

2. Dividing

Before we can introduce the general notion of dimension, we must introduce a basic model theoretic definition.

DEFINITION 2.1. Suppose that $(I, <)$ is an infinite linear order and A is a subset of a model M. A sequence $\{a_i : i \in I\}$ of tuples from M is said to be *indiscernible over A* if, whenever $i_1 < \cdots < i_n$ and $j_1 < \cdots < j_n$ are sequences from I, $\text{tp}(a_{i_1} \ldots a_{i_n}/A) = \text{tp}(a_{j_1} \ldots a_{j_n}/A)$.

EXAMPLE 2.2. 1. A transcendence basis in an algebraically closed field, ordered in any way, is an example of an indiscernible sequence over the empty set.
2. $(Q, <)$ with the usual order is also an example of an indiscernible sequence over the empty set.

We now introduce the most important definition on our way towards a general dimension theory. This definition is due to Shelah [1978].

DEFINITION 2.3. 1. A formula $\varphi(x, a)$ (and the set it defines) is said to *divide* over A if there is an sequence I, indiscernible over A, with $a \in I$ such that $\{\varphi(x, b) : b \in I\}$ is inconsistent.
2. A definable set X (and the formula which defines it) is said to *fork* over A if X is contained in the union of finitely many definable sets, each of which divides over A.
3. A type p divides (forks) over A if there is $\varphi \in p$ which divides (forks) over A.

We introduce a ternary relation symbol $\perp$ between subsets of M as follows:

$$A \perp_C B \text{ if and only if } \text{tp}(A/B \cup C) \text{ does not divide over } C.$$

We will say that A and B are independent over C if $A \perp_C B$ although the justification for this terminology will only come later.

EXAMPLE 2.4. 1. Suppose that X is a strongly minimal set defined over A. Then any infinite, definable subset of X does not divide over A.

To see this, by compactness, it is enough to see that any finitely many A-conjugates of X have a nonempty intersection. But X, and each of its conjugates, is cofinite and so the intersection is nonempty.

2. In an ω-stable theory, a type $p \in S(B)$ does not divide over $A \subseteq B$ if $\mathrm{RM}(p) = \mathrm{RM}(p|A)$.

To see this, suppose that p divides over A. Then there is a sequence $\{p_i : i \in \omega\}$ of A-conjugates of p so that for some n, any n distinct p_i's are mutually inconsistent. We can assume that $p_i \in S(B_i)$ and that the sequence $\{B_i : i \in \omega\}$ is indiscernible over A. Choose $\varphi \in p|A$ so that $\mathrm{RM}(\varphi) = \mathrm{RM}(p|A)$. Now since the p_i's are n-inconsistent, we can find $\psi_i(x, b_i) \in p_i$ and $m \leq n$ with the following properties:

1. $\psi(x, b_i)$ strengthens φ for every i;
2. if we let $\theta_k = \bigwedge_{mk \leq i < m(k+1)} \psi_i(x, b_i)$ then $\mathrm{RM}(\theta_k) = \mathrm{RM}(\varphi)$ for all k;
3. $\mathrm{RM}(\theta_k \wedge \theta_l) < \mathrm{RM}(\varphi)$ for all $k \neq l$

From the properties of Morley rank, we see that the formulas

$$\{\theta_k \wedge \bigwedge_{l<k} \neg\theta_l : k \in \omega\}$$

are pairwise inconsistent, strengthen φ and have the same Morley rank as φ which is a contradiction.

3. Suppose our model is an infinite vector space over a finite field with a non-degenerate symplectic bilinear form. Then any nonalgebraic one-type does not divide over the empty set. For suppose that p is a nonalgebraic 1-type over a subspace A (in general, it is enough to check that all 1-types do not divide over small sets). Let $\{A_i : i \in \omega\}$ be any indiscernible sequence with $A_0 = A$ and let B be their common intersection. Then it follows that $\{A_i : i \in \omega\}$ is linearly disjoint over B. Now p is determined by the linear map f it defines on A (for all $a \in A$, $f(a) = \alpha$ if and only if $\langle x, a \rangle = \alpha \in p$). If we let p_i be the conjugate of p over A_i and f_i be the corresponding conjugate of f then the consistency of $\bigcup_{i \in \omega} p_i$ is equivalent to the ability to extend all the maps f_i to a linear map on the subspace generated by A_i's. The latter is clear by the linear disjointness of the A_i's over B.

Dividing and forking satisfy many properties in all theories.

1. Both dividing and forking are invariant under automorphisms of the large, saturated model that we are working in.
2. If $X \subseteq Y$ are definable sets and Y divides (forks) over A then so does X.
3. (Extension) If $A \subseteq B \subseteq C$ and $p \in S(B)$ does not fork over A then p has an extension in $S(C)$ which does not fork over A.
4. (Finite Character) $A \downarrow_C B$ if and only if $a \downarrow_c b$ for every finite $a \in A$, $b \in B$ and $c \in C$.
5. If $A \subseteq B$, $a \in \mathrm{acl}(B)$ and $a \downarrow_A B$ then $a \in \mathrm{acl}(A)$.
6. (Weak transitivity) If $B \subseteq C \subseteq D$ and $A \downarrow_B D$ then $A \downarrow_B C$ and $A \downarrow_C D$.
7. (Left transitivity) If $C \subseteq B \subseteq A$, $A \downarrow_B D$ and $B \downarrow_C D$ then $A \downarrow_C D$.

The first two properties listed above follow immediately from the definitions. The extension property is the entire reason for defining forking and is easily

verified by compactness. Both finite character and weak transitivity are easily
verified. Left transitivity is extremely useful (see [Hart et al. 1998], section 4); a
proof can be found in [Shelah 1980] and [Kim 1998].

DEFINITION 2.5. 1. A sequence $\{a_i : i < \alpha\}$ is *independent over* A if, for every
j, $a_j \downarrow_A \{a_i : i < j\}$.
2. A *Morley sequence* for a type $p \in S(A)$ is a sequence of realizations of p which
is both independent and indiscernible over A.

Key general assumptions

1. (Forking equals dividing) Whenever a definable set forks over a set A, it
divides over A.
2. (The Kim property, KP) A formula $\varphi(x, a)$ does not divide over A if and
only if there is a Morley sequence I in $\mathrm{tp}(a/A)$ such that $\{\varphi(x, b) : b \in I\}$ is
consistent.

In a theory which satisfies the key general assumptions the following properties
hold, trivially, dividing satisfies the extension property. Far less trivially, we
have

THEOREM 2.6. *In a theory which satisfies the key general assumptions, dividing
is symmetric, that is,* $A \downarrow_C B$ *if and only if* $B \downarrow_C A$.

As a corollary of the left transitivity property,

COROLLARY 2.7. *In a theory which satisfies the key general assumptions, dividing is transitive; that is, if* $C \subseteq B \subseteq A$ *and* $D \downarrow_B A$ *and* $D \downarrow_C B$ *then*
$D \downarrow_C A$.

In fact recently in [Kim 1999], Kim has shown the the key general assumptions
are equivalent to dividing being symmetric.

The most important of the properties that dividing satisfies in certain theories
is the following (proved in [Kim and Pillay 1997]):

THEOREM 2.8. *(Type amalgamation over a model; also known as the Independence Theorem) Fix a theory which satisfies the key general assumptions.
Suppose that* A *and* B *both contain a model* M *and are independent over* M. *If
p and q are types over A and B respectively and both are nonforking extensions
of a common type over* M *then p and q have a common nonforking extension.*

Simple theories

DEFINITION 2.9. A theory T is simple if every type does not divide over a set
of size at most $|T|$. We say that in such a theory, dividing (or forking) satisfies
local character.

THEOREM 2.10. *Simple theories satisfy the key general assumptions; that is,
forking equals dividing and the Kim property holds.*

EXAMPLE 2.11. 1. Any ω-stable theory is simple. This follows from Example 2.4.2.

2. The theory of an infinite vector space over a finite field with a nondegenerate symplectic bilinear form is simple. This follows from Example 2.4.3.

3. The theory of the random graph is simple. One can show this in a manner similar to (but easier than) Example 2.4.3; see the remarks after the next theorem for an alternative approach.

4. The generic triangle-free graph is **not** an example of a simple theory. There are several ways to see this; the one I present will be an application of type amalgamation.

The first observation is that the only indiscernible sequence of singletons in any model of this theory has no edges between the points. Otherwise, any three points will form a triangle.

So suppose that the theory of the generic triangle-free graph is simple. Fix a model M (we do this only to match the form of the type amalgamation theorem). I claim that any two points a and b, not in M, with an edge between them are independent over M. This follows from the first observation since if we consider an M-indiscernible sequence $\{b_i : i \in \omega\}$ starting with b then there are no edges between the b_i's so there is nothing inconsistent about the type which contains the type of a over M and the statement that x is connected to each of the b_i's. But then if we let p_a be the type of an element connected to a but not to any element of M and p_b be the similar type, connected to b but nothing in M, then by what we just said, p_a and p_b are nondividing extensions of their common restriction to M. However, any point which would realize p_a and p_b would form a triangle. Since type amalgamation fails, this theory cannot be simple.

5. The theory of the real field is **not** simple; the theory of $(Q, <)$ is **not** simple.

Let's show that $(Q, <)$ is not simple (real closed fields can be done in a similar manner). We will show that dividing is not symmetric. Consider the type $p(x; y, z)$ determined by "$y < x < z$". For any a it is fairly clear that $p(a; y, z)$ does not divide over the empty set (for any sequence of singletons in any model of this theory it is consistent that there is something bigger than and something smaller than them all). However, if $a < b$ then $p(x; a, b)$ does divide over the empty set. For we can choose an indiscernible sequence $a = a_0 < b = b_0 < a_1 < b_1 < a_2 < b_2 < \cdots$ for which even $p(x; a, b) \cup p(x, a_1, b_1)$ is not consistent.

The following characterization theorem shows the connection between the existence of an independence relation satisfying many of the properties we have mentioned up until now and simplicity. Its proof is due to Kim and Pillay and appears in [Kim and Pillay 1997].

THEOREM 2.12. *A theory is simple if and only if there is an invariant ternary relation on sets which has finite and local character, is symmetric and transitive and satisfies extension and type amalgamation over models. If such a ternary relation exists then it must be nondividing.*

The usefulness of this theorem cannot be overstated. In practice, when one encounters a theory "in nature", it often comes with a suggestion for an independence relation. A case in point is the theory of algebraically closed fields with a generic automorphism (ACFA, see [Chatzidakis 2000]). It is frequently easier to check that the example satisfies the properties listed above then it is to work through the definition of dividing. For instance, for the random graph, suppose A is a subgraph of B and C, and say that B and C are independent over A if and only if $B \cap C = A$. Then it is straightforward to check that all the properties listed in the characterization theorem are satisfied and so the random graph is simple (and this independence relation is the dividing relation).

Orthogonality, supersimplicity and regular types. Any sufficiently general notion of independence leads to a derived notion of orthogonality; see the almost axiomatic treatments in [Makkai 1984] and [Shelah 1978]. In the case of simple theories, the details are worked out in [Hart et al. 1998] where the following definitions appear.

DEFINITION 2.13.   1. If $p, q \in S(A)$ then p and q are *almost orthogonal* if whenever a and b realize p and q respectively then a and b are independent over A.

2. p and q are said to be *orthogonal* if all their nonforking extensions to common domains are almost orthogonal.

3. p is regular if it is orthogonal to all its forking extensions.

It is shown in [Hart et al. 1998] that regular types are the dimension carrying objects in simple theories.

PROPOSITION 2.14.   *If $p \in S(A)$ is regular then independence over A is a pregeometry on the realizations of p.*

EXAMPLE 2.15.   1. The nonalgebraic type over the empty set in a strongly minimal set is a regular type.

2. The unique rank ω 1-type over the empty set in the theory of differentially closed fields of characteristic zero is a regular type. (See [Marker 2000] in this volume.)

There is a rank which is more general than Morley rank called SU-rank; it is also defined inductively:

DEFINITION 2.16.   For a type p,

1. $\mathrm{SU}(p) \geq 0$.

2. $\mathrm{SU}(p) \geq \alpha + 1$ if and only if there is q, a forking extension of p, such that $\mathrm{SU}(q) \geq \alpha$.

3. For a limit ordinal δ, $\mathrm{SU}(p) \geq \delta$ if and only if $\mathrm{SU}(p) \geq \alpha$ for all $\alpha < \delta$.

4. $\mathrm{SU}(p) = \alpha$ if $\mathrm{SU}(p) \geq \alpha$ and $\mathrm{SU}(p) \not\geq \alpha + 1$; $\mathrm{SU}(p) = \infty$ if $\mathrm{SU}(p) \geq \alpha$ for all α.

A theory for which $\mathrm{SU}(p) < \infty$ for all p is called *supersimple*. The following appears in [Hart et al. 1998] and shows that it is supersimple theories that have "enough" regular types.

PROPOSITION 2.17. *If T is simple and $\mathrm{SU}(p) < \infty$ then p is nonorthogonal to a regular type.*

3. Stability

The rationale for this section stems from the need for many of concepts in [Chatzidakis 2000]. On the other hand, many discussions of dimension theory in a model theoretic context will revolve around the notions here; we start with one which predates simplicity.

DEFINITION 3.1. A partial type p is stable if there is a λ such that, for any set A of size λ, the number of extensions of p over A is of size at most λ.

EXAMPLE 3.2. 1. The partial type $x = x$ is stable in all of the following theories: algebraically closed fields of any fixed characteristic, differentially closed fields of characteristic zero, any strongly minimal set. In such a case, one says that the theory is stable.
2. Consider a theory whose universe is partitioned by two unary predicates U and V; U contains a an algebraically closed field and V contains a copy of a real closed field. It is easy to see that the partial type $U(x)$ is stable.
3. Consider a theory again with two unary predicates which partition the universe, call them V and V^*. The model (V, V^*) will be an infinite vector space over a finite field, V and its dual V^*. Neither of these predicates is stable even though the theory of a vector space over any field is stable in its own right.

Definability of types. Independence on realizations of stable types has a more intrinsic definition than the one found in the last section which depends on the following definition.

DEFINITION 3.3. A type $p \in S(A)$ is definable if, for every formula $\varphi(x, y)$, there is a formula $d_\varphi(y, a)$ with $a \in A$ such that, for every $b \in A$, $\varphi(x, b) \in p$ if and only if $d_\varphi(b, a)$ holds.

EXAMPLE 3.4. Suppose that M is a left R-module for some ring R. M is considered a structure in the language with $0, +, -$ and a unary function symbol for every $r \in R$. The following is a theorem of Baur ([Baur 1976]).

THEOREM 3.5. *If $T = \mathrm{Th}(M)$ then any definable set is a boolean combination of cosets of definable subgroups. In fact, each coset is an instance of a formula*

$\varphi(x, y)$ *of the form* $\exists z(Axz = y)$ *where* A *is an* R-*matrix and the subgroup is defined by* $\varphi(x, 0)$.

Now if p is a type over M and $\varphi(x, y) = \exists z(Axz = y)$ then the φ-definition d_φ is

1. false, if p does not contain an instance of φ and
2. $\varphi(b, y)$, if $\varphi(x, c) \in p$ for some $c \in M$ and $\varphi(b, c)$ holds for some $b \in M$.

Here are two key equivalences to the notion of a stable type.

THEOREM 3.6. *The following are equivalent:*

1. p *is a stable type.*
2. *There is no formula* $\varphi(x, y)$ *and realizations* a_i *of* p *and elements* b_i *for* $i \in \omega$ *such that* $\varphi(a_i, b_j)$ *holds if and only if* $i < j$.
3. *Every extension of* p *is definable.*

In a stable theory or for a stable type, nonforking and the definability of types are very closely related.

FACT 3.7. *If* $q \in S(B)$ *is a stable type and* $A \subseteq B$ *then* q *does not divide over* A *if and only if* q *is defined almost over* A.

COROLLARY 3.8. *If* q *is a stable type then* q *does not divide over a set of size at most* $|T|$.

COROLLARY 3.9. *Stable theories are simple.*

Stability has added advantages over simplicity and this is no more evident than in the notion of multipicity.

DEFINITION 3.10. *If* $p \in S(A)$ *then the multiplicity of* p *is the supremum, if it exists, over all* B, $A \subseteq B$ *of* $|\{q \in S(B) : q$ *is a nonforking extension of* $p\}|$.

FACT 3.11. *A type* p *is stable if and only if every extension of* p *has bounded multiplicity and does not divide over a set of size* $|T|$.

A stable type with multiplicity one is called *stationary*. A stable type over a model is stationary.

　　For the rest of this paper, I will assume that we are working inside a stable type p. All of the forking technology discussed in the previous section goes through in this context. If the reader likes, there is no real loss in assuming that the ambient theory is simple.

The canonical base

DEFINITION 3.12. 1. If $\varphi(x, a)$ defines X then the canonical parameter of X, written $\lceil \varphi(x, a) \rceil$ or just $\lceil \varphi \rceil$, is the element of M^{eq}, a/E_φ, where $E_\varphi(x, y) := \forall z(\varphi(z, x) \leftrightarrow \varphi(z, y))$.
2. For a stationary type p, the canonical base of p, $Cb(p) = \text{dcl}\{\lceil d_\varphi \rceil : \varphi\}$.

The canonical base of a stationary type p has many properties.

FACT 3.13. *Suppose that p is a stationary type and $\mathbf{p}$ is the nonforking extension of p to a large saturated model M.*

1. *p does not fork over $Cb(p)$.*
2. *$Cb(p)$ is the unique subset C of M^{eq} such that for all automorphisms σ of M, σ fixes C if and only if σ fixes $\mathbf{p}$.*
3. *$Cb(p)$ is contained in the definable closure of any Morley sequence in p.*

It is immediate that if T has elimination of imaginaries and is stable then the canonical base of any stationary type lies in M not M^{eq}.

1-based theories and types. In this last subsection, we make an attempt to tie together many of the concepts from this paper. Additionally, the Theorem below is used critically in [Chatzidakis 2000] where the concepts are discussed in more detail.

DEFINITION 3.14. 1. If $p \in S(C)$ and A is a set of realizations of p we will write $\bar{A}$ for $\mathrm{acl}^{\mathrm{eq}}(A \cup C)$.
2. $p \in S(C)$ is *1-based* if for every set A of realizations of p and every model M containing A, $Cb(A/M) \subseteq \bar{A}$.

FACT 3.15. *$p \in S(C)$ is 1-based if and only if for every pair A and B of sets of realizations of p, $\bar{A} \underset{\bar{A} \cap \bar{B}}{\downarrow} \bar{B}$.*

REMARKS 3.16. As the previous fact points out, 1-based types have an independence relation which is as simple as possible. The terminology "1-based" stems from the fact that in general, for a stationary type, the canonical base lies in the closure of any Morley sequence in the type; for 1-based types, the canonical base of any extension lies in the closure of a single realization of the type.

EXAMPLE 3.17. 1. A modular, strongly minimal set is 1-based.
2. Any complete theory of modules is 1-based.

DEFINITION 3.18. If G is the set of realizations of a partial type p then we say G is a *stable group* if there is a definable, binary function $*$ so that $(G, *)$ is a group.

REMARKS 3.19. A stable group may have more structure than just a group structure; for example, by the previous definition, a field is a stable group.

The following Theorem, found in [Hrushovski and Pillay 1987], shows how the strong model theoretic assumptions we have mentioned in this section impact on the definable sets in a stable group.

THEOREM 3.20. *A type-definable group G is stable and 1-based if and only if every definable subset is equivalent to a boolean combination of cosets of definable subgroups of G^n for some n.*

References

[Baldwin and Lachlan 1971] J. T. Baldwin and A. H. Lachlan, "On strongly minimal sets", *J. Symbolic Logic* **36** (1971), 79–96.

[Baur 1976] W. Baur, "Elimination of quantifiers for modules", *Israel J. Math.* **25**:1-2 (1976), 64–70.

[Chatzidakis 2000] Z. Chatzidakis, "A survey on the model theory of difference fields", pp. 65–96 in *Model theory, algebra and geometry*, edited by D. Haskell et al., Math. Sci. Res. Inst. Publ. **39**, Cambridge Univ. Press, New York, 2000.

[Cherlin et al. 1985] G. Cherlin, L. Harrington, and A. H. Lachlan, "$\aleph_0$-categorical, $\aleph_0$-stable structures", *Ann. Pure Appl. Logic* **28**:2 (1985), 103–135.

[Hart et al. 1998] B. Hart, B. Kim, and A. Pillay, "Coordinatisation and canonical bases in simple theories", preprint, 1998. To appear in *J. Symb. Logic*.

[Hrushovski 1989] E. Hrushovski, "Totally categorical structures", *Trans. Amer. Math. Soc.* **313**:1 (1989), 131–159.

[Hrushovski and Pillay 1987] U. Hrushovski and A. Pillay, "Weakly normal groups", pp. 233–244 in *Logic colloquium '85* (Orsay, 1985), edited by the Paris Logic Group, Stud. Logic Found. Math. **122**, North-Holland, Amsterdam, 1987.

[Kim 1998] B. Kim, "Forking in simple unstable theories", *J. London Math. Soc.* (2) **57**:2 (1998), 257–267.

[Kim 1999] B. Kim, "Simplicity and stability in there", preprint, 1999.

[Kim and Pillay 1997] B. Kim and A. Pillay, "Simple theories", *Ann. Pure Appl. Logic* **88**:2-3 (1997), 149–164.

[Makkai 1984] M. Makkai, "A survey of basic stability theory, with particular emphasis on orthogonality and regular types", *Israel J. Math.* **49**:1-3 (1984), 181–238.

[Marker 2000] D. Marker, "Introduction to model theory", pp. 15–35 in *Model theory, algebra and geometry*, edited by D. Haskell et al., Math. Sci. Res. Inst. Publ. **39**, Cambridge Univ. Press, New York, 2000.

[Peterzil and Starchenko 1996] Y. Peterzil and S. Starchenko, "Geometry, calculus and Zil'ber's conjecture", *Bull. Symbolic Logic* **2**:1 (1996), 72–83.

[Pillay 1996] A. Pillay, *Geometric stability theory*, Oxford Science Publications, Oxford Univ. Press, New York, 1996.

[Shelah 1978] S. Shelah, *Classification theory and the number of non-isomorphic models*, Stud. Logic Found. Math. **92**, North-Holland, Amsterdam, 1978.

[Shelah 1980] S. Shelah, "Simple unstable theories", *Ann. Math. Logic* **19**:3 (1980), 177–203.

[Zil'ber 1984] B. I. Zil'ber, "Strongly minimal countably categorical theories, III", *Sibirsk. Mat. Zh.* **25**:4 (1984), 63–77.

BRADD HART
DEPARTMENT OF MATHEMATICS AND STATISTICS
MCMASTER UNIVERSITY
1280 MAIN ST.
HAMILTON, ONTARIO L8S 4K1
CANADA
  hartb@mcmaster.ca

Model Theory, Algebra, and Geometry
MSRI Publications
Volume **39**, 2000

Subanalytic Geometry

EDWARD BIERSTONE AND PIERRE D. MILMAN

ABSTRACT. Lou van den Dries has suggested that the o-minimal structure
of the classes of semialgebraic or subanalytic sets makes precise Grothendieck's idea of a "tame topology" based on stratification. These notes
present another viewpoint (with intriguing possible relationships): we describe a range of classes of spaces between semialgebraic and subanalytic,
that do not necessarily fit into the o-minimal framework, but that are
"tame" from algebraic or analytic perspectives.

1. Introduction

Semialgebraic and subanalytic sets capture ideas in several areas: In model
theory, they express properties of quantifier elimination. In geometry and analysis, they provide a language for questions about the local behaviour of algebraic and analytic mappings. Lou van den Dries has suggested that the o-
minimal structure of the classes of semialgebraic or subanalytic sets makes precise Grothendieck's vision of a "tame topology". (In his provocative *Esquisse
d'un programme*, Grothendieck [1984] proposes an axiomatic development of a
topology based on ideas of stratification in order to study, for example, singularities that arise in compactifications of moduli spaces.) These notes present
another point of view (with intriguing possible relationships): we will describe a
range of geometric classes of spaces between semialgebraic and subanalytic, that
do not necessarily fit into the o-minimal framework, but that are "tame" from
algebraic or analytic perspectives. The questions we discuss are in directions
pioneered by Whitney, Thom, Łojasiewicz, Gabrielov and Hironaka. (We will

1991 *Mathematics Subject Classification*. Primary 14P10, 32B20; Secondary 03C10, 32S10,
32S60.

Lectures of E. Bierstone in the Introductory Workshop on Model Theory of Fields, MSRI,
January 1998.

Research partially supported by NSERC grants OGP0009070 and OGP0008949, and by the
Connaught Fund.

not try to give a general survey of recent results in the area of semialgebraic and subanalytic sets.)

Semialgebraic and semianalytic sets. A *semialgebraic subset of* $\mathbb{R}^n$ is a subset of the form

$$X = \bigcup_{i=1}^{p} \bigcap_{j=1}^{q} X_{ij}, \tag{1.1}$$

where each X_{ij} is of the form $\{f_{ij}(x) = 0\}$ or $\{f_{ij}(x) > 0\}$, with $f_{ij}(x) = f_{ij}(x_1, \ldots, x_n)$ a polynomial. For example, Figure 1 shows an *algebraic* subset X of $\mathbb{R}^3$ defined by the equation $z^2 - xy^2 = 0$ ("Whitney's umbrella"). Figure 1 illustrates a stratification of X. A *stratification* means a finite (or locally finite) partition into connected smooth manifolds (*strata*) within the class (here, semialgebraic), such that the frontier of each stratum is a union of strata of lower dimension.

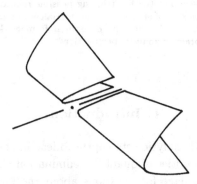

Figure 1. Stratification of Whitney's umbrella $z^2 - xy^2 = 0$.

According to the *Tarski–Seidenberg theorem*, the image of a semialgebraic subset X of $\mathbb{R}^{m+n}$ by a projection $\mathbb{R}^{m+n} \to \mathbb{R}^n$ is semialgebraic. (For the basic properties of semialgebraic or subanalytic sets, see [Bierstone and Milman 1988], for example.) The Tarski–Seidenberg theorem is an assertion about *elimination of quantifiers*; it says that any formula obtained using a finite number of "and", "or", negations, existential and universal quantifiers, from formulas of the form $f(x) = 0$ or $f(x) > 0$, where $f(x) = f(x_1, \ldots, x_n)$ is a polynomial, describes the same set as a formula without quantifiers.

A *semianalytic* subset of $\mathbb{R}^n$ is a subset that is defined *locally* (i.e., in some neighbourhood of any point of $\mathbb{R}^n$) by an expression of the form (1.1), but where the functions f_{ij} are real-analytic. A projection of even a compact semianalytic set need not be semianalytic. (See Example 2.3 below.)

Subanalytic sets. A subset X of $\mathbb{R}^n$ is *subanalytic* if, locally, X is a projection of a relatively compact semianalytic set. The class of semianalytic sets was enlarged to include projections in this way by Łojasiewicz [1964], although the term "subanalytic" is due to Hironaka [1973]. Gabrielov's *theorem of the complement* asserts that the complement of any subanalytic set is subanalytic [Gabrielov 1968]. This is also an assertion about "quantifier simplification": The complement of a subanalytic set is defined locally by a formula involving real-analytic equations and inequalities, with existential and universal quantifiers; Gabrielov's theorem says that the complement can be defined locally by an existential formula.

The uniformization theorem. A closed subanalytic subset X of $\mathbb{R}^n$ is locally a projection of a compact semianalytic set. In fact, closed subanalytic sets are precisely the images of proper real-analytic mappings (from manifolds), according to the following *uniformization theorem*:

THEOREM 1.2. *Let X be a closed subanalytic subset of $\mathbb{R}^n$. Then X is the image of a proper real-analytic mapping $\varphi\colon M \to \mathbb{R}^n$, where M is a real-analytic manifold of the same dimension as X.*

The uniformization theorem is a consequence of resolution of singularities [Hironaka 1964; 1974; Bierstone and Milman 1997]; but see [Bierstone and Milman 1988, Sections 4 and 5] for a short elementary proof. From the point of view of the uniformization theorem, subanalytic sets can be viewed as real analogues of complex-analytic sets, or analytic analogues of semialgebraic sets. (See the table *Images of proper mappings* on the next page.) These classes share many properties of a "tame topology". For example, any set in the class (locally) has finitely many connected components, each in the class; the components, boundary and interior of a set in the class are in the class; a set in the class can be stratified or even triangulated by subsets in the class.

But there are crucial distinctions between the behaviour of semialgebraic and general subanalytic sets. An important example that we will not deal with explicitly is their behaviour at infinity: A semialgebraic subset of $\mathbb{R}^n$ remains semialgebraic at infinity (i.e., when $\mathbb{R}^n$ is compactified to real projective space $\mathbb{P}^n(\mathbb{R})$). This is false for subanalytic sets, in general. An understanding of the behaviour at infinity of certain important classes of subanalytic sets (as in [Wilkie 1996]) represents the most striking success of the model-theoretic point of view in subanalytic geometry.

Grothendieck, in *Esquisse d'un programme*, suggests that "tame" should reflect not only conditions on strata, but also the way that the strata fit together. (For example, semialgebraic and subanalytic sets admit stratifications that are *Lipschitz locally trivial* [Mostowski 1985; Parusiński 1994]). The way that strata are attached to each other is closely related to the way that the local behaviour of X varies along a given stratum S, or as we approach $\bar{S} \setminus S$. Grothendieck envisaged a hierarchy of tame geometric categories from semialgebraic to subanalytic.

Section 2 below includes a sequence of examples illustrating differences in the local behaviour of semialgebraic and general subanalytic sets. The results described in the following sections are directed towards understanding these phenomena; Theorems 3.1 and 4.4 characterize certain subclasses of subanalytic sets that are tame from algebraic or or analytic perspectives, although they do not necessarily fit into an o-minimal framework. The uniformization theorem provides the point of view toward subanalytic geometry that is taken here: On the one hand, subanalytic sets provide a natural language for questions about the local behaviour of analytic mappings, and, on the other, local invariants of analytic mappings can be used to characterize a hierarchy of "tame" classes of sets (*Nash-subanalytic, semicoherent, ...*) intermediate between semialgebraic and subanalytic.

The phenomena studied in these notes concern not peculiarities of the reals, but rather the local behaviour of analytic mappings whether real or complex. Although it is true, for example, that the image of an arbitrary complex-analytic mapping is a closed analytic set (by the theorem of Remmert [1957]), a complex-analytic set X can be realized, more precisely, as the image of a proper complex-analytic mapping φ that is *relatively algebraic* over any sufficiently small open subset V of the target; this means there is a closed embedding $\iota : \varphi^{-1}(V) \to V \times \mathbb{P}^k(\mathbb{C})$ commuting with the projections to V, whose image is defined by homogeneous polynomial equations (in terms of the homogeneous coordinates of $\mathbb{P}^k(\mathbb{C})$) with coefficients analytic functions on V (by resolution of singularities [Hironaka 1964; 1974; Bierstone and Milman 1997]). The image of a proper real-analytic mapping satisfying the analogous condition is semianalytic, by Łojasiewicz's generalization of the Tarski–Seidenberg theorem [Łojasiewicz 1964; Bierstone and Milman 1988, Theorem 2.2]. Subanalytic sets, on the other hand, provide a natural setting for questions about the local behaviour of analytic mappings in general.

Images of proper mappings.

	Algebraic	Relatively algebraic	Analytic
$\mathbb{C}$	closed algebraic sets	closed analytic sets $\rightarrow$	
$\mathbb{R}$	closed semialgebraic sets	closed semianalytic sets	closed subanalytic sets

(In the real case, "proper" imposes no restriction on local behaviour.)

Uniformization and rectilinearization. The uniformization theorem above is closely related to the following *rectilinearization theorem* for subanalytic functions. Let N denote a real-analytic manifold; e.g., $\mathbb{R}^n$. (A function $f\colon X \to \mathbb{R}$, where $X \subset N$, is called *subanalytic* if the graph of f is subanalytic as a subset of $N \times \mathbb{R}$.)

THEOREM 1.3 [Bierstone and Milman 1988, § 5]. *Let $f\colon N \to \mathbb{R}$ be a continuous subanalytic function. Then there is a proper analytic surjection $\varphi\colon M \to N$, where $\dim M = \dim N$, such that $f \circ \varphi$ is analytic and locally has only normal crossings.*

The latter condition means that each point of M admits a neighbourhood with a coordinate system $x = (x_1, \ldots, x_n)$ in which $f(\varphi(x)) = x_1^{\alpha_1} \cdots x_n^{\alpha_n} u(x)$ and $u(x)$ does not vanish.

It may be interesting to ask whether the uniformization and rectilinearization properties of semialgebraic or subanalytic sets have reasonable analogues for a given o-minimal structure (or *geometric category* in the sense of [van den Dries and Miller 1996]). This is true, for example, for *restricted subpfaffian sets* (projections of relatively compact semianalytic sets that are defined using Pfaffian functions in the sense of [Khovanskiĭ 1991]). The point is that [Bierstone and Milman 1988, § 4], on "transforming an analytic function to normal crossings by blowings-up", preserves subalgebras of analytic functions that are closed under composition by polynomial mappings, differentiation and division (when the quotient is analytic).

2. Examples

In this section, we describe a range of phenomena that distinguish between the behaviour or algebraic and general analytic mappings. Given a subset X of $\mathbb{R}^n$ (or $\mathbb{C}^n$), let $\mathcal{A}_a(X)$ denote the ideal of germs of analytic functions at a that vanish on X.

Coherence. Every complex-analytic set is *coherent* (according to the theory of Oka and Cartan). This means that, if X is a closed complex-analytic subset of $\mathbb{C}^n$ and $a \in X$, then $\mathcal{A}_a(X)$ generates $\mathcal{A}_b(X)$, for all $b \in X$ in some neighbourhood of a. (See, for example, [Łojasiewicz 1991, § VI.1].)

Real-algebraic sets already need not be coherent [Whitney 1965]; for example, Whitney's umbrella (Figure 1) is not coherent at the origin. Here are two more examples:

EXAMPLE 2.1 [Hironaka 1974]. Let X be the closed algebraic subset of $\mathbb{R}^3$ defined by $z^3 - x^2 y^3 = 0$ (Figure 2).

In this example, $\mathcal{A}_0(X) = (z^3 - x^2 y^3)$, the ideal of germs of real-analytic functions at 0 generated by $z^3 - x^2 y^3$, but, at a nonzero point b of the x-axis,

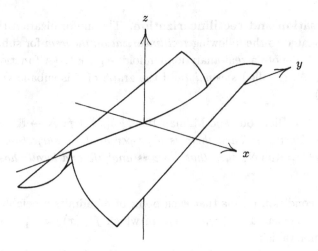

Figure 2. $z^3 - x^2 y^3 = 0$.

$z^3 - x^2 y^3$ factors non-trivially,

$$z^3 - x^2 y^3 = (z - x^{2/3}y)(z^2 + x^{2/3}yz + x^{4/3}y^2),$$

and $\mathcal{A}_b(X) = (z - x^{2/3}y)$. The analytic function $t = z - x^{2/3}y$ defined for $x \neq 0$, is called a *Nash function*: t satisfies a nontrivial polynomial equation $P(x, y, z, t) = 0$. (We can take $P = (t - z)^3 + x^2 y^3$.)

EXAMPLE 2.2 [Bierstone and Milman 1988]. Let X be the real-algebraic subset $z^3 - x^2 yz - x^4 = 0$ of $\mathbb{R}^3$ (Figure 3).

The *singularities* of X form a half-line $\{x = z = 0, \ y \geq 0\}$. In particular, $z^3 - x^2 yz - x^4$ does not generate $\mathcal{A}_b(X)$, $b \in \{x = z = 0, \ y < 0\}$, so that X is not coherent. In fact, over $\{(x, y) : \ y < 0\}$, $z^3 - x^2 yz - x^4 = 0$ can be solved uniquely as $z = g(x, y)$, where g is analytic. This can be seen by transforming the given equation by the quadratic mapping

$$\sigma : \ x = u, \ y = v, \ z = uw.$$

(σ is the *blowing-up* of $\mathbb{R}^3$ with *centre* $\{x = z = 0\}$ (restricted to a local coordinate chart)). Then $\sigma^{-1}(X)$ is given by $u^3(w^3 - vw - u) = 0$; i.e., $\sigma^{-1}(X) = E' \cup X'$, where E' is the coordinate plane $\{u = 0\}$ and X' is the smooth hypersurface $\{u = w^3 - vw\}$, which is transverse to E' when $v < 0$.

Local dimensions of a subanalytic set. At each point of a subanalytic set, we can consider its local topological dimension, as well as the dimensions of its local analytic and formal closures (in a sense we will make precise below). These three local dimensions coincide for analytic or semianalytic sets, but they may all be distinct for subanalytic sets in general [Gabrielov 1971]. Gabrielov's construction (Example 2.5) is based on the following classical example of Osgood (1920's).

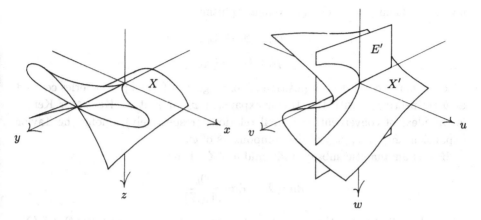

Figure 3. Blowing-up of $X = \{z^3 - x^2 yz - x^4 = 0\}$.

EXAMPLE 2.3. Let φ denote the analytic mapping

$$\varphi(x_1, x_2) = (x_1, x_1 x_2, x_1 x_2 e^{x_2}).$$

Then there are no (nonzero) formal relations (among the components of φ) at the origin $(x_1, x_2) = (0, 0)$; i.e., if $G(y_1, y_2, y_3)$ is a nonzero formal power series and $G(x_1, x_1 x_2, x_1 x_2 e^{x_2}) = 0$, then $G = 0$. Indeed, writing $G = \sum_{j=0}^{\infty} G_j$, where $G_j(y_1, y_2, y_3)$ denotes the homogeneous part of G of order k, we have

$$0 = G(x_1, x_1 x_2, x_1 x_2 e^{x_2}) = \sum_{j=0}^{\infty} x_1^j G_j(1, x_2, x_2 e^{x_2}),$$

so that all $G_j(1, x_2, x_2 e^{x_2}) = 0$, and therefore all polynomials G_j are zero because e^{x_2} is transcendental.

Relations. If $y = \varphi(x)$, where $x = (x_1, \ldots, x_m)$ and $y = (y_1, \ldots, y_n)$, is a mapping (in some given class), we let φ^* denoted the homomorphism of rings of functions given by composition with φ; i.e., $\varphi^* : g(y) \mapsto (g \circ \varphi)(x)$. Then

$$\mathrm{Ker}\, \varphi^* = \{g(y_1, \ldots, y_n) : g(\varphi_1(x), \ldots, \varphi_n(x)) = 0\}$$

is, by definition, the ideal of *relations* among the components $\varphi_1, \ldots, \varphi_n$ of φ.

Let $\mathbb{K} = \mathbb{R}$ or $\mathbb{C}$. Let U be an open subset of $\mathbb{K}^m$ and let $a \in U$. We write $\mathbb{K}\{x - a\} = \mathbb{K}\{x_1 - a_1, \ldots, x_m - a_m\}$ or $\mathbb{K}[\![x - a]\!] = \mathbb{K}[\![x_1 - a_1, \ldots, x_m - a_m]\!]$ for the rings of convergent or formal power series (respectively) centred at a. The ring $\mathbb{K}\{x - a\}$ can be identified with the local ring $\mathcal{O}_a$ of germs of analytic functions at a, and $\mathbb{K}[\![x - a]\!]$ with the completion $\widehat{\mathcal{O}}_a$ of $\mathcal{O}_a$. (These notions and remarks make sense, more generally, using local coordinates on an m-dimensional $\mathbb{K}$-analytic manifold U.) We will write $\underline{m}_a$ or $\widehat{\underline{m}}_a$ for the maximal ideal of $\mathcal{O}_a$ or $\widehat{\mathcal{O}}_a$ (respectively). Suppose that φ is an analytic mapping $\varphi : U \to \mathbb{K}^n$, and let

$b = \varphi(a)$. Then φ induces ring homomorphisms

$$\varphi_a^* : \mathcal{O}_b \to \mathcal{O}_a,$$

$$\hat{\varphi}_a^* : \widehat{\mathcal{O}}_b \to \widehat{\mathcal{O}}_a;$$

φ_a^* or $\hat{\varphi}_a^*$ corresponds to composition of convergent or formal power series centred at b (respectively) with the Taylor expansion $\hat{\varphi}_a$ of φ at a. $\operatorname{Ker} \varphi_a^*$ or $\operatorname{Ker} \hat{\varphi}_a^*$ is the ideal of convergent or formal relations (respectively) among the Taylor expansions $\hat{\varphi}_{1,a}, \ldots, \hat{\varphi}_{n,a}$ of the components of φ.

If X is an analytic subset of $\mathbb{K}^n$ and $b \in X$, then

$$\dim_b X = \dim \frac{\mathcal{O}_b}{\mathcal{A}_b(X)},$$

where $\dim_b X$ denotes the geometric dimension of X at a, and $\dim \mathcal{O}_b/\mathcal{A}_b(X)$ denotes the *Krull dimension* of the local ring $\mathcal{O}_b/\mathcal{A}_b(X)$; i.e., the length of a longest chain of prime ideals in this ring. (See [Łojasiewicz 1991, § IV.4.3].) Let φ be an analytic mapping and $b = \varphi(a)$, as above. Then there is a smallest (germ of an) analytic set Y_b at b, such that Y_b contains $\varphi(V)$, for a sufficiently small neighbourhood V of a. Clearly,

$$\mathcal{A}_b(Y_b) = \operatorname{Ker} \varphi_a^*.$$

Ranks of Gabrielov. Gabrielov introduced the following three *ranks* associated to an analytic mapping φ at a point a of its source:

$$r_a(\varphi) := \text{generic rank of } \varphi \text{ at } a,$$

$$r_a^{\mathcal{F}}(\varphi) := \dim \frac{\mathbb{K}[\![y - \varphi(a)]\!]}{\operatorname{Ker} \hat{\varphi}_a^*},$$

$$r_a^{\mathcal{A}}(\varphi) := \dim \frac{\mathbb{K}\{y - \varphi(a)\}}{\operatorname{Ker} \varphi_a^*}.$$

(The "generic rank" of φ at a is the largest rank of the tangent mapping of φ in a small neighbourhood of a.) It is not difficult to see that

$$r_a(\varphi) \leq r_a^{\mathcal{F}}(\varphi) \leq r_a^{\mathcal{A}}(\varphi).$$

(We have $r_a^{\mathcal{F}}(\varphi) \leq r_a^{\mathcal{A}}(\varphi)$ because $\operatorname{Ker} \varphi_a^* \subset \operatorname{Ker} \hat{\varphi}_a^*$. On the other hand, $r_x(\varphi)$ is constant in a neighbourhood of a, and at a point x where $r_x(\varphi)$ equals the rank of the tangent mapping of φ, all three ranks of Gabrielov coincide (by the implicit function theorem) and $r_x^{\mathcal{F}}(\varphi) \leq r_a^{\mathcal{F}}(\varphi)$ (for example, by [Bierstone and Milman 1987a, Prop. 8.3.7]). Therefore, $r_a(\varphi) \leq r_a^{\mathcal{F}}(\varphi)$. See also [Milman 1978].)

In the 1960's, Artin and Grothendieck asked: Is $\operatorname{Ker} \hat{\varphi}_a^*$ generated by $\operatorname{Ker} \varphi_a^*$? In other words, is $r_a^{\mathcal{F}}(\varphi) = r_a^{\mathcal{A}}(\varphi)$? Gabrielov [1971] showed that the answer is "no". (See Example 2.5 below.) Of course, if φ is a proper complex-analytic mapping, then all three ranks above coincide, by Remmert's proper mapping theorem [Remmert 1957].

DEFINITION 2.4. We say that φ is *regular at a* if $r_a(\varphi) = r_a^A(\varphi)$. We say that φ is *regular* if it is regular at every point of the source.

Local dimensions. The ranks of Gabrielov have counterparts for a subanalytic set. Let X denote a closed subanalytic subset of $\mathbb{R}^n$. If $b \in X$, then, by definition, $\mathcal{A}_b(X) = \mathcal{A}_b(Y_b)$, where Y_b is the smallest germ of an analytic set at b containing the germ of X. Suppose that $\varphi : M \to \mathbb{R}^n$ is a proper real-analytic mapping from a manifold M, such that $\varphi(M) = X$. Clearly,

$$\mathcal{A}_b(X) = \bigcap_{a \in \varphi^{-1}(b)} \operatorname{Ker} \varphi_a^*.$$

This suggests that we define the *formal local ideal* $\mathcal{F}_b(X)$ of X at b as

$$\mathcal{F}_b(X) = \bigcap_{a \in \varphi^{-1}(b)} \operatorname{Ker} \hat{\varphi}_a^*.$$

(The preceding intersections are finite: $\operatorname{Ker} \varphi_a^*$ and $\operatorname{Ker} \hat{\varphi}_a^*$ are constant on connected components of the fibre $\varphi^{-1}(b)$ [Bierstone and Milman 1998, Lemma 5.1].) The ideal $\mathcal{F}_b(X)$ does not depend on the mapping φ: There are equivalent ways to define it using X alone [Bierstone and Milman 1998, Lemma 6.1]; for example, $\mathcal{F}_b(X) = \{G \in \hat{\mathcal{O}}_b : (G \circ \gamma)(t) \equiv 0$ for every real-analytic arc $\gamma(t)$ in X such that $\gamma(0) = b\}$.

We define

$$d_b(X) := \dim_b X,$$

$$d_b^A(X) := \dim Y_b = \dim \frac{\mathcal{O}_b}{\mathcal{A}_b(X)},$$

$$d_b^{\mathcal{F}}(X) := \dim \frac{\hat{\mathcal{O}}_b}{\mathcal{F}_b(X)}.$$

Then

$$d_b(X) \leq d_b^{\mathcal{F}}(X) \leq d_b^A(X),$$

by the corresponding inequalities among the ranks of Gabrielov. If X is semianalytic, then $\mathcal{F}_b(X)$ is generated by $\mathcal{A}_b(X)$, and all three local dimensions coincide [Lojasiewicz 1964; Bierstone and Milman 1988, Theorem 2.13].

EXAMPLE 2.5 [Gabrielov 1971]. Let $\varphi(x) = (\varphi_1(x), \varphi_2(x), \varphi_3(x))$ denote Osgood's mapping (Example 2.3). Then there is a *divergent* power series $G(y) = G(y_1, y_2, y_3)$ such that $\varphi_4(x) := G(\varphi(x))$ *converges*: Write

$$\varphi_3(x) = x_1 x_2 + x_1 x_2^2 + \frac{x_1 x_2^3}{2!} + \cdots.$$

We construct a sequence of polynomials $g_k(y)$, for $k = 1, 2, \ldots$, to kill terms of higher and higher order in the expansion of $\varphi_3(x)$:

$$g_1(y) := y_3 - y_2, \qquad\qquad g_1 \circ \varphi = x_1 x_2^2 + \cdots,$$
$$g_2(y) := 2(y_1(y_3 - y_2) - y_2^2), \qquad g_2 \circ \varphi = x_1^2 x_2^3 + \cdots,$$
$$\vdots \qquad\qquad\qquad\qquad\qquad \vdots$$
$$g_k(y) := k(y_1 g_{k-1}(y) - y_2^k), \qquad g_k \circ \varphi = x_1^k x_2^{k+1} + \cdots.$$

Note that the maximum absolute value of a coefficient of $g_k(y)$ is $k!$, while the maximum absolute value of a coefficient in the power series expansion of $g_k(\varphi(x))$ is 1. It follows that the power series $G(y) := \sum_{k=1}^{\infty} g_k(y)$ diverges, while $\varphi_4(x) := G(\varphi(x))$ converges.

Set $\psi(x) := (\varphi_1(x), \varphi_2(x), \varphi_3(x), \varphi_4(x))$. It is not difficult to see that

$$r_0(\psi) = 2, \quad r_0^{\mathcal{F}}(\psi) = 3, \quad r_0^{\mathcal{A}}(\psi) = 4.$$

The images of proper real-analytic mappings that are regular form an important subclass of closed subanalytic sets, called *Nash-subanalytic*. Regular mappings (and therefore Nash-subanalytic sets) are characterized by a theorem of Gabrielov described in Section 3 below. Although none of Examples 2.1, 2.2, 2.3 or 2.5 above is coherent, it is clear that each satisfies a stratified version of the property of coherence, in some reasonable sense. (For example, the images of Osgood's and Gabrielov's mappings — Examples 2.3 and 2.5 — are coherent outside the origin.) We will describe a larger class of "semicoherent" subanalytic sets that captures such an idea. In Section 4, we will see that semicoherent sets are "tame" from an analytic viewpoint.

Images of proper mappings.

	Algebraic	Relatively algebraic	Regular	
$\mathbb{C}$	closed algebraic sets	closed analytic sets $\longrightarrow$	$\longrightarrow$	
$\mathbb{R}$	closed semialgebraic sets	closed semianalytic sets	closed Nash-subanalytic sets	semicoherent sets

Semicoherence. We will say that a subanalytic set X is semicoherent if it has a stratification such that the formal local ideals $\mathcal{F}_b(X)$ of X are generated over each stratum by finitely many subanalytically parametrized formal power series. More precisely:

DEFINITION 2.6. Let $X \supset Z$ denote closed subanalytic subsets of $\mathbb{R}^n$. We say that X is *(formally) semicoherent rel Z (relative to Z)* if X has a (locally finite) subanalytic stratification $X = \bigcup X_i$ such that Z is a union of strata and X satisfies the following *formal semicoherence* property *along* every stratum X_i outside Z: For every point of $\overline{X}_i$, there is a neighbourhood V and there are finitely many formal power series

$$f_{ij}(\,\cdot\,,Y) = \sum_{\alpha \in \mathbb{N}^n} f_{ij,\alpha}(\,\cdot\,)Y_1^{\alpha_1} \cdots Y_n^{\alpha_n}$$

whose coefficients $f_{ij,\alpha}$ are analytic functions on $X_i \cap V$ that are subanalytic (i.e., their graphs are subanalytic as subsets of $V \times \mathbb{R}$), such that, for all $b \in X_i \cap V$, $\mathcal{F}_b(X)$ is generated by the elements $\sum_\alpha f_{ij,\alpha}(b)(y-b)^\alpha \in \mathbb{R}[\![y-b]\!]$. $((y-b)^\alpha$ means $(y_1 - b_1)^{\alpha_1} \cdots (y_n - b_n)^{\alpha_n}$, where $\alpha = (\alpha_1, \ldots, \alpha_n)$.)

Subanalyticity of the coefficients, $f_{ij,\alpha}$ is a natural restriction on their growth at the boundary of a stratum (as expressed by a *Łojasiewicz inequality*; compare [Łojasiewicz 1964; Bierstone and Milman 1988, §6]). We can formulate an (analytic) semicoherence condition analogous to that above by using the ideals $\mathcal{A}_b(X)$ in place of $\mathcal{F}_b(X)$, but the formal condition seems to be the more useful (as in Theorem 4.4 below). The formal and analytic semicoherence conditions are equivalent if each $\mathcal{F}_b(X)$ is generated by $\mathcal{A}_b(X)$ (as in the case of real- or complex-analytic sets). Semicoherence of semi-algebraic sets was proved by Tougeron and Merrien [Merrien 1980], and of Nash-subanalytic sets by Bierstone and Milman [1987a; 1987b]; in these cases, each $\mathcal{F}_b(X)$ is generated by $\mathcal{A}_b(X)$.

Pawłucki has proved that, if Z denotes the set of *non-Nash points* of X (i.e., the points that do not admit Nash-subanalytic neighbourhoods in X), then Z is subanalytic [1990] and X is semicoherent rel Z [1992]. Pawłucki's theorem implies the analogous statement for the non-semianalytic points of X.

QUESTION. Pawłucki's result suggests the following general question about o-minimal structures on $\mathbb{R}$ (probably easier than Pawłucki's theorem; Nash-subanalytic sets do not correspond to an o-minimal structure): Let $\mathcal{S}_1 \subset \mathcal{S}_2$ be o-minimal structures on $\mathbb{R}$. If X is $\mathcal{S}_2$-definable and $Y \subset X$ denotes the points of X that do not admit $\mathcal{S}_1$-definable neighbourhoods, then is Y $\mathcal{S}_2$-definable? (Piękosz [1998] gives a result in this direction.)

In 1986, Hironaka announced that every subanalytic set X is both $\mathcal{F}$- and $\mathcal{A}$-semicoherent (and, as a consequence, that X admits a subanalytic stratification with the local dimensions $d_b^\mathcal{F}(X)$ and $d_b^A(X)$ constant on every stratum) [Hironaka 1986]. But Pawłucki has given a counterexample!

EXAMPLE 2.7 [Pawłucki 1989]. Let $\{a_n\}$ be any sequence of points in an open interval $I = (-\delta, \delta)$ of $\mathbb{R}$ (where $\delta > 0$). Pawłucki constructs an analytic mapping $\varphi \colon I^3 \to \mathbb{R}^5$ of the form

$$\varphi(u, w, t) = \big(u,\, t,\, tw,\, t\Phi(u,w),\, t\Psi(u,w,t)\big)$$

such that, in $I = I \times \{0\} \times \{0\}$, φ admits no nonzero formal relation (i.e., $\operatorname{Ker} \hat{\varphi}_a^* = 0$) precisely at the points a of $\{a_n\}$, but φ has a nonzero convergent relation (i.e., $\operatorname{Ker} \varphi_a^* \neq 0$) throughout any open interval in $I \setminus \{a_n\}$.

Pawłucki's idea is based on Gabrielov's example 2.5. Take $\delta = \frac{1}{2}$. We can define

$$\Phi(u,w) := \sum_{n=1}^{\infty} \big((u - a_1) \cdots (u - a_n)\big)^{r(n)} w^n,$$

where $\{r(n)\}$ is an increasing sequence of positive integers with $\limsup r(n)/n = \infty$. Write $p_n(u) := \big((u - a_1) \cdots (u - a_n)\big)^{r(n)}$. We define a sequence of rational functions $f_n(u, t, x, y)$ by

$$f_1(u, t, x, y) := \frac{y}{p_1(u)}$$

and, for $n > 1$,

$$f_n(u, t, x, y) := \frac{p_{n-1}(u)}{p_n(u)} t\big(f_{n-1}(u, t, x, y) - x^{n-1}\big)$$

$$= \frac{t^{n-1} y}{p_n(u)} - \frac{1}{p_n(u)} \sum_{k=1}^{n-1} p_k(u) t^{n-k} x^k.$$

Then each

$$f_n\big(u, t, tw, t\Phi(u,w)\big) = t^n \sum_{k=n}^{\infty} \frac{p_k(u)}{p_n(u)} w^k,$$

and we can set

$$t\Psi(u, w, t) := \sum_{n=1}^{\infty} f_n\big(u, t, tw, t\Phi(u,w)\big).$$

(See [Pawłucki 1989] for details.)

Pawłucki's construction provides examples of a variety of interesting phenomena, depending on the choice of the sequence $\{a_n\}$; for instance:

(1) If $\lim_{n \to \infty} a_n = 0$ but $a_n \neq 0$ for all n, then there is no (nonzero) relation at each a_n, a divergent relation (but no convergent relation) at 0, and a convergent relation at any other point of I. Suppose that $X = \varphi(K)$, where K is a compact subanalytic neighbourhood of 0 in I^3. Clearly, X is neither $\mathcal{F}$- nor $\mathcal{A}$-semicoherent.

(2) If $\{a_n\}$ is dense in I, then there is no convergent relation at any point of I, but there is a formal relation at every point of $I \setminus \{a_n\}$. Therefore, $\mathcal{A}$-semicoherent $\nRightarrow$ $\mathcal{F}$-semicoherent.

We believe it is not known whether $\mathcal{F}$-semicoherent $\Rightarrow$ $\mathcal{A}$-semicoherent.

(3) If the accumulation points of $\{a_n\}$ themselves form a convergent sequence $\{c_k\}$, then $X = \varphi(K)$ is not semicoherent precisely at the points $\varphi(c_k)$ and $\varphi(\lim c_k)$ (i.e., these points do not admit semicoherent neighbourhoods in X). In other words, the points at which X is not semicoherent do not necessarily form a subanalytic subset!

The phenomena above show that subanalytic sets in general can be wild indeed. The class of semicoherent sets, on the other hand, can be characterized by several (remarkably equivalent) "tameness" properties (to be described in Section 4). For example, let X be a closed subanalytic subset of $\mathbb{R}^n$, and let $\mathcal{C}^k(X)$ denote the ring of restrictions to X of $\mathcal{C}^k$ (i.e., k times continuously differentiable) functions on $\mathbb{R}^n$, where $k \in \mathbb{N} \cup \{\infty\}$. Then X is $(\mathcal{F}\text{-})$ semicoherent if and only if $\mathcal{C}^\infty(X)$ is the intersection $\bigcap_{k \in \mathbb{N}} \mathcal{C}^k(X)$ of all finite differentiability classes [Bierstone and Milman 1998; Bierstone et al. 1996]).

QUESTION. Are restricted subpfaffian sets semicoherent? (A closed *restricted subpfaffian set* is a proper projection of a semianalytic set that is defined using *Pfaffian functions* in the sense of Khovanskii [1991]; compare [Wilkie 1996].)

3. Gabrielov's Theorem

THEOREM 3.1. *Let* $y = \varphi(x)$, *where* $x = (x_1, \ldots, x_m)$ *and* $y = (y_1, \ldots, y_n)$, *denote a real-analytic (or complex-analytic) mapping defined in a neighbourhood of a point* a. *Set* $b = \varphi(a)$. *Then the following conditions are equivalent:*

(1) $r_a(\varphi) = r_a^{\mathcal{F}}(\varphi)$; *i.e., there are "sufficiently many" formal relations.*

(2) $r_a(\varphi) = r_a^{\mathcal{F}}(\varphi) = r_a^{\mathcal{A}}(\varphi)$; *i.e.,* φ *is regular at* a.

(3) *Composite function property:*

$$\mathcal{O}_a \cap \hat{\varphi}_a^*(\widehat{\mathcal{O}}_b) = \varphi_a^*(\mathcal{O}_b).$$

(4) *Linear equivalence of topologies. Let* $R := \widehat{\mathcal{O}}_b / \operatorname{Ker} \hat{\varphi}_a^*$ *and* $R' := \widehat{\mathcal{O}}_a$, *so there is a natural inclusion of local rings* $R \hookrightarrow R'$. *Let* $\underline{m}$ *and* $\underline{m}'$ *denote the maximal ideals of* R *and* R', *respectively. Then there exist* $\alpha, \beta \in \mathbb{N}$ *such that, for all* k,*

$$(\underline{m}')^{\alpha k + \beta} \cap R \subset \underline{m}^k.$$

The composite function property (3) concerns the solution of an equation $f(x) = g(\varphi(x))$, where f is a given analytic function at a and g is the unknown; (3) says that if there is a formal power series solution g, then there is also an analytic solution. Condition (4) concerns the $\underline{m}$-adic (or *Krull*) *topologies* of the local rings. (The powers $\underline{m}^k$ of the maximal ideal $\underline{m}$ of R form a fundamental system of neighbourhoods of 0 for the $\underline{m}$-adic topology.) Clearly, $\underline{m}^k \subset (\underline{m}')^k \cap R$ for all k, so (4) implies that the $\underline{m}$-adic topology of R coincides with its $\underline{m}'$-adic topology as a subspace of R'.

Gabrielov [1973] proved that (1) $\Longleftrightarrow$ (2) and (2) $\Rightarrow$ (3). The implication (3) $\Rightarrow$ (2) follows from [Becker and Zame 1979] and [Milman 1978], and (2) $\Longleftrightarrow$ (4) is due to Izumi [1986], Rees [1989] and Spivakovsky [1990].

Chevalley estimate. Condition (4) is related to the following elementary lemma of Chevalley [1943, § II, Lemma 7] (compare [Bierstone and Milman 1998, Lemma 5.2]).

LEMMA 3.2. (We use the notation of Theorem 3.1.) *For all $k \in \mathbb{N}$, there exists $l \in \mathbb{N}$ such that if $G \in \widehat{\mathbb{O}}_b$ and $\hat{\varphi}_a^*(G) \in \hat{\underline{m}}_a^{l+1}$, then $G \in \operatorname{Ker} \hat{\varphi}_a^* + \hat{\underline{m}}_b^{k+1}$.*

Given $k \in \mathbb{N}$, let $l_{\varphi^*}(a, k)$ denote the least l satisfying Chevalley's lemma. We call $l_{\varphi^*}(a, k)$ a *Chevalley estimate*. Condition (4) of Theorem 3.1 means there is a *linear Chevalley estimate* $l_{\varphi^*}(a, k) \leq \alpha k + \beta'$ (where $\beta' = \alpha + \beta - 1$).

QUESTION. Suppose that $\varphi \colon M \to \mathbb{R}^n$ is a regular mapping (Definition 2.4). Is there a uniform linear Chevalley estimate $l_{\varphi^*}(a, k) \leq \alpha_L k + \beta_L$, where $L \subset M$ is compact and $a \in L$? This question is equivalent to a uniform version of a product theorem of Izumi [1985] and D. Rees [1989]; see [Wang 1995], where the question also is answered positively in a special case.

4. Semicoherent Sets

In this section, we characterize the class of semicoherent subanalytic sets: We describe several metric, algebro-geometric and differential properties of subanalytic sets that might seem of quite different natures, but that turn out each to be equivalent to semicoherence. The ideas and results here come from [Bierstone and Milman 1998; Bierstone et al. 1996]. Theorem 4.4 below can be viewed as a parallel to Gabrielov's theorem 3.1, but is expressed in terms of properties of a closed subanalytic set X (i.e., the image $X = \varphi(M)$ of a proper real-analytic mapping $\varphi \colon M \to \mathbb{R}^n$) rather than in terms of properties of φ. The composite function property (3) of Theorem 3.1 is replaced by an analogous property concerning composite differentiable functions. According to Theorem 4.4, the $\mathbb{C}^\infty$ composite function property depends on the way that the formal local ideals $\mathcal{F}_b(X) = \bigcap_{a \in \varphi^{-1}(b)} \operatorname{Ker} \hat{\varphi}_a^*$ vary with respect to $b \in X$. The analytic composite function property (Theorem 3.1(3)), on the other hand, depends on the relationship between the convergent and formal ideals $\operatorname{Ker} \varphi_a^*$ and $\operatorname{Ker} \hat{\varphi}_a^*$. Theorem 4.4 shows that spaces of differentiable functions are natural function spaces on subanalytic sets.

$\mathbb{C}^\infty$ **composite function problem** (Thom, Glaeser). Let M denote a real-analytic manifold and $\varphi \colon M \to \mathbb{R}^n$ a proper real-analytic mapping. Suppose that $f \colon M \to \mathbb{R}$ is a $\mathbb{C}^\infty$ function. Under what conditions is f a composite $f(x) = g(\varphi(x))$, where g is a $\mathbb{C}^\infty$ function on $\mathbb{R}^n$?

An obvious necessary condition is that f be constant on the fibres $\varphi^{-1}(b)$, where $b \in X := \varphi(M)$.

EXAMPLES 4.1 ($\mathbb{C}^\infty$ INVARIANTS OF A GROUP ACTION). In the early 1940's, Whitney proved that every $\mathbb{C}^\infty$ even function $f(x)$ (of one variable) can be written $f(x) = g(x^2)$, where g is $\mathcal{C}^\infty$ [Whitney 1943]. (See Example 4.2 below.) Whitney's result is the earliest version of the $\mathcal{C}^\infty$ composite function theorem. About twenty years later, Glaeser (answering a question posed by Thom in connection with the $\mathcal{C}^\infty$ preparation theorem) showed that a $\mathcal{C}^\infty$ function $f(x_1, \ldots, x_n)$ which is invariant under permutation of the coordinates can be expressed $f(x) = g(\sigma_1(x), \ldots, \sigma_n(x))$, where g is $\mathcal{C}^\infty$ and the $\sigma_i(x)$ are the elementary symmetric polynomials [Glaeser 1963]. G. W. Schwarz [1975] extended these results to a $\mathcal{C}^\infty$ analogue of Hilbert's classical theorem on polynomial invariants: Hilbert's theorem says that, on a linear representation of a compact Lie group, the algebra of invariant polynomials is finitely generated; i.e., there are finitely many invariant polynomials $p_1(x), \ldots, p_r(x)$ such that any invariant polynomial $f(x)$ can be written $f(x) = g(p_1(x), \ldots, p_r(x))$, where g is a polynomial. Schwarz's theorem asserts that a $\mathcal{C}^\infty$ invariant function $f(x)$ can be expressed in the same way, with g $\mathcal{C}^\infty$.

Formal composition. In each of Examples 4.1, f is constant on the fibres of the mapping (given by the basic invariant polynomials). In general, however, not every $\mathcal{C}^\infty$ function f that is constant on the fibres of a proper real-analytic mapping $\varphi \colon M \to \mathbb{R}^n$ can be expressed as a composite $f = g \circ \varphi$, where g is $\mathcal{C}^\infty$. (Consider, for example, $\varphi(x) = x^3$, $f(x) = x$.) There is a necessary formal condition [Glaeser 1963]: The Taylor expansions of f along any fibre $\varphi^{-1}(b)$ are the pull-backs of a formal power series centred at b; i.e., $f \in \big(\varphi^* C^\infty(\mathbb{R}^n)\big)\widehat{}$, where

$$\big(\varphi^* \mathcal{C}^\infty(\mathbb{R}^n)\big)\widehat{} := \left\{ \begin{array}{l} f \in \mathcal{C}^\infty(M) : \text{for all } b \in \varphi(M), \text{ there exists } G_b \in \\ \widehat{\mathcal{O}}_b = \mathbb{R}[\![y-b]\!] \text{ such that } \hat{f}_a = \hat{\varphi}_a^*(G_b), \text{ for all } a \in \varphi^{-1}(b) \end{array} \right\}.$$

(Here $\hat{f}_a$ denotes the element of $\widehat{\mathcal{O}}_a$ induced by f: the formal Taylor expansion of f at a, with respect to any local coordinate system.) The functions in $\big(\varphi^* \mathcal{C}^\infty(\mathbb{R}^n)\big)\widehat{}$ are "formally composites with φ". (In each of Examples 4.1, the hypothesis implies formal composition.)

It is easy to see that $\big(\varphi^* \mathcal{C}^\infty(\mathbb{R}^n)\big)\widehat{}$ contains the closure of $\varphi^* \mathcal{C}^\infty(\mathbb{R}^n)$ in $\mathcal{C}^\infty(M)$ (with respect to the $\mathcal{C}^\infty$ topology); in fact, $\big(\varphi^* \mathcal{C}^\infty(\mathbb{R}^n)\big)\widehat{}$ is closed [Bierstone et al. 1996, Corollary 1.4]. (For a definition of the $\mathcal{C}^\infty$ topology, see Question (2) following Theorem 4.4 below.) The *composite function property*

$$\varphi^* \mathcal{C}^\infty(\mathbb{R}^n) = \big(\varphi^* \mathcal{C}^\infty(\mathbb{R}^n)\big)\widehat{}$$

depends only on the image $X = \varphi(M)$ and holds, for example, if X is Nash-subanalytic [Bierstone and Milman 1982]. (The latter and [Bierstone and Milman 1987a; 1987b] are the sources of the ideas involved in Theorem 4.4.)

EXAMPLE 4.2 (PROOF OF WHITNEY'S THEOREM ON $\mathcal{C}^\infty$ EVEN FUNCTIONS). Suppose that $f(x)$ is a $\mathcal{C}^\infty$ function that is even; equivalently, f is formally a composite with $y = x^2$. We can assume that f is *flat at* 0 (i.e., f vanishes at 0 together with its derivatives of all orders) as follows: $\hat{f}_0(x) = H(x^2)$, where

$H \in \mathbb{R}[\![y]\!]$. By a classical lemma of E. Borel, there exists $h(y)$ of class $\mathcal{C}^\infty$ such that $\hat{h}_0 = H$. We can replace $f(x)$ by $f(x) - h(x^2)$.

Now let $g(y) = f(\sqrt{y})$, $y > 0$. Differentiating repeatedly, we have

$$
\begin{pmatrix} f(x) \\ f'(x) \\ f''(x) \\ \vdots \end{pmatrix} = \begin{pmatrix} 1 & 0 & 0 & \cdots \\ 0 & 2x & 0 & \cdots \\ 0 & 2 & 4x^2 & \cdots \\ \vdots & \vdots & \vdots & \ddots \end{pmatrix} \begin{pmatrix} g(x^2) \\ g'(x^2) \\ g''(x^2) \\ \vdots \end{pmatrix}.
$$

For any l, the determinant of the $l \times l$ upper left-hand block of the matrix is $c_l x^{l(l-1)/2}$, where $c_l > 0$. By Cramer's rule, since f is flat at 0, we have $\lim_{y \to 0+} g^{(k)}(y) = 0$, for all k. By L'Hôpital's rule (compare [Spivak 1994, Chapter 11, Theorem 7]), g extends to a $\mathcal{C}^\infty$ function that is flat at 0.

Approach to the composite function problem. Our point of view is similar to that in Example 4.2, and shows how the various properties of semicoherent sets enter into the composite function problem. Let $\varphi \colon M \to \mathbb{R}^n$ be a proper real-analytic mapping. Suppose $f \in \widehat{\left(\varphi^* \mathcal{C}^\infty(\mathbb{R}^n)\right)}$. Then, for every $b \in X = \varphi(M)$, there exists $G_b \in \widehat{\mathcal{O}}_b$ such that $\hat{f}_a = G_b \circ \hat{\varphi}_a$, for all $a \in \varphi^{-1}(b)$. But in general G_b is uniquely determined only modulo $\mathcal{F}_b(X)$. A choice of a complementary subspace V_b to $\mathcal{F}_b(X)$ (i.e., $\widehat{\mathcal{O}}_b = \mathcal{F}_b(X) \oplus V_b$) provides a unique determination of G_b.

The equation $\hat{f}_a = G \circ \hat{\varphi}_a$, where $a \in \varphi^{-1}(b)$, implies that $\hat{f}_{a'} = G \circ \hat{\varphi}_{a'}$, for all a' in the same connected component of $\varphi^{-1}(b)$ as a [Bierstone and Milman 1998, Lemma 5.1]. Therefore, to find G_b as above, it is enough to solve the system of equations

$$
\hat{f}_{a^i} = G_b \circ \hat{\varphi}_{a^i}, \quad \text{for } i = 1, \ldots, s, \tag{4.3}
$$

where there is at least one a^i in every component of $\varphi^{-1}(b)$. Since φ is a proper analytic mapping, there is a uniform bound on the number of connected components of a fibre $\varphi^{-1}(b)$, over any compact subset of the target.

We argue by successively flattening over a stratification $X = \cup X_j$ (as in Example 4.2). To guarantee that the G_b are Taylor expansions of a $\mathcal{C}^\infty$ function (at least along a stratum), we need to stratify so that the V_b can be chosen independent of $b \in X_j$, and invariant under formal differentiation in $\widehat{\mathcal{O}}_b = \mathbb{R}[\![y - b]\!]$. These properties hold on a stratification by the "diagram of initial exponents" $\mathcal{N}\left(\mathcal{F}_b(X)\right)$ (to be described below).

In general, it is not true that in (4.3) the coefficients of G_b of order $\leq k$ are determined (modulo $\mathcal{F}_b(X)$) by the coefficients of the $\hat{f}_{a^i}$ of order $\leq k$. (Even in Example 4.2 above, $l = 2k$ derivatives of f at 0 are needed to determine k derivatives of a formal solution H_0.) A uniform Chevalley estimate provides a uniform bound $l = l(k, K)$ on the number of formal derivatives of the $\hat{f}_{a^i}$, for $i = 1, \ldots, s$, that are needed to determine the derivatives of order $\leq k$ of $G_b \bmod \mathcal{F}_b(X)$, for b in a compact subset K of $\mathbb{R}^n$.

We can then solve the composite function problem inductively over a stratification by the diagram, using Cramer's rule and a subanalytic version of L'Hôpital's rule or Hestenes's Lemma [Bierstone and Milman 1982, Corollary 8.2; Bierstone et al. 1996, Proposition 3.4], in a manner similar to Example 4.2.

Chevalley estimate. Let $\varphi \colon M \to \mathbb{R}^n$ be a proper real-analytic mapping and let $X = \varphi(M)$. Theorem 4.4 involves a variant of the Chevalley estimate (as defined in Section 3) for a fibre, or for the image X. For all $b \in X$ and $k \in \mathbb{N}$, we define

$$l_{\varphi^*}(b,k) := \min \left\{ \begin{array}{l} l \in \mathbb{N} : \text{if } G \in \widehat{\mathcal{O}}_b \text{ and } \hat{\varphi}_a^*(G) \in \hat{m}_a^{l+1} \text{ for} \\ \text{all } a \in \varphi^{-1}(b), \text{ then } G \in \hat{m}_b^{k+1} + \mathcal{F}_b(X) \end{array} \right\},$$

$$l_X(b,k) := \min \left\{ \begin{array}{l} l \in \mathbb{N} : \text{if } G \in \widehat{\mathcal{O}}_b \text{ and } |T_b^l G(y)| = o(|y-b|^l), \\ \text{where } y \in X, \text{ then } G \in \hat{m}_b^{k+1} + \mathcal{F}_b(X) \end{array} \right\},$$

where $T_b^l G(y)$ denotes the Taylor polynomial of order l of G. Then $l_{\varphi^*}(b,k) < \infty$ because, if $\underline{a} = (a^1, \ldots, a^s)$, where each $a^i \in \varphi^{-1}(b)$ and some a^i belongs to each component of $\varphi^{-1}(b)$, then $l_{\varphi^*}(b,k) < l_{\varphi^*}(\underline{a},k)$, where

$$l_{\varphi^*}(\underline{a},k) := \min \left\{ \begin{array}{l} l \in \mathbb{N} : \text{if } G \in \widehat{\mathcal{O}}_b \text{ and } \hat{\varphi}_{a^i}^*(G) \in \hat{m}_{a^i}^{l+1} \text{ for} \\ i = 1, \ldots, s, \text{ then } G \in \bigcap_i \operatorname{Ker} \hat{\varphi}_{a^i}^* + \hat{m}_b^{l+1} \end{array} \right\},$$

and $l_{\varphi^*}(\underline{a},k) < \infty$ as in Lemma 3.2. On the other hand, l_{φ^*} and l_X are equivalent on compact subsets of X in the sense that, for every compact $K \subset X$, there exists r_K ($r_K \geq 1$) such that

$$l_X(b, \cdot) \leq l_{\varphi^*}(b, \cdot) \leq r_K l_X(b, \cdot),$$

$b \in K$. These inequalities are consequences of the two metric inequalities

$$|\varphi(x) - b| \leq c_\varphi(K) d(x, a), \qquad \text{for } b \in K,$$
$$d\big(x, \varphi^{-1}(b)\big)^r \leq c_\varphi(b, K) |\varphi(x) - b|, \qquad \text{for } b \in K,$$

where $r \geq 1$ and $d(\cdot, \cdot)$ denotes a locally Euclidean metric on M [Bierstone and Milman 1998, Lemma 6.5]. The first of these metric inequalities is simple; the second is an important estimate of Tougeron [1971].

Diagram of initial exponents. Let I be an ideal in the ring of formal power series $\mathbb{R}[[y - b]] = \mathbb{R}[[y_1 - b_1, \ldots, y_n - b_n]]$. The *diagram of initial exponents* $\mathcal{N}(I) \subset \mathbb{N}^n$ is a combinatorial representation of I, in the spirit of the classical Newton diagram of a formal power series;

$$\mathcal{N}(I) := \{ \exp G : \ G \in I \setminus \{0\} \},$$

where $\exp G$ denotes the smallest exponent β of a monomial

$$(y - b)^\beta = (y_1 - b_1)^{\beta_1} \cdots (y_n - b_n)^{\beta_n}$$

with nonzero coefficient in the expansion of G ("smallest" with respect to the lexicographic order of $(|\beta|, \beta_1, \ldots, \beta_n)$, where $|\beta| = \beta_1 + \cdots + \beta_n$). The diagram $\mathcal{N} = \mathcal{N}(I)$ has the form $\mathcal{N} = \mathcal{N} + \mathbb{N}^n$ (since I is an ideal); therefore, there is a smallest finite subset $\mathcal{V}$ of $\mathbb{N}^n$ such that $\mathcal{N} = \mathcal{V} + \mathbb{N}^n$. We call the elements of $\mathcal{V}$ the *vertices* α^j of $\mathcal{N}$.

Set $\operatorname{supp} G := \{\beta : G_\beta \neq 0\}$, where $G = \sum G_\beta (y - b)^\beta$. Hironaka's *formal division algorithm* [Hironaka 1964; Bierstone and Milman 1987a, Theorem 6.2; 1998, Theorem 3.1] shows that

$$\mathbb{R}[\![y - b]\!] = I \oplus \mathbb{R}[\![y - b]\!]^{\mathcal{N}(I)},$$

where

$$\mathbb{R}[\![y - b]\!]^{\mathcal{N}(I)} := \{G : \operatorname{supp} G \cap \mathcal{N}(I) = \varnothing\},$$

and that, if we write

$$(y - b)^{\alpha_j} = F^j(y) + R^j(y),$$

where $F^j(y) \in I$ and $R^j(y) \in \mathbb{R}[\![y-b]\!]^{\mathcal{N}(I)}$, for every vertex α^j, then $\{F^j\}$ is a set of generators of I [Bierstone and Milman 1987a, Corollary 6.8; 1998, Corollary 3.2]. We call $\{F^j\}$ the *standard basis* of I. (It is uniquely determined by the condition that $F^j - (y-b)^{\alpha^j} \in \mathbb{R}[\![y-b]\!]^{\mathcal{N}(I)}$, for each j.) Since $\mathcal{N}(I) + \mathbb{N}^n = \mathcal{N}(I)$, it follows that $\mathbb{R}[\![y - b]\!]^{\mathcal{N}(I)}$ is stable with respect to formal differentiation.

The diagram of initial exponents determines many important algebraic invariants of the ring $\mathbb{R}[\![y - b]\!]/I$; for example, the *Hilbert–Samuel function*, defined for $k \in \mathbb{N}$ by

$$H_I(k) := \dim_\mathbb{R} \frac{\mathbb{R}[\![y - b]\!]}{I + (y - b)^{k+1}} = \#\{\beta \in \mathbb{N}^n \setminus \mathcal{N}(I) : |\beta| \leq k\},$$

where $(y - b)$ here denotes the maximal ideal of $\mathbb{R}[\![y - b]\!]$.

Characterization of semicoherent sets. Suppose that $Z \subset X$ are closed analytic subsets of $\mathbb{R}^n$. If $k \in \mathbb{N} \cup \{\infty\}$, $\mathcal{C}^k(X; Z)$ denotes the algebra of restrictions to X of $\mathcal{C}^k$ functions on $\mathbb{R}^n$ that are k-*flat* on Z (i.e., that vanish on Z together with all partial derivatives of orders at most k). If $\varphi : M \to \mathbb{R}^n$ is a proper real-analytic mapping such that $\varphi(M) = X$, we set

$$(\varphi^* \mathcal{C}^\infty(\mathbb{R}^n; Z))\hat{} := (\varphi^* \mathcal{C}^\infty(\mathbb{R}^n))\hat{} \cap \mathcal{C}^\infty(M; \varphi^{-1}(Z)).$$

THEOREM 4.4 [Bierstone and Milman 1998; Bierstone et al. 1996]. *The following conditions are equivalent:*

(1) X *is semicoherent rel* Z.

(2) *Composite function property. If* $\varphi : M \to \mathbb{R}^n$ *is a proper real-analytic mapping such that* $X = \varphi(M)$, *then*

$$\varphi^* \mathcal{C}^\infty(\mathbb{R}^n; Z) = (\varphi^* \mathcal{C}^\infty(\mathbb{R}^n; Z))\hat{}.$$

(3) $\mathcal{C}^\infty(X; Z) = \bigcap_{k \in \mathbb{N}} \mathcal{C}^k(X, Z)$.

(4) *Uniform Chevalley estimate.* For every compact subset K of X, there is a function $l_K: \mathbb{N} \to \mathbb{N}$ such that

$$l_X(b, k) \leq l_K(k), \quad \text{for all } b \in K \cap (X \setminus Z).$$

(5) *Stratification by the diagram of initial exponents.* X has a (locally finite) subanalytic stratification $X = \bigcup X_i$ such that Z is a union of strata and $\mathcal{N}_b := \mathcal{N}(\mathcal{F}_b(X))$ is constant on every stratum outside Z.

(6) *Stratification by the Hilbert–Samuel function.*

The conditions of Theorem 4.4 are satisfied, for example, if X is a closed subanalytic set and Z is the set of non-Nash points of X.

Conditions (5) and (6) of the theorem can be replaced by conditions of *subanalytic semicontinuity* that are *a priori* stronger. Subanalytic semicontinuity of $\mathcal{N}_b$, for example, means adding to (5) the condition that, if $X_j \subset \overline{X}_i$, $X_j \not\subset Z$, then $\mathcal{N}_j \geq \mathcal{N}_i$, where, for each i, $\mathcal{N}_i$ denotes the value of $\mathcal{N}_b$ on X_i, and $\geq$ is a natural ordering on the set of all possible diagrams. See [Bierstone and Milman 1998].

If $X = \bigcup X_i$ is a stratification by the diagram, as in (5), then X satisfies the formal semicoherence property along every stratum X_i outside Z; in fact, the standard basis

$$F^{ij}(b, y) = (y - b)^{\alpha^{ij}} + \sum_{\beta \in \mathbb{N}^n \setminus \mathcal{N}_i} f_{ij,b}(b)(y - b)^{\beta}$$

(where $\{\alpha^{ij}\}$ denotes the vertices of $\mathcal{N}_i$) provides a semicoherent structure [Bierstone and Milman 1998, §9].

QUESTION 4.5. Suppose that X is a Nash-subanalytic set. Is there a uniform linear Chevalley estimate $l_X(b, k) \leq \alpha_K k + \beta_K$, where $K \subset X$ is compact and $b \in K$? (A variant of the question in Section 3 above.)

QUESTION 4.6. Functional-analytic characterization of "tame"; for example, characterization of semicoherent sets by the extension property. Let X be a closed subanalytic subset of $\mathbb{R}^n$. We say that X has the *extension property* if the restriction mapping $\mathcal{C}^\infty(\mathbb{R}^n) \to \mathcal{C}^\infty(X)$ has a continuous linear splitting (or right inverse) E. If X is semicoherent, then there is an extension operator E [Bierstone and Milman 1998, Theorem 1.23].

The topology of $\mathcal{C}^\infty(\mathbb{R}^n)$ is defined by a system of seminorms

$$\|f\|_k^K := \sup_{\substack{y \in K \\ |\beta| \leq k}} \left| \frac{\partial^{|\beta|} f(y)}{\partial y^{\beta}} \right|,$$

where $k \in \mathbb{N}$ and $K \subset \mathbb{R}^n$ is compact. The topology of $\mathcal{C}^\infty(X)$ is defined by the induced quotient seminorms $\|g\|_l^L := \inf\{\|f\|_l^L : f \in \mathcal{C}^\infty(\mathbb{R}^n), f|X = g\}$. If X is semicoherent and $E: \mathcal{C}^\infty(X) \to \mathcal{C}^\infty(\mathbb{R}^n)$ is an extension operator, then for every $k \in \mathbb{N}$ and $K \subset \mathbb{R}^n$ compact, there exist $l = l(K, k) \in \mathbb{N}$, $L = L(K, k)$

compact, and a constant $c = c(K, k)$ such that $\|E(g)\|_k^K \leq c\|g\|_l^L$, for all $g \in$ $\mathcal{C}^\infty(X)$. We do not have a precise estimate on $l = l(K, k)$, in general. But, for example, if $X = \overline{\text{int} X}$, then there is an extension operator with a linear estimate $l(K, k) = \lambda k$, where $\lambda = \lambda(K)$ [Bierstone 1978]. In the direction converse to "semicoherence implies the extension property", we can prove that if X has an extension operator with an estimate $l(0, K) = 0$ on the zeroth seminorms for every compact K, then there is a uniform Chevalley estimate $l_X(b, k) \leq l(K, 2k)$, $b \in K$ [Bierstone and Milman 1998, Proposition 1.24]. In view of Theorem 4.4, it is therefore interesting to ask: Does semicoherence imply the extension property with $l(\,\cdot\,, 0) = 0$?

Acknowledgement. We are happy to thank Paul Centore for drawing Figure 3 on page 157.

References

[Becker and Zame 1979] J. Becker and W. R. Zame, "Applications of functional analysis to the solution of power series equations", *Math. Ann.* **243** (1979), 37–54.

[Bierstone 1978] E. Bierstone, "Extension of Whitney fields from subanalytic sets", *Invent. Math.* **46** (1978), 277–300.

[Bierstone and Milman 1982] E. Bierstone and P. D. Milman, "Composite differentiable functions", *Ann. of Math.* (2) **116** (1982), 541–558.

[Bierstone and Milman 1987a] E. Bierstone and P. D. Milman, "Relations among analytic functions, I", *Ann. Inst. Fourier (Grenoble)* **37**:1 (1987), 187–239.

[Bierstone and Milman 1987b] E. Bierstone and P. D. Milman, "Relations among analytic functions, II", *Ann. Inst. Fourier (Grenoble)* **37**:2 (1987), 49–77.

[Bierstone and Milman 1988] E. Bierstone and P. D. Milman, "Semianalytic and subanalytic sets", *Inst. Hautes Études Sci. Publ. Math.* **67** (1988), 5–42.

[Bierstone and Milman 1997] E. Bierstone and P. D. Milman, "Canonical desingularization in characteristic zero by blowing up the maximum strata of a local invariant", *Invent. Math.* **128** (1997), 207–302.

[Bierstone and Milman 1998] E. Bierstone and P. D. Milman, "Geometric and differential properties of subanalytic sets", *Ann. of Math.* (2) **147** (1998), 731–785.

[Bierstone et al. 1996] E. Bierstone, P. D. Milman, and W. Pawłucki, "Composite differentiable functions", *Duke Math. J.* **83** (1996), 607–620.

[Chevalley 1943] C. Chevalley, "On the theory of local rings", *Ann. of Math.* (2) **44** (1943), 690–708.

[van den Dries and Miller 1996] L. van den Dries and C. Miller, "Geometric categories and o-minimal structures", *Duke Math. J.* **84** (1996), 497–540.

[Gabrielov 1968] A. M. Gabrielov, "Projections of semi-analytic sets", *Funkcional. Anal. i Priložen.* **2**:4 (1968), 18–30. In Russian; translated in *Functional Anal. Appl.* **2** (1968), 282–291.

[Gabrielov 1971] A. M. Gabrielov, "Formal relations between analytic functions", *Funktsional. Anal. i Priložen.* **5**:4 (1971), 64–65. In Russian; translated in *Functional Anal. Appl.* **5** (1971), 318–319.

[Gabrielov 1973] A. M. Gabrielov, "Formal relations between analytic functions", *Izv. Akad. Nauk SSSR Ser. Mat.* **37** (1973), 1056–1090. In Russian; translated in *Math. USSR Izv.* **7** (1973), 1056–1088.

[Glaeser 1963] G. Glaeser, "Fonctions composées différentiables", *Ann. of Math.* (2) **77** (1963), 193–209.

[Grothendieck 1984] A. Grothendieck, "Esquisse d'un programme", research proposal, 1984. Reprinted as pp. 5–48 of *Geometric Galois actions*, vol. 1, edited by Leila Schneps and Pierre Lochak, London Math. Soc. Lecture Note Series **242**, Cambridge Univ. Press, Cambridge, 1997; translated on pp. 243–283 of same volume.

[Hironaka 1964] H. Hironaka, "Resolution of singularities of an algebraic variety over a field of characteristic zero", *Ann. of Math.* (2) **79** (1964), 109–203, 205–326.

[Hironaka 1973] H. Hironaka, "Subanalytic sets", pp. 453–493 in *Number theory, algebraic geometry and commutative algebra: in honor of Yasuo Akizuki*, edited by Y. Kusunoki et al., Kinokuniya, Tokyo, 1973.

[Hironaka 1974] H. Hironaka, *Introduction to the theory of infinitely near singular points*, Mem. Mat. Instituto Jorge Juan **28**, Consejo Superior de Investigaciones Científicas, Madrid, 1974.

[Hironaka 1986] H. Hironaka, "Local analytic dimensions of a subanalytic set", *Proc. Japan Acad. Ser. A Math. Sci.* **62** (1986), 73–75.

[Izumi 1985] S. Izumi, "A measure of integrity for local analytic algebras", *Publ. Res. Inst. Math. Sci.* **21** (1985), 719–735.

[Izumi 1986] S. Izumi, "Gabrielov's rank condition is equivalent to an inequality of reduced orders", *Math. Ann.* **276** (1986), 81–89.

[Khovanskiĭ 1991] A. G. Khovanskiĭ, *Fewnomials*, Translations of mathematical monographs **88**, Amer. Math. Soc., Providence, RI, 1991.

[Lojasiewicz 1964] S. Lojasiewicz, "Ensembles semi-analytiques", Notes, Inst. Hautes Études Sci., Bures-sur-Yvette, 1964.

[Lojasiewicz 1991] S. Lojasiewicz, *Introduction to complex analytic geometry*, Birkhäuser, Basel, 1991.

[Merrien 1980] J. Merrien, "Faisceaux analytiques semi-cohérents", *Ann. Inst. Fourier (Grenoble)* **30**:4 (1980), 165–219.

[Milman 1978] P. D. Milman, "Analytic and polynomial homomorphisms of analytic rings", *Math. Ann.* **232** (1978), 247–253.

[Mostowski 1985] T. Mostowski, *Lipschitz equisingularity*, Rozprawy Matematyczne (Dissertationes Math.) **243**, Panstwowe Wydawn. Naukowe, Warsaw, 1985.

[Parusiński 1994] A. Parusiński, "Lipschitz stratification of subanalytic sets", *Ann. Sci. École Norm. Sup.* (4) **27** (1994), 661–696.

[Pawłucki 1989] W. Pawłucki, "On relations among analytic functions and geometry of subanalytic sets", *Bull. Polish Acad. Sci. Math.* **37** (1989), 117–125.

[Pawłucki 1990] W. Pawłucki, *Points de Nash des ensembles sous-analytiques*, Mem. Amer. Math. Soc. **425**, Amer. Math. Soc., Providence, RI, 1990.

[Pawłucki 1992] W. Pawłucki, "On Gabrielov's regularity condition for analytic mappings", *Duke Math. J.* **65** (1992), 299–311.

[Piękosz 1998] A. Piękosz, "On semialgebraic points of definable sets", pp. 189–193 in *Singularities symposium Łojasiewicz 70*, edited by B. Jakubczyk et al., Banach Center Publ. **44**, Polish Academy of Sciences, Warsaw, 1998.

[Rees 1989] D. Rees, "Izumi's theorem", pp. 407–416 in *Commutative algebra* (Berkeley, CA, 1987), edited by M. Hochster et al., Math. Sci. Res. Inst. Publ. **15**, Springer, New York, 1989.

[Remmert 1957] R. Remmert, "Holomorphe und meromorphe Abbildungen komplexer Räume", *Math. Ann.* **133** (1957), 328–370.

[Schwarz 1975] G. W. Schwarz, "Smooth functions invariant under the action of a compact Lie group", *Topology* **14** (1975), 63–68.

[Spivak 1994] M. Spivak, *Calculus*, 3rd ed., Publish or Perish, Houston, 1994.

[Spivakovsky 1990] M. Spivakovsky, "On convergence of formal functions: a simple algebraic proof of Gabrielov's theorem", pp. 69–77 in *Seminario di Geometria Reale a.a. 1989/90*, Dipartimento di Matematica, Universitá di Pisa, 1990.

[Tougeron 1971] J.-C. Tougeron, "An extension of Whitney's spectral theorem", *Inst. Hautes Études Sci. Publ. Math.* **40** (1971), 139–148.

[Wang 1995] T. Wang, "Linear Chevalley estimates", *Trans. Amer. Math. Soc.* **347** (1995), 4877–4898.

[Whitney 1943] H. Whitney, "Differentiable even functions", *Duke Math. J.* **10** (1943), 159–160.

[Whitney 1965] H. Whitney, "Local properties of analytic varieties", pp. 205–244 in *Differential and Combinatorial Topology: A symposium in honor of Marston Morse*, edited by S. S. Cairns, Princeton Math. Series **27**, Princeton Univ. Press, Princeton, NJ, 1965.

[Wilkie 1996] A. J. Wilkie, "Model completeness results for expansions of the ordered field of real numbers by restricted Pfaffian functions and the exponential function", *J. Amer. Math. Soc.* **9** (1996), 1051–1094.

EDWARD BIERSTONE
DEPARTMENT OF MATHEMATICS
UNIVERSITY OF TORONTO
TORONTO, ONTARIO M5S 3G3
CANADA
    bierston@math.toronto.edu

PIERRE D. MILMAN
DEPARTMENT OF MATHEMATICS
UNIVERSITY OF TORONTO
TORONTO, ONTARIO M5S 3G3
CANADA
    milman@math.toronto.edu

Model Theory, Algebra, and Geometry
MSRI Publications
Volume **39**, 2000

Arithmetic and Geometric Applications of Quantifier Elimination for Valued Fields

JAN DENEF

ABSTRACT. We survey applications of quantifier elimination to number theory and algebraic geometry, focusing on results of the last 15 years. We start with the applications of p-adic quantifier elimination to p-adic integration and the rationality of several Poincar series related to congruences $f(x) = 0$ modulo a prime power, where f is a polynomial in several variables. We emphasize the importance of p-adic cell decomposition, not only to avoid resolution of singularities, but especially to obtain much stronger arithmetical results. We survey the theory of p-adic subanalytic sets, which is needed when f is a power series instead of a polynomial. Next we explain the fundamental results of Lipshitz–Robinson and Gardener–Schoutens on subanalytic sets over algebraically closed complete valued fields, and the connection with rigid analytic geometry. Finally we discuss recent geometric applications of quantifier elimination over $\mathbb{C}((t))$, related to the arc space of an algebraic variety.

One of the most striking applications of the model theory of valued fields to arithmetic is the work of Ax and Kochen [1965a; 1965b; 1966; Kochen 1975], and of Ershov [1965; 1966; 1967], which provided for example the first quantifier elimination results for discrete valued fields [Ax and Kochen 1966], and the decidability of the field $\mathbb{Q}_p$ of p-adic numbers. As a corollary of their work, Ax and Kochen [1965a] proved the following celebrated result: For each prime number p, big enough with respect to d, any homogeneous polynomial of degree d over $\mathbb{Q}_p$ in $d^2 + 1$ variables has a nontrivial zero in $\mathbb{Q}_p$. However in the present survey we will not discuss this work, but focus on results of the last 15 years.

In Section 1 we explain the applications of p-adic quantifier elimination to p-adic integration and the rationality of several Poincaré series related to a congruence $f(x) \equiv 0 \bmod p^m$, where $f(x)$ is a polynomial in several variables with integer coefficients. We emphasize the importance of p-adic cell decomposition, not only to avoid resolution of singularities, but especially to obtain much stronger results (for example, on local singular series in several variables).

To obtain results similar to those in Section 1, but when f is a power series instead of a polynomial, one needs the theory of p-adic subanalytic sets which we survey in Section 2.

In Section 3 we explain the fundamental results of Lipshitz–Robinson and Gardener–Schoutens on subanalytic sets over algebraically closed nonarchimedean complete valued fields and the connection with rigid analytic geometry. Finally, in Section 4 we discuss recent geometric applications of quantifier elimination over the field $\mathbb{C}((t))$ of Laurent series over $\mathbb{C}$. Here p-adic integration is replaced by "motivic integration", a notion recently introduced by Kontsevich.

1. Integration on Semi-Algebraic Subsets over $\mathbb{Q}_p$

1.1. Motivating Problem. Let $f(x) \in \mathbb{Z}[x], x = (x_1, \ldots, x_n)$. Let p be a prime number and $m \in \mathbb{N}$. Denote the ring of p-adic integers by $\mathbb{Z}_p$ and the field of p-adic numbers by $\mathbb{Q}_p$; see [Koblitz 1977], for example. For $a \in \mathbb{Z}_p$, we denote the image of a in $\mathbb{Z}/p^m\mathbb{Z}$ by $a \bmod p^m$. We use the notations

$$N_m := \text{number of elements in } \{x \in (\mathbb{Z}/p^m\mathbb{Z})^n \mid f(x) \equiv 0 \equiv p^m\},$$

$$\tilde{N}_m := \text{number of elements in } \{x \equiv p^m \mid x \in \mathbb{Z}_p^n, f(x) = 0\},$$

$$P(T) := \sum_{m \in \mathbb{N}} N_m T^m, \qquad \tilde{P}(T) := \sum_{m \in \mathbb{N}} \tilde{N}_m T^m.$$

Borevich and Shafarevich conjectured that $P(T)$ is a rational function of T. This was proved by Igusa [1974; 1975; 1978] using Hironaka's resolution of singularities. Serre [1981, §3] and Oesterlé [1982] investigated the behaviour of $\tilde{N}_m$ for $m \to \infty$, and they asked the question whether $\tilde{P}(T)$ is a rational function of T. This was proved by Denef [1984] using resolution of singularities and Macintyre's Theorem [Macintyre 1976] on quantifier elimination for $\mathbb{Q}_p$. Denef [1984] also gave an alternative proof of the rationality of $P(T)$ and $\tilde{P}(T)$, avoiding the use of resolution of singularities, using instead Macintyre's Theorem and a cell decomposition theorem. We will briefly explain these proofs below.

1.2.1. The p-adic measure. There exists a unique ($\mathbb{R}$-valued Borel) measure on $\mathbb{Q}_p^n$ which is invariant under translation such that $\mathbb{Z}_p^n$ has measure 1. We denote this Haar measure by $|dx| = |dx_1| \cdots |dx_n|$. The measure of $a + p^m\mathbb{Z}_p^n$ equals p^{-mn}, for each $a \in \mathbb{Q}_p^n$, because these sets have the same measure (being translates of $p^m\mathbb{Z}_p^n$) and p^{nm} of them form a partition of $\mathbb{Z}_p^n$. For any measurable $A \subset \mathbb{Q}_p^n$ and $\lambda \in \mathbb{Q}_p$, the measure of $\lambda A = \{\lambda a \mid a \in A\}$ equals the measure of A times $|\lambda|^n$, where $|\lambda|$ denotes the p-adic absolute value $|\lambda| := p^{-\operatorname{ord}\lambda}$, with $\operatorname{ord} : \mathbb{Q}_p \to \mathbb{Z} \cup \{+\infty\}$ the p-adic valuation. We recall that each λ in $\mathbb{Q}_p \setminus \{0\}$ can be written as $\lambda = up^{\operatorname{ord}\lambda}$ with u a unit in the ring $\mathbb{Z}_p$. Integration of (integrable) real valued functions on $\mathbb{Q}_p^n$ is defined in the standard way. As an example we calculate the following integral for $n = 1$:

$$\int_{x \in \mathbb{Z}_p, \, \text{ord} \, x \geq m} |x|^s \, |dx| = \sum_{j \geq m} p^{-sj} \int_{\text{ord} \, x = j} |dx| = \sum_{j \geq m} p^{-sj}(p^{-j} - p^{-j-1})$$

$$= (1 - p^{-1}) p^{-(s+1)m} / (1 - p^{-s-1}),$$

for any nonnegative $s \in \mathbb{R}$.

1.2.2. Rationality of $P(T)$ and $\tilde{P}(T)$. The proof of the rationality of $P(T)$ and $\tilde{P}(T)$ is based on the simple formulas

$$N_m = p^{mn} \text{ measure } (\{x \in \mathbb{Z}_p^n \mid \text{ord} \, f(x) \geq m\}),$$

$$\tilde{N}_m = p^{mn} \text{ measure } (\{x \in \mathbb{Z}_p^n \mid \exists y \in \mathbb{Z}_p^n : f(y) = 0, y \equiv x \equiv p^m\}),$$

which are justified by observing that the set in the right-hand side is a union of respectively N_m and $\tilde{N}_m$ residue classes mod p^m, each having measure p^{-nm}.

The set in the first formula is of a very simple type, but the set in the second is more complicated, involving an existential quantifier. We need Macintyre's Theorem (see Section 1.3 below) on elimination of quantifiers to see that this set is not too complicated, so that its measure (as a function of m) can be controlled. To prove the rationality of $P(T)$ and $\tilde{P}(T)$ one has to know how the measures of the above sets vary with m. This is provided by the Basic Theorem 1.5 below.

1.3. Definable Subsets of $\mathbb{Q}_p$. Let $\mathcal{L}_{\text{Pres}}$ be the (first order) language (in the sense of logic) whose variables run over $\mathbb{Z}$ and with symbols to denote $+, \leq$ $, 0, 1$ and with for each $d = 2, 3, 4, \ldots$ a symbol to denote the binary relation $x \equiv y \equiv d$. Note that in $\mathcal{L}_{\text{Pres}}$ there is no symbol for multiplication. As for any (first order) language, the formulas of $\mathcal{L}_{\text{Pres}}$ are built up in the obvious way from the above specified symbols and variables, together with the logical connectives $\wedge$ (and), $\vee$ (or), $\neg$, the quantifiers $\exists, \forall$, brackets, and $=$. A well-known result of Presburger [1930] states that $\mathbb{Z}$ has elimination of quantifiers in the language $\mathcal{L}_{\text{Pres}}$, meaning that each formula in that language is equivalent (in $\mathbb{Z}$) to a formula without quantifiers. (For readers who are not familiar with this terminology from logic, we refer to [Denef and van den Dries 1988, § 0], where these notions are explained for non-logicians.)

Let $\mathcal{L}_{\text{Mac}}$ be the (first order) language whose variables run over $\mathbb{Q}_p$ and with symbols to denote $+, -, \times, 0, 1$ and with for each $d = 2, 3, 4, \ldots$ a symbol P_d to denote the predicate "x is a d-th power in $\mathbb{Q}_p$". Moreover for each element in $\mathbb{Z}_p$ there is a symbol to denote that element. Macintyre's theorem [1976] states that $\mathbb{Q}_p$ has elimination of quantifiers in the language $\mathcal{L}_{\text{Mac}}$, meaning that each formula in that language is equivalent (in $\mathbb{Q}_p$) to a formula without quantifiers.

Let $\mathcal{L}$ be the (first order) language with two sorts of variables: A first sort of variables running over $\mathbb{Q}_p$, and a second sort of variable running over $\mathbb{Z}$. The symbols of $\mathcal{L}$ consist of the symbols of $\mathcal{L}_{\text{Mac}}$ (for the first sort), the symbols of $\mathcal{L}_{\text{Pres}}$ (for the second sort), and a symbol to denote the valuation function ord : $\mathbb{Q}_p \setminus \{0\} \to \mathbb{Z}$ (from the first sort to the second sort). (We use the convention that $\mathrm{ord}\, 0 = +\infty, (+\infty) + l = +\infty$ and $+\infty \equiv l \bmod d$, for all l in $\mathbb{Z} \cup \{+\infty\}$.) An easy adaptation of Macintyre's proof yields that $\mathbb{Q}_p$ has elimination of quantifiers in the language $\mathcal{L}$; see [Denef 1984, Remark 6.4].

A subset of $\mathbb{Q}_p^n$ is called *semi-algebraic* if it is definable by a quantifier-free formula of $\mathcal{L}_{\mathrm{Mac}}$ (that is, a formula without quantifiers). Every subset of $\mathbb{Q}_p^n$ which is definable in $\mathcal{L}$ is semi-algebraic. This follows from quantifier elimination for $\mathcal{L}$ and the fact that the relation "$\mathrm{ord}\, x \le \mathrm{ord}\, y$" can be expressed in terms of the predicate P_2; see [Denef 1984, Lemma 2.1].

1.4. The Cell Decomposition Theorem

THEOREM [Denef 1984; 1986]. *Let $f_i(x,t) \in \mathbb{Q}_p[x,t]$, where $i = 1,\ldots,m$, $x = (x_1,\ldots,x_{n-1})$, and t is one variable. Fix $d \in \mathbb{N}$ with $d \ge 2$. Then there exists a finite partition of $\mathbb{Q}_p^n$ into subsets (called cells) of the form*

$$A = \left\{(x,t) \in \mathbb{Q}_p^n \mid x \in C \text{ and } |a_1(x)| \,\square_1\, |t - c(x)| \,\square_2\, |a_2(x)| \right\},$$

where C is an $\mathcal{L}$-definable subset of $\mathbb{Q}_p^{n-1}$, each of $\square_1$ and $\square_2$ denotes either $\le$, $<$, or no condition, and $a_1(x), a_2(x), c(x)$ are $\mathcal{L}$-definable functions from $\mathbb{Q}_p^{n-1}$ to $\mathbb{Q}_p$, such that, for all $(x,t) \in A$,

$$f_i(x,t) = u_i(x,t)^d h_i(x)(t - c(x))^{\nu_i}, \text{ for } i = 1,\ldots,m,$$

with $u_i(x,t)$ a unit in $\mathbb{Z}_p$ for all (x,t) in A, $h_i(x)$ an $\mathcal{L}$-definable function from $\mathbb{Q}_p^{n-1}$ to $\mathbb{Q}_p$, and $\nu_i \in \mathbb{N}$.

We recall that a function is called $\mathcal{L}$-definable if its graph is $\mathcal{L}$-definable, meaning that it can be expressed by a formula in the language $\mathcal{L}$.

REMARK. This was first proved in [Denef 1984] using Macintyre's Theorem. Conversely Macintyre's Theorem follows easily from The Cell Decomposition Theorem which can be proved directly using a method due to Cohen [1969]; see [Denef 1986].

1.5. Basic Theorem on p-adic Integration

THEOREM [Denef 1985]. *Let $(A_{\lambda,l})_{\lambda \in \mathbb{Q}_p^k, l \in \mathbb{Z}^r}$ be an $\mathcal{L}$-definable family of bounded subsets of $\mathbb{Q}_p^n$. Then*

$$I(\lambda, l) := \text{measure of } A_{\lambda,l} := \int_{A_{\lambda,l}} |dx|$$

is a $\mathbb{Q}$-valued function of λ, l belonging to the $\mathbb{Q}$-algebra generated by the functions

$$\theta(\lambda, l) \quad \text{and} \quad p^{\theta(\lambda,l)},$$

where θ is $\mathbb{Z}$-valued $\mathcal{L}$-definable.

(Saying that $(A_{\lambda,l})$ is $\mathcal{L}$-definable means that the relation $x \in A_{\lambda,l}$ can be expressed by a formula in the language $\mathcal{L}$ where x, λ are variables running over $\mathbb{Q}_p$ and l are variables running over $\mathbb{Z}$. Saying that θ is $\mathbb{Z}$-valued $\mathcal{L}$-definable means that the relation $z = \theta(\lambda, l)$ can be expressed by a formula in $\mathcal{L}$, where λ are variables running over $\mathbb{Q}_p$ and z, λ are variables running over $\mathbb{Z}$.)

We call the elements of the algebra mentioned in the theorem $\mathcal{L}$-*simple p-exponential functions*, and if there are no variables λ involved we call them $\mathcal{L}_{\mathrm{Pres}}$-*simple p-exponential functions*. The Basic Theorem and its proof also hold for integrals of the form $\int_{A_{\lambda,l}} p^{-\alpha(x,\lambda,l)} |dx|$, with α a *positive* $\mathbb{Z}$-valued $\mathcal{L}$-definable function.

PROOF OF THE BASIC THEOREM. By quantifier elimination $A_{\lambda,l}$ is given by a quantifier-free formula Ψ of $\mathcal{L}$. Let $f_1, f_2, \ldots, f_m$ be the polynomials (in variables of the first sort) which appear in this formula Ψ. We now apply the Cell Decomposition Theorem 1.4 to $f_1, \ldots, f_m$. This enables us to separate off the last variable and integrate first with respect to that variable. The Basic Theorem is obtained by iterating this procedure. For the details we refer to [Denef 1985, § 3], where a similar result is proved.   □

1.6. Meaning of the Basic Theorem with No λ. If in Theorem 1.5 there are no variables λ, then the function $I(l)$ is built from Presburger functions (that is, $\mathcal{L}_{\mathrm{Pres}}$-definable functions from $\mathbb{Z}^r$ to $\mathbb{Z}$) by multiplication, exponentiation, and $\mathbb{Q}$-linear combinations. Such functions $I(l)$ are easy to understand because any Presburger function is piecewise $\mathbb{Q}$-linear, the pieces being Presburger subsets of $\mathbb{Z}^r$ (that is, $\mathcal{L}_{\mathrm{Pres}}$-definable subsets). But Presburger subsets are finite unions of convex polyhedrons intersected with residue classes. A completely elementary argument now yields:

THEOREM 1.6.1. *Assume the notation of Theorem 1.5 with no λ involved. Let* $T = (T_1, \ldots, T_r)$. *Then*

$$\sum_{l \in \mathbb{N}^r} I(l) T^l \in \mathbb{Q}[[T_1, \ldots, T_r]]$$

is a rational function of T.

Actually this holds for any $\mathcal{L}_{\mathrm{Pres}}$-simple p-exponential function $I(l)$.

COROLLARY 1.6.2. *The series $P(T)$ and $\tilde{P}(T)$ from Section 1.1 are rational.*

PROOF. Direct consequence of 1.6.1 and 1.2.2.   □

COROLLARY 1.6.3. *Assume the notation of Section 1.1 and let $N_{m,r}$ be the number of solutions in $\mathbb{Z}/p^m\mathbb{Z}$ of $f(x) \equiv 0 \bmod p^m$ that can be lifted to a solution of $f(x) \equiv 0 \bmod p^{m+r}$ in $\mathbb{Z}/p^{m+r}\mathbb{Z}$. Then $\sum_{m,r \in \mathbb{N}} N_{m,r} T^m U^r$ is a rational function of T, U.*

PROOF. This is a direct consequence of 1.6.1 and the obvious fact that $p^{-mn}N_{m,r}$
equals the measure of the set

$$\{x \in \mathbb{Z}_p^n \mid \exists y \in \mathbb{Z}_p^n : f(y) \equiv 0 \bmod p^{m+r}, \, y \equiv x \bmod p^m\}. \qquad \square$$

The Basic Theorem 1.5 with no λ involved can also be proved without using the
Cell Decomposition Theorem, using instead resolution of singularities. Indeed,
by the p-adic Analytic Resolution Theorem of Section 1.7 below (applied to the
polynomials $f_1, \ldots, f_m$ appearing in a quantifier-free formula Ψ describing A_l),
we can pull back the integral $I(l)$ to the p-adic manifold M. The so obtained
integral on M can be easily evaluated by an elementary local calculation, using
[Denef 1985, Lemma 3.2]. A special case of such a calculation is given in the example of Section 1.2.1. However when there are at least two variables λ involved
(meaning that $k \geq 2$) then I do not know how to prove the Basic Theorem 1.5
without using the Cell Decomposition Theorem (even when $r = 0$).

1.7. Resolution of Singularities (p-adic Analytic Case). Let $U \subset \mathbb{Q}_p^n$ be
open and $f : U \to \mathbb{Q}_p$ a map. We call f analytic if each $a \in U$ has an open
neighbourhood V_a on which f can be written as a power series in $x - a$, with
coefficients in $\mathbb{Q}_p$, which converges for all $x \in V_a$.

By a *p-adic manifold* we mean a p-adic analytic manifold (defined in the same
way as a complex analytic manifold) which is Hausdorff and everywhere of the
same dimension (see [Bourbaki 1967], for example). *Analytic functions* from
a p-adic manifold M_1 to a p-adic manifold M_2 are defined in the obvious way
by working locally. Also the notion of *isomorphic* p-adic manifolds is defined
straightforwardly.

It is easy to verify that each compact p-adic manifold of dimension n is a
disjoint union of a finite number of open compact submanifolds which are isomorphic to $\mathbb{Z}_p^n$.

Let M be a compact p-adic manifold and C a closed submanifold of codimension r at least 2. We refer to [Denef and van den Dries 1988, § 2.1] for the
definition of the *blowing-up of M with respect to C*. This is an analytic map
$h : \tilde{M} \to M$, with $\tilde{M}$ a compact p-adic manifold of the same dimension as M,
such that the restriction $\tilde{M} \setminus h^{-1}(C) \to M \setminus C$ of h is an isomorphism, and which
is constructed in a special way (well-known to geometers). In particular, using
suitable local coordinates, the map h is locally given by

$$(x_1, \ldots, x_n) \mapsto (x_1 x_r, x_2 x_r, \ldots, x_{r-1} x_r, x_r, \ldots, x_n).$$

(In these local coordinates, the submanifold C is locally given by $x_1 = x_2 = \cdots = x_r = 0$.)

p-ADIC ANALYTIC RESOLUTION THEOREM. *Let $f_1, \ldots, f_m : \mathbb{Z}_p^n \to \mathbb{Q}_p$ be analytic functions. Then there exists a compact p-adic manifold M of dimension n
and an analytic map $\pi : M \to \mathbb{Z}_p^n$ such that*

(i) M is the disjoint union of a finite number of clopens $U_i = \mathbb{Z}_p^n$, such that on each U_i, the jacobian of π and all $f_j \circ \pi$ are monomials times analytic functions with constant absolute value.

(ii) π is a composition of finitely many blowing-up maps with respect to closed submanifolds of codimension ≥ 2. In particular π is an isomorphism outside closed sets of measure zero.

This is an easy consequence of Hironaka's embedded resolution of singularities [Hironaka 1964]; see, for example, [Denef and van den Dries 1988, Theorem 2.2].

1.8. Meaning of the Basic Theorem with No l. If in Theorem 1.5 there are no variables l, then the function $I(\lambda)$ is built from $\mathcal{L}$-definable functions $\mathbb{Q}_p^k \to \mathbb{Z}$, by multiplication, exponentiation and $\mathbb{Q}$-linear combinations. Such functions $I(\lambda)$ are easy to understand. Indeed, by [Denef 1984, Theorem 6.3], for any $\mathcal{L}$-definable function $\theta : \mathbb{Q}_p^k \to \mathbb{Z}$ there exists a finite partition of $\mathbb{Q}_p^k$ in semi-algebraic subsets S such that on each such S the function θ is a $\mathbb{Q}$-linear combination of the ord of polynomials over $\mathbb{Q}_p$ with no zeros on S. Applying the Analytic Resolution Theorem (Section 1.7) to the polynomials appearing in the linear combinations and formulas for S above mentioned, and expressing any locally constant function on $\mathbb{Z}_p^\times$ as a $\mathbb{C}$-linear combination of characters (i.e., homomorphisms $\chi : \mathbb{Z}_p^\times \to \mathbb{C}^\times$ with finite image, where $\mathbb{Z}_p^\times$ and $\mathbb{C}^\times$ are the groups of units in the rings $\mathbb{Z}_p$ and $\mathbb{C}$), we obtain:

THEOREM 1.8.1. *Let $I : \mathbb{Z}_p^k \to \mathbb{Q}$ be an $\mathcal{L}$-simple p-exponential function (e.g., the function I (restricted to $\mathbb{Z}_p^k$) in the Basic Theorem 1.5, when there is no l involved). Then there exists a p-adic manifold M of dimension k and an analytic map $\pi : M \to \mathbb{Z}_p^k$, which is the composition of finitely many blowing-up maps with respect to closed submanifolds of codimension ≥ 2, such that locally at each $b \in M$ there exist local coordinates $y_1, \ldots, y_k$ centered at b such that $I \circ \pi$ is a finite $\mathbb{C}$-linear combination of functions of the form*

$$\prod_{i=1}^{k} \chi_i(\mathrm{ac}(y_i))(\mathrm{ord}\, y_i)^{n_i} |y_i|^{\gamma_i}, \qquad (*)$$

where the χ_i are characters on $\mathbb{Z}_p^\times$, $\mathrm{ac}(y_i) := y_i p^{-\,\mathrm{ord}\, y_i}$ denotes the angular component of $y_i \in \mathbb{Q}_p$, the n_i are in $\mathbb{N}$, and the γ_i are in $\mathbb{C}$. (Here we use the following conventions: $\chi(\mathrm{ac}(0)) = 0$ if χ is a nontrivial character, $\chi(\mathrm{ac}(0)) = 1$ if χ is the trivial character 1; and $(\mathrm{ord}\, 0)^{n_i} |0|^{\gamma_i} = 0$, unless $n_i = \gamma_i = 0$ in which case it equals 1.)

REMARK. Working with complex exponents in $(*)$ we are able to express, for example, the function $g : \mathbb{Z}_p \setminus \{0\} \to \mathbb{Q}$ with $g(x) = 1$ if $(\mathrm{ord}\, x) \equiv 0 \equiv d$ and $g(x) = 0$ otherwise.

Application to the Local Singular Series in Several Variables. Let

$$f = (f_1, \ldots, f_k) \in (\mathbb{Z}_p[x])^k,$$

with $x = (x_1, \ldots, x_n)$. Let $a = (a_1, \ldots, a_k) \in \mathbb{Z}_p^k$ be a regular value of $f : \mathbb{Z}_p^n \to \mathbb{Z}_p^k$, this means that a belongs to the image $f(\mathbb{Z}_p^n)$ but is not the image of any point in $\mathbb{Z}_p^n$ where the Jacobian of f has rank $< k$. Then it is known [Igusa 1978] that

$$p^{-m(n-k)} \#\{x \in (\mathbb{Z}/p^m)^n \mid f_i(x) \equiv a_i \bmod p^m \text{ for } i = 1, \ldots, k\}$$

is constant for m big enough. (Here $\#$ stands for the number of elements.) We denote this constant value by $F(a)$. The function $\lambda \mapsto F(\lambda)$, for λ a regular value of f, is called the local singular series of $f : \mathbb{Z}_p^n \to \mathbb{Z}_p^k$ and plays an important role in number theory (for example, for the circle method). We put $F(\lambda) = 0$ if λ is not a regular value; thus F is a $\mathbb{Q}$-valued function on $\mathbb{Z}_p^k$. It is easy to see that $F(\lambda)$ is a locally constant function in the neighbourhood of any regular value a of f. But if λ tends to a nonregular value c, then $F(\lambda)$ has a nontrivial singular behavior. For $k = 1$, this has been studied in depth by Igusa [1974; 1975; 1978], who obtained an asymptotic expansion of $F(\lambda)$ for $\lambda \to c$. His work is based on Mellin inversion over p-adic fields and the study of local zeta functions using resolution of singularities. Igusa [1978, p. 32] asked how one could extend his result to the general case $k > 1$. A contribution to Igusa's question is given by

COROLLARY 1.8.2. *The local singular series $F(\lambda)$ is an $\mathcal{L}$-simple p-exponential function of $\lambda = (\lambda_1, \ldots, \lambda_k) \in \mathbb{Z}_p^k$. Hence Theorem 1.8.1 applies to $I(\lambda) := F(\lambda)$.*

PROOF. This follows from Theorem 1.5 and the simple fact that

$$F(\lambda) = \int_{x \in \mathbb{Z}_p^n,\, f(x) = \lambda} |dx / (df_1 \wedge \cdots \wedge df_k)|,$$

whenever λ is a regular value of f. $\square$

Some first results on local singular series in several variables were obtained by Loeser [1989], who conjectured that Theorem 1.8.1 holds for $I(\lambda) := F(\lambda)$ with π being an isomorphism above the set of regular values, when $f = (f_1, \ldots, f_k)$ satisfies some nice geometric conditions (for example when the fibers of $f : \bar{\mathbb{Q}}_p^n \to \bar{\mathbb{Q}}_p^k$ are $(n-k)$-dimensional complete intersections with only isolated singularities, where $\bar{\mathbb{Q}}_p$ denotes the algebraic closure of $\mathbb{Q}_p$). Loeser's conjecture has several important implications and is still wide open. Indeed Corollary 1.8.2 does not yield any information about where π is locally an isomorphism. Very recently Lichtin [$\geq$ 2000a; $\geq$ 2000b] obtained explicit results assuming $k = 2$ together with some other hypothesises. It was only after seeing Lichtin's results that I obtained Theorem 1.8.1 and Corollary 1.8.2. I do not know how to prove Corollary 1.8.2 (for $k \geq 2$) without using the Cell Decomposition Theorem. The problem of relating the γ_i in Theorem 1.8.1 to geometric invariants remains open, although Lichtin [$\geq$ 2000a; $\geq$ 2000b] achieved a first breakthrough. Much remains to be done. Moreover Lichtin's method also has important applications in analysis and geometry.

Applications to Ax–Kochen-Definable Subsets. Let A be an $\mathcal{L}$-definable subset of $\mathbb{Q}_p^n$, then

$$\tilde{P}_A(T) := \sum_{m \in \mathbb{N}} (\#\{x \bmod p^m \mid x \in A\}) T^m$$

is a rational function of T, the proof being the same as for $\tilde{P}(T)$. This can be proved without the Cell Decomposition Theorem (using instead resolution of singularities and quantifier elimination; compare Section 1.6). By contrast, it was proved in [Denef 1985] that, if we take for A a subset definable in the language of Ax and Kochen [1966], then $\tilde{P}_A(T)$ is still rational, but in this case the Cell Decomposition Theorem seems to be essential. (The language of Ax and Kochen is equivalent to the language obtained from $\mathcal{L}$ by adjoining a symbol for the function $\mathbb{Z} \to \mathbb{Q}_p : m \mapsto p^m$ from the second sort to the first sort.)

1.9. Dependence on p. It is well kown that $\mathbb{Q}_p$ does not have a quantifier elimination in $\mathcal{L}_{\mathrm{Mac}}$ or $\mathcal{L}$ which holds for all p (or for almost all p). To have a uniform quantifier elimination one has to work in a more complicated language (and here it becomes tedious to avoid the logical terminology of languages.) For such a quantifier elimination and its applications to integration we refer to [Pas 1989; 1990; 1991; Macintyre 1990].

1.10. Igusa's Local Zeta Function. Let $f(x) \in \mathbb{Z}[x], x = (x_1, \dots, x_n)$. Igusa's local zeta function (for the trivial character) is the function

$$Z(s) := \int_{\mathbb{Z}_p^n} |f(x)|^s \, |dx|,$$

for $s \in \mathbb{C}$ with $\mathrm{Re}(s) \geq 0$. It is an easy excercise to verify that $P(p^{-n-s}) = (1 - p^{-s} Z(s))/(1 - p^{-s})$. The rationality of $P(T)$ is equivalent to $Z(s)$ being a rational function of p^{-s}. It was in this way that Igusa [1974; 1975; 1978] proved that rationality of $P(T)$, by applying a resolution of singularities $\pi : M \to \mathbb{Z}_p^n$ as in Section 1.7, and pulling back the integral $Z(s)$ through π, so obtaining a very simple integral on M whose calculation is a straightforward exercise (compare the example in Section 1.2.1). There are fascinating conjectures about $Z(s)$, such as the monodromy and holomorphy conjectures, which relate the poles of $Z(s)$ (and hence the poles of $P(T)$) to topological invariants of the singularities of $\{x \in \mathbb{C}^n \mid f(x) = 0\}$. For all these and the many geometric and arithmetic results related to this we refer to the survey papers [Denef 1991; Igusa 1987; 1996; Veys 1996], and to the articles [Veys 1993; 1997].

1.11. Integration on Orbits. Let G be an algebraic group (defined over $\mathbb{Q}_p$) acting (algebraically) on the affine n-space (over $\mathbb{Q}_p$). Let $U \subset \mathbb{Q}_p^n$ be a $G(\mathbb{Q}_p)$-orbit (where $G(\mathbb{Q}_p)$ denotes the group of $\mathbb{Q}_p$-rational points on G). Igusa [1984] considered the orbital integral $Z_U(s) = \int_{U \cap \mathbb{Z}_p^n} |f(x)|^s \, |dx|$ which plays an essential role in several investigations (for example, study of the Γ-matrix of a prehomogeneous vectorspace [Sato 1989]). For this work it is essential to know that $Z_U(s)$ is a rational function of p^{-s}. The rationality is proved by using quantifier elimination: Indeed,

$$Z_U(s) = \sum_{m \in \mathbb{N}} \left(\int_{\substack{U \cap \mathbb{Z}_p^n \\ \mathrm{ord}\, f(x)=m}} |dx| \right) (p^{-s})^m,$$

so that we can apply Theorem 1.6.1, since the orbit U is definable by an existential $\mathcal{L}$-formula.

2. Integration on Subanalytic Sets over $\mathbb{Q}_p$

2.1. Motivating Problem. Let $P(T)$ and $\tilde{P}(T)$ be as in Section 1.1, but now with $f(x)$ a power series over $\mathbb{Z}_p$ which converges on $\mathbb{Z}_p^n$. Again we can ask whether $P(T)$ and $\tilde{P}(T)$ are rational. And indeed they are rational. For $P(T)$ this can be proved by adapting Igusa's method in a straightforward way; compare Section 1.10. Concerning $\tilde{P}(T)$, we have a problem in adapting the proof in §1: the set $\{x \in \mathbb{Z}_p^n \mid \exists y \in \mathbb{Z}_p^n : f(y) = 0, y \equiv x \equiv p^m\}$ is in general not $\mathcal{L}$-definable when f is a power series. For this reason we have to introduce analytic functions in our language.

2.2. The Languages $\mathcal{L}_{\mathrm{an}}$ and $\mathcal{L}_{\mathrm{an}}^D$. We continue to use the language $\mathcal{L}$ introduced in 1.3, but from now on the variables of the first sort will run over $\mathbb{Z}_p$ (instead of over $\mathbb{Q}_p$ in §1). Thus quantifiers with respect to variables of the first sort will always run over $\mathbb{Z}_p$ instead of over $\mathbb{Q}_p$. (Otherwise existential formulas in $\mathcal{L}_{\mathrm{an}}$ could define very pathological sets, if we also allowed symbols for analytic functions in these variables.)

Let $\mathcal{L}_{\mathrm{an}}$ be the (first order) language (in the sense of logic) obtained from $\mathcal{L}$ by adding a symbol for each analytic function $g : \mathbb{Z}_p^n \to \mathbb{Z}_p$.

Let $\mathcal{L}_{\mathrm{an}}^D$ be the language obtained from $\mathcal{L}_{\mathrm{an}}$ by adding a symbol D for the function (truncated division)

$$D : \mathbb{Z}_p \times \mathbb{Z}_p \to \mathbb{Z}_p : (x,y) \mapsto \begin{cases} x/y & \text{if } y \neq 0 \text{ and } |x| \leq |y|, \\ 0 & \text{otherwise.} \end{cases}$$

Let S be a subset of $\mathbb{Z}_p^n$. We call S *semi-analytic in* $\mathbb{Z}_p^n$, if S is definable by a quantifier-free formula of $\mathcal{L}_{\mathrm{an}}$. We say that S is *D-semi-analytic in* $\mathbb{Z}_p^n$ if it is definable by a quantifier-free formula of $\mathcal{L}_{\mathrm{an}}^D$. Finally, we call S *subanalytic in* $\mathbb{Z}_p^n$ if it is definable by an existential formula of $\mathcal{L}_{\mathrm{an}}$. (A formula is called existential if it is obtained from a quantifier-free formula by putting some existential quantifiers in front of it.)

Let S be a subset of a p-adic manifold M, and $a \in M$. We say that S is *blue in M at a*, where "blue" is one of the three above properties, if a has an open neighbourhood $U \cong \mathbb{Z}_p^n$ in M such that $S \cap U$ is blue. We call S *blue in M* if S is blue in M at each $a \in M$. Note that the subanalytic subsets of M are precisely the images of semi-analytic sets under proper analytic maps.

2.3. The p-adic Analytic Elimination Theorem

THEOREM [Denef and van den Dries 1988; Denef 1988]. $\mathbb{Z}_p$ *has elimination of quantifiers in* $\mathcal{L}_{\mathrm{an}}^D$.

Easy examples show that $\mathbb{Z}_p$ has no quantifier elimination in $\mathcal{L}_{\mathrm{an}}$.

2.4. COROLLARY. (i) *A subset of* $\mathbb{Z}_p^n$ *is subanalytic in* $\mathbb{Z}_p^n$ *if and only if it is*
   D-semi-analytic in $\mathbb{Z}_p^n$.
(ii) *Each* $\mathcal{L}_{\mathrm{an}}^D$-*definable subset of* $\mathbb{Z}_p^n$ *is subanalytic in* $\mathbb{Z}_p^n$.
(iii) *The complement and the closure of a subanalytic subset in a p-adic manifold*
   are again subanalytic.

2.5. About the Proof of Theorem 2.3. It suffices to prove that every subanalytic subset of $\mathbb{Z}_p^n$ is D-semi-analytic. Consider for *example* a subanalytic set $S \subset \mathbb{Z}_p^n$ of the form

$$S = \{x = (x_1, \ldots, x_n) \in \mathbb{Z}_p^n \mid \exists y = (y_1, \ldots, y_m) \in \mathbb{Z}_p^m : f(x,y) = 0\},$$

with $f = \sum_{i \in \mathbb{N}^m} a_i(x) y^i$ a power series over $\mathbb{Z}_p$ which converges on $\mathbb{Z}_p^{n+m}$, and $f \not\equiv 0 \bmod p$. If f where regular in y_m (meaning that $f \equiv p$ is a monic polynomial in y_m over $\mathbb{Z}/p\mathbb{Z}[[x, y_1, \ldots, y_{m-1}]]$), then, by a well-known p-adic version of the Weierstrass Preparation Theorem, we could write $f = ug$, with u having no zeros in $\mathbb{Z}_p^{n+m}$ and g a polynomial with respect to the last variable y_m. (Both u and g are power series over $\mathbb{Z}_p$ which converge on $\mathbb{Z}_p^{n+m}$.) Hence we could apply quantifier elimination in the language $\mathcal{L}_{\mathrm{Mac}}$ to get rid of the quantifier $\exists y_m$. Although there exists an invertible change of the variables (x, y) which makes f regular in y_m, this is of no help because we are not allowed to mix the variables x and y. However, dividing f by a coefficient $a_j(x)$ with maximal absolute value (depending on $x \in \mathbb{Z}_p^n$, using case distinction), we can nevertheless apply the Weierstrass Preparation Theorem after an invertible transformation of only the y variables (which is certainly permitted). Divisions by $a_j(x)$ introduce the D-functions. In order to apply the D-function only a finite number of times, one has to express all the $a_i(x)$ as linear combinations of only finitely many of them, which is possible by Noetherianness. See [Denef and van den Dries 1988] for the details of the proof, which are somewhat lengthy. $\square$

2.6. Basic Theorem on p-adic integration (analytic case)

THEOREM. *Theorem 1.5 with no λ involved (and hence also Theorem 1.6.1) still holds if we replace $\mathbb{Q}_p$ by $\mathbb{Z}_p$ and "$\mathcal{L}$-definable" by "$\mathcal{L}_{\mathrm{an}}$-definable".*

An easy adaptation of the proof of this theorem shows that $\int_{A_l} p^{-\theta(x,l)} |dx|$ is an $\mathcal{L}_{\mathrm{Pres}}$-simple p-exponential function, whenever $A_l \subset \mathbb{Z}_p^n$ and $\theta : \mathbb{Z}_p^n \times \mathbb{Z}^r \to \mathbb{N}$ are $\mathcal{L}_{\mathrm{an}}$-definable. (Here $l = (l_1, \ldots, l_r)$ are $\mathbb{Z}$-variables).

COROLLARY [Denef and van den Dries 1988]. $\tilde{P}(T)$ *is rational.*

PROOF OF THEOREM 2.6. The next Theorem reduces it to Theorem 1.5, by pulling back the integral through π. □

REMARK. We expect that Theorem 2.6 remains true when there are variables λ involved as in 1.5, but the above proof collapses in this case. Probably a proof can be obtained using the Cell Decomposition Theorem 1.5 and the method in Section 2.5 of [van den Dries 1992].

2.7. Uniformization Theorem for Subanalytic Sets

THEOREM [Denef and van den Dries 1988]. *Let $A \subset \mathbb{Z}_p^n$ be subanalytic in $\mathbb{Z}_p^n$. Then there exists a compact p-adic manifold M of dimension n and an analytic map $\pi : M \to \mathbb{Z}_p^n$ satisfying these conditions:*

(i) *$\pi^{-1}(A)$ is semi-analytic, and actually semi-algebraic on each $U_i = \mathbb{Z}_p^n$ in a suitable decomposition of M as disjoint union of compact open subsets U_i.*

(ii) *π is a composition of finitely many blowing-up maps with respect to closed submanifolds of codimension ≥ 2. In particular π is an isomorphism outside closed sets of measure zero.*

Moreover the same holds if A depends in an $\mathcal{L}_{an}$-definable way on a parameter $l \in \mathbb{Z}^r$ (replacing "semi-analytic", resp. "semi-algebraic", by "definable by a quantifier-free formula in $\mathcal{L}_{an}$, resp. $\mathcal{L}$, involving the parameter l). We can also require that on each U_i the Jacobian of π equals a monomial times an analytic function with constant absolute value.

The proof of Theorem 2.7 is based on the fact that A is D-semi-analytic and on an induction on the number of occurences of D in the description of A, using p-adic analytic resolution (Section 1.7).

2.8. THEOREM [Denef and van den Dries 1988]. *A subanalytic subset of $\mathbb{Z}_p^2$ is semi-analytic.*

PROOF. Follows from Theorem 2.7 taking advantage of the simple nature of blowing-ups of $\mathbb{Z}_p^2$. □

2.9. Further Results. Using the above theorems one can prove (see [Denef and van den Dries 1988]) that subanalytic sets have many good properties: finite stratification in subanalytic manifolds, good dimension theory, Łojasiewicz inequalities, rationality of Łojasiewicz exponents, existence of a uniform bound for the cardinality of the finite members of a subanalytic family of subanalytic sets, semi-analytic nature of one-dimensional subanalytic sets, etc. Finally we mention the result of Z. Robinson [1993] that the singular locus of a subanalytic set is subanalytic.

To make the Analytic Elimination Theorem 2.3 uniform in p, one has to work in a more complicated language; see [van den Dries 1992].

2.10. Application to Counting Subgroups. For a group G and an integer $n \geq 1$, let $a_n(G)$ be the number of subgroups of index n in G. For a finitely generated group or for compact p-adic analytic group this number $a_n(G)$ is always finite (see [Grunewald, Segal, and Smith 1988; du Sautoy 1993], for example).

THEOREM 2.10.1 [Grunewald, Segal, and Smith 1988].   *If G is a torsion-free finitely generated nilpotent group, then $\sum_m a_{p^m}(G)T^m$ is rational, for each prime number p.*

THEOREM 2.10.2 [du Sautoy 1993].   *If G is a compact p-adic analytic group then $\sum_m a_{p^m}(G)T^m$ is rational.*

Theorem 2.10.1 is proved by expressing $a_{p^m}(G)$ in terms of a p-adic integral $\int_{A_m} p^{-\theta(x)} |dx|$ with $(A_m)_{m \in \mathbb{N}}$ and θ definable in $\mathcal{L}$. The proof of 2.10.2 is based on the same idea, with $\mathcal{L}_{an}$ replacing $\mathcal{L}$.

3. Subanalytic Sets over $\mathbb{C}_p$ and Rigid Analytic Geometry

3.1. Definition of $\mathbb{C}_p$. $\mathbb{C}_p$ is the completion of the algebraic closure $\bar{\mathbb{Q}}_p$ of $\mathbb{Q}_p$: The valuation ord on $\mathbb{Q}_p$ extends to a valuation ord on $\bar{\mathbb{Q}}_p$, taking values in $\mathbb{Q}$. This yields a norm $|\cdot| = p^{-\operatorname{ord}(\cdot)}$ on $\bar{\mathbb{Q}}_p$, and we can take the completion $\mathbb{C}_p$ of $\bar{\mathbb{Q}}_p$ with respect to this norm. One verifies that $\mathbb{C}_p$ is a nonarchimedean normed field and that $\mathbb{C}_p$ is algebraically closed. Most of what follows holds for any algebraically closed nonarchimedean complete normed field, except possibly Theorem 3.9 where we have to assume at this moment that the characteristic is zero to apply resolution of singularities.

NOTATION. Put $R = \{x \in \mathbb{C}_p \mid \operatorname{ord} x \geq 0\}$.

3.2. Motivating Problem. Let $f : R^m \to R^n$ be "analytic" (we will discuss in 3.3 below what we mean by "analytic"). What can be said about the image $f(R^m)$ of f? Can one make $f(R^m)$ semi-analytic by blowing-ups? The work of Lipshitz, Robinson, Gardener and Schoutens yields analogiess over $\mathbb{C}_p$ for most of the p-adic results in § 2, but the proofs are much more complicated.

3.3. First Motivation for Rigid Analysis. If in 3.2 we define "analytic" in the local sense (namely, that each point $a \in R^m$ has an open neighbourhood U_a in R^m on which f can be written as a converging power series), then any non-empty countable subset of R^n can be obtained as the image of a suitable "analytic" map $f : R^m \to R^n$. (The reason is that R^m is the disjoint union of infinitely many clopen subsets.) With this definition of "analytic" we obtain very "pathological" sets as images. To avoid this we will require that f is rigid analytic.

A *rigid analytic* function $h : R^m \to \mathbb{C}_p$ is a function which is given by a power series over $\mathbb{C}_p$ which converges on R^m. We denote the ring consisting of these functions by

$$\mathbb{C}_p\langle X_1, \ldots, X_m\rangle := \{h : R^m \to \mathbb{C}_p \mid h \text{ is rigid analytic}\}.$$

This is called a Tate algebra, and is a Noetherian unique factorization domain (see [Bosch et al. 1984], for example).

3.4. The Languages L_{an} and L_{an}^D. Let L be the (first order) language (in the sense of logic) whose variables run over R, and with symbols to denote $+, -, \times, 0, 1$ and the binary relation $|x| \leq |y|$. It follows from a well-known result of A. Robinson [1956] that R has quantifier elimination in the language L. (In that paper Robinson only proves model completeness for the theory of algebraically closed valued fields. But since this theory satisfies the prime extension property, its model completeness actually implies elimination of quantifiers; see [van den Dries 1978], for example.)

Let L_{an} be the (first order) language obtained from L by adding a symbol for each rigid analytic function $f : R^m \to R$. Easy examples show that R has no quantifier elimination in L_{an}.

Let L_{an}^D be the language obtained from L_{an} by adding a symbol D for the function (truncated division)

$$D : R \times R \to R : (x, y) \mapsto \begin{cases} x/y & \text{if } y \neq 0 \text{ and } |x| \leq |y|, \\ 0 & \text{otherwise.} \end{cases}$$

Let A be a subset of R^n. We call A *globally semi-analytic in R^n*, resp. *D-semi-analytic in R^n*, if A is definable by a quantifier-free formula of L_{an}, resp. L_{an}^D. We call A *(rigid) subanalytic in R^n*, if it is definable by an existential formula of L_{an}.

3.5. The Main Theorems

THEOREM 3.5.1 (MODEL COMPLETENESS THEOREM [Lipshitz and Robinson 1996a]). *R is model complete in L_{an}, meaning that any formula in L_{an} is equivalent (for R) to an existential formula in L_{an}.*

Some of the ingredients in the proof of this theorem are discussed in Section 3.10 below.

COROLLARY 3.5.2. (i) *Each L_{an}-definable subset of R^n is subanalytic in R^n.*
(ii) *The complement and the closure (with respect to the norm topology) of a subanalytic subset of R^n are again subanalytic.*

THEOREM 3.5.3 (RIGID ANALYTIC ELIMINATION THEOREM [Gardener and Schoutens $\geq$ 2000]). *R has quantifier elimination in L_{an}^D.*

Some of the ideas in the proof of this theorem are discussed in Section 3.11 below.

REMARKS 3.5.4. Theorem 3.5.1 is a direct consequence of Theorem 3.5.3, but 3.5.3 uses 3.5.1 in its proof. Lipshitz [1993] proved already much earlier that R

has quantifier elimination in the language L_{sep}^D (see Section 3.10 below), which is richer than L_{an}^D. This important result of Lipshitz is at the same time stronger and weaker than Theorem 3.5.3.

COROLLARY 3.5.5. (i) *A subset of R^n is subanalytic in R^n if and only if it is D-semi-analytic in R^n.*

(ii) *The image of a rigid analytic map $R^m \to R^n$ is D-semi-analytic.*

REMARK 3.5.6. Using the theorems above one proves [Lipshitz 1993; Lipshitz and Robinson 1996a; Lipshitz and Robinson 1999] that subanalytic sets in R^n have many good properties. In particular all the results mentioned in Section 2.9 remain valid.

3.6. Further Motivation for Rigid Analysis. In the p-adic case any subanalytic subset of $\mathbb{Z}_p^2$ is semi-analytic. It is not true that any subanalytic subset of R^2 is globally semi-analytic. The reason is that the definition of "global semi-analytic" is too rigid. We need a more local definition. If we make the definition completely local, then we lose information and projections of semi-analytic sets would become pathological in some cases. Therefore we define a subset A of R^n to be (rigid) semi-analytic if there exists a finite covering of R^n by *admissible* open sets U in R^n, such that on each such U, $A \cap U$ is a finite boolean combination of sets of the form $\{x \in R^n \mid |f(x)| \le |g(x)|\}$ with f, g rigid analytic on U. We still have to define the notions "admissible open in R^n" and "rigid analytic function on an admissible open". We give these definitions in Section 3.7 below. They are the key notions of rigid analysis and rigid analytic geometry. With these definitions, subanalytic subsets of R^2 are indeed semi-analytic; see Theorem 3.8.

3.7. First Steps in Rigid Analysis. A (reduced) *affinoid variety V* is a subset of some R^n of the form

$$V = \{x \in R^n \mid f_1(x) = \cdots = f_r(x) = 0\},$$

where the f_i are rigid analytic functions on R^n. The elements of V are in one-one correspondence with the maximal ideals of the affinoid algebra $A :=$ $\mathbb{C}_p\langle X_1, \ldots, X_n\rangle/(f_1, \ldots, f_r)$.

A *rigid analytic function* on V is the restriction to V of a rigid analytic function on R^n. A *morphism $f : W \to V$* of affinoid varieties is a map $f = (f_1, \ldots, f_n)$ with each f_i rigid analytic.

A *rational subdomain U* of an affinoid variety V is a subset of V of the form

$$U = \{x \in V \mid |\, p_i(x)| \le |p_0(x)|, \quad \text{for } i = 1, \cdots, s\}, \qquad (*)$$

where $p_0, p_1, \ldots, p_s$ are rigid analytic functions on V with no common zero in V. Note that U is open and closed in V. Moreover U is actually an affinoid variety, its points being in 1-1 correspondence with $\{(x, t_1, \cdots, t_s) \mid x \in V, p_i - p_0 t_i = 0\}$.

Note that a "closed" disc $\{x \in V | \ |x - a| \leq |r|\}$, and the complement $\{x \in V \ | \ |x - a| \geq |r|\}$ of an "open" disc, with $a, r \in \mathbb{C}_p$, are rational subdomains of V. Moreover the intersection of two rational subdomains is again a rational subdomain.

A *rigid analytic function* on the rational subdomain U of V is a function of the form

$$f(x, p_1/p_0, \ldots, p_s/p_0),$$

with $f(x, t_1, \ldots, t_s) \in \mathbb{C}_p\langle x, t_1, \ldots, t_s \rangle$ and $p_0, p_1, \ldots, p_s$ as in $(*)$.

An *admissible open* of an affinoid variety V is a rational subdomain of V, and an *admissible covering* of an admissible open U is a finite covering of U consisting of rational subdomains of V.

THEOREM (TATE). *Let $U_1, U_2, \ldots, U_k$ be an admissible cover of an affinoid variety V. Let $f : V \to \mathbb{C}_p$ be a function whose restriction to each U_i is rigid analytic. Then f is rigid analytic.*

A *quasi-compact rigid analytic variety* is obtained by "gluing together" a finite number of affinoid varieties (see [Bosch et al. 1984] for the details).

The preceding notions are the cornerstones of rigid analysis and rigid analytic geometry, founded by J. Tate. Basic references are [Bosch et al. 1984; Fresnel and van der Put 1981].

The definition of semi-analytic subsets of R^n given in Section 3.6 (based on the notion of admissible open given above) extend in the obvious way to the notion of semi-analytic subsets of an affinoid variety V.

3.8. THEOREM [Gardener and Schoutens $\geq$ 2000]. *Let $A \subset R^2$ be subanalytic in R^2. Then A is semi-analytic.*

The proof is based on the following theorem of Gardener and Schoutens [$\geq$ 2000], and on the simple nature of blowing-ups of R^2.

3.9. Uniformization Theorem for Rigid Subanalytic Sets

THEOREM. *Let $A \subset R^n$ be subanalytic in R^n. Then there exist a finite number of morphisms $f_i : V_i \to R^n$ with the following properties:*

(i) V_i *is an affinoid variety and f_i is a composition of smooth local blowing-up maps. (By a smooth local blowing-up map we mean the restriction to an open affinoid subvariety of a blowing-up map (in the sense of rigid analytic geometry) with respect to a smooth center of codimension at least 2.)*

(ii) $\bigcup_i f_i(V_i) = R^n$.

(iii) $f_i^{-1}(A)$ *is semi-analytic in V_i.*

The proof (see [Gardener and Schoutens $\geq$ 2000]) is not difficult, since we know already by Corollary 3.5.5 that A is D-semi-analytic, and by resolution of singularities it can be proved that D-semi-analytic sets can be made semi-analytic by smooth local blowing-ups (compare the proof of Theorem 2.7).

3.10. Ideas in the Proof of the Model Completeness Theorem 3.5.1

3.10.1. The Languages L_{sep} and L_{sep}^D. Let L_{sep} be the (first order) language obtained from L by introducing a second sort of variables running over $\mathcal{P} := \{x \in R \mid |x| < 1\}$ and by adding a symbol for each function $f : R^n \times \mathcal{P}^m \to R$ with $f \in \mathbb{C}_p\langle X_1, \ldots, X_n\rangle[[Y_1, \ldots, Y_m]]_s$. Here $\mathbb{C}_p\langle X\rangle[[Y]]_s$ is the ring of *separated power series*, which is a Noetherian subring of $R < X > [[Y]] \otimes_R \mathbb{C}_p$, where $R\langle X\rangle$ denotes the ring of power series over R which converge on R^n. We refer to [Lipshitz 1993; Lipshitz and Robinson $\geq$ 2000] for the exact definition. The restriction to separated power series is essential to avoid pathologies. At any rate we have $\mathbb{Z}_p\langle X\rangle[[Y]] \subset \mathbb{C}_p\langle X\rangle[[Y]]_s$. A nonzero separated power series $f(Y_1)$ in one variable has only a finite number of zeroes in R. (This can fail when $f(Y_1)$ is not separated.) A systematic study of the rings of separated power series has been made by Lipshitz and Robinson in their fundamental paper [Lipshitz and Robinson $\geq$ 2000].

Let L_{sep}^D be the (first order) language obtained from L_{sep} by adding a symbol for the function D defined in 3.4, and a symbol for the function

$$D_0 : R \times R \to \mathcal{P} : (x, y) \mapsto \begin{cases} x/y & \text{if } y \neq 0 \text{ and } |x| \leq |y|, \\ 0 & \text{otherwise.} \end{cases}$$

3.10.2. THEOREM [Lipshitz 1993]. *R has elimination of quantifiers in L_{sep}^D.*

The proof uses ideas from the proof of the p-adic Analytic Elimination Theorem 2.3, but is much more complicated. Variants of the Weierstrass Preparation Theorem play an important role.

3.10.3. The language L_E^D. Let L_E^D be the sublanguage of L_{sep}^D having a symbol for $f : R^n \times \mathcal{P}^m \to R$ only if f and all its partial derivatives are definable by existential formulas of L_{an}. The set of these functions is denoted by E. To be fully correct one should include more local functions as well, which only converge on $U \times \mathcal{P}^m$, with U a rational subdomain of R^n. (When $\mathbb{C}_p$ is replaced by an algebraically closed nonarchimedean normed field of nonzero characteristic one has to modify the definition of L_E^D slightly.)

3.10.4. THEOREM [Lipshitz and Robinson $\geq$ 2000; 1996a]. *R has quantifier elimination in L_E^D.*

The proof is based on the verification that in the proof of the quantifier elimination for L_{sep}^D one only needs functions in E. For this, one has (among other things) to prove a Weierstrass Preparation Theorem for E.

3.10.5.   Note now that the Model Completeness Theorem 3.5.1 is a direct consequence of the above Theorem 3.10.4.

3.11. Some ideas in the proof of the Rigid Analytic Elimination Theorem 3.5.3. We have to prove that any subanalytic subset A of R^n is D-semi-analytic. By Corollary 3.5.2 of the Model Completeness Theorem, we may

suppose that A is closed in R^n. Indeed the closure $\bar{A}$ of A and $\bar{A} \setminus A$ are suban-alytic and $\dim(\bar{A} \setminus A) < \dim A$ (see [Lipshitz and Robinson 1996a]), so that we can use induction.

An easy argument shows that a closed subanalytic subset of R^n is "almost" the image $f(X)$ of a morphism $f : X \to R^n$ with X an affinoid variety.

Recall that a morphism $f : X \to Y$ is called flat if, for each point x in X, the local ring of X at x is flat over the local ring of Y at $f(x)$. When $f : X \to Y$ is a flat morphism of affinoid varieties, a theorem of Raynaud and Mehlmann states that $f(X)$ is a finite union of rational subdomains of Y, hence D-semi-analytic.

When f is not flat, Gardener and Schoutens [$\geq$ 2000] proved using results from [Gardener 2000; Schoutens 1999] that one can make f flat by taking its strict transform under a suitable finite sequence of local blowing-ups. This Flattening Theorem is an analogy of a difficult result of Hironaka in real analytic geometry. The adaptation to the rigid analytic case is difficult and is based on Berkovich's approach [1990] to rigid analytic geometry. Since the image of a D-semi-analytic set under a local blowing-up map is D-semi-analytic (up to a subanalytic subset in an affinoid variety of smaller dimension), the Flattening Theorem (and some extra work) reduces us to the case that f itself is flat, which we considered already.

4. Semi-algebraic Sets over $\mathbb{C}((t))$ and Motivic Integration

4.1. Motivating Problem. Let $f(x) \in \mathbb{C}[x]$, with $x = (x_1, \ldots, x_n)$. We use the notations

$X := \{x \in \mathbb{C}^n \mid f(x) = 0\}$, the hypersurface defined by f,

$\mathcal{A} := \mathcal{A}(X) := \{\gamma \in (\mathbb{C}[[t]])^n \mid f(\gamma) = 0\} = $ the arc space of X,

$\mathcal{A}_m := \mathcal{A}_m(X) := \{\gamma \in (\mathbb{C}[t]/t^m)^n \mid f(\gamma) \equiv 0\rangle(t^m)\}$,

$\pi_m : \mathcal{A} \to \mathcal{A}_m$ the natural projection,

$\tilde{\mathcal{A}}_m := \tilde{\mathcal{A}}_m(X) := \pi_m(\mathcal{A}) = $ the set of truncations mod t^m of arcs on X.

Note that $\mathcal{A}_m$ is an algebraic variety over $\mathbb{C}$ in a natural way. Indeed, we identify it with

$$\{(a_{1,0}, a_{1,1}, \ldots, a_{1,m-1}, a_{2,0}, \ldots, a_{n,m-1}) \in \mathbb{C}^{nm} \mid$$
$$f(a_{1,0} + a_{1,1}t + \cdots, \ldots, a_{n,0} + a_{n,1}t + \cdots) \equiv 0\rangle(t^m)\}.$$

PROPOSITION 4.1.1 [Nash 1995]. *$\tilde{\mathcal{A}}_m$ is a constructible subset of the algebraic variety $\mathcal{A}_m$, meaning that it is a finite union of (Zariski) locally closed subvarieties of $\mathcal{A}_m$.*

PROOF. By a theorem of Greenberg [Greenberg 1966], for each m there exists $m' \geq m$ such that $\tilde{\mathcal{A}}_m$ equals the image of $\mathcal{A}_{m'}$, under the natural map $\mathcal{A}_{m'} \to \mathcal{A}_m$. The Proposition follows now from quantifier elimination for $\mathbb{C}$ (Chevalley's

Theorem asserting that the image of a constructible subset under a morphism of algebraic varieties is again constructible).   □

REMARK. The notions above can be defined for any algebraic variety over $\mathbb{C}$, and all results of the present § 4 hold in this more general case.

REMARK. The $\tilde{A}_m$ were first studied by J. Nash [1995], in relation with Hironaka's resolution of singularities. In the same paper, Nash formulated a very intriging conjecture about the $\tilde{A}_m$, which is still open. For related work see [Gonzalez-Sprinberg and Lejeune Jalabert 1996; Lejeune-Jalabert 1990].

FORMULATION OF THE PROBLEM. How does $\tilde{A}_m$ vary with m? We will give an answer (Theorem 4.3) to this problem, modulo the equivalence relation which calls two algebraic varieties equivalent if they can be cut in a finite number of (Zariski locally closed) pieces, the pieces of the first variety being isomorphic (as algebraic varieties) with the pieces of the second, or if this can be done after replacing the two varieties by the disjoint union with a third variety. The set of all varieties modulo this equivalence relation generates a ring:

4.2. The Grothendieck Ring $\mathcal{M}$ of Algebraic Varieties over $\mathbb{C}$. This ring $\mathcal{M}$ is generated by symbols $[V]$, for V running over all algebraic varieties $\mathbb{C}$ (reduced and separated schemes of finite type over $\mathbb{C}$), with relations

$$[V] = [V'] \quad \text{if } V \text{ is isomorphic with } V',$$
$$[V] = [V \setminus V'] + [V'] \quad \text{if } V' \text{ is Zariski closed in } V,$$
$$[V \times V'] = [V][V'].$$

Note that for V any algebraic variety over $\mathbb{C}$, the map $V' \mapsto [V']$, for V' Zariski locally closed in V, extends uniquely to the map $W \mapsto [W]$, for W any constructible subset of V, satisfying $[W \cup W'] = [W] + [W'] - [W \cap W']$.

Set $\mathbb{L} := [\mathbb{A}^1] \in \mathcal{M}$, where $\mathbb{A}^1$ denotes the affine line over $\mathbb{C}$.

Set $\mathcal{M}_{\mathrm{loc}} := M[\mathbb{L}^{-1}]$, the localization of the ring M obtained by inverting $\mathbb{L}$.

4.3. Rationality Theorem

THEOREM [Denef and Loeser 1999a]. $\tilde{P}(T) := \sum_{m=1}^{\infty} [\tilde{A}_m] T^m$, *considered as a power series over* $\mathcal{M}_{\mathrm{loc}}$, *is rational and belongs to the subring of* $\mathcal{M}_{\mathrm{loc}}[[T]]$ *generated by* $\mathcal{M}_{\mathrm{loc}}[T]$ *and the series* $(1 - \mathbb{L}^a T^b)^{-1}$ *with* $a \in \mathbb{Z}$ *and* $b \in \mathbb{N} \setminus \{0\}$.

4.4. Analogy with the p-adic Case. Note the analogy with the series $\tilde{P}(T)$ in Section 1.1, considering $\mathbb{Z}_p$ as an analogue of $\mathbb{C}[[t]]$. The proof of the rationality Theorem 4.3 runs along the same lines as in Section 1 and is based on the quantifier elimination for $\mathbb{C}((t))$ due to Pas and integration on the arc space $(\mathbb{C}[[t]])^n$ of the affine n-space $\mathbb{A}^n$. Integration on $(\mathbb{C}[[t]])^n$ is called motivic integration and was recently introduced by Kontsevich [1995] and refined by Denef and Loeser [1999a]. We briefly discuss motivic integration in Section 4.7 and in 4.8 we present some ideas of the proof of Theorem 4.3. These integrals take values in a certain completion $\hat{\mathcal{M}}$ of $\mathcal{M}$, unlike the p-adic integrals of Section 1 which take values in $\mathbb{R}$.

4.5. The Completion $\hat{\mathcal{M}}$ of $\mathcal{M}_{\text{loc}}$. Define $F^m(\mathcal{M}_{\text{loc}})$ as the subgroup of $\mathcal{M}_{\text{loc}}$ generated by

$$\{[V]\mathbb{L}^{-i} \mid V \text{ is algebraic variety and } i \geq m + \dim V\}.$$

These form a filtration of $\mathcal{M}_{\text{loc}}$. Let $\hat{\mathcal{M}}$ be the completion of $\mathcal{M}_{\text{loc}}$ with respect to this filtration. (An element of $\mathcal{M}_{\text{loc}}$ is "small" if it belongs to $F^m(\mathcal{M}_{\text{loc}})$ for m big.) In comparison with p-adic integration, consider $\hat{\mathcal{M}}$ as the analogue of $\mathbb{R}$ (the target of integration), and $\mathbb{L}$ as the analogue of $p \in \mathbb{R}$. The ring structure on $\mathcal{M}_{\text{loc}}$ induces a ring structure on $\hat{\mathcal{M}}$. The ring $\hat{\mathcal{M}}$ was first introduced by Kontsevich [Kontsevich 1995].

REMARK. We do not know whether the natural map $\mathcal{M}_{\text{loc}} \to \hat{\mathcal{M}}$ is injective. But many geometric invariants, such as the topological Euler characteristic factor through the image $\overline{\mathcal{M}}_{\text{loc}}$ of $\mathcal{M}_{\text{loc}}$ in $\hat{\mathcal{M}}$.

4.6. Semi-Algebraic Sets over $\mathbb{C}[[t]]$. Let $\mathcal{L}_{\text{Pas}}$ be the (first order) language (in the sense of logic) with three sorts of variables: variables running over the valued field $\mathbb{C}((t))$ (= the fraction field of $\mathbb{C}[[t]]$), variables running over the value group $\mathbb{Z}$, and variables running over the residue field $\mathbb{C}$. The symbols of $\mathcal{L}_{\text{Pas}}$ consist of the symbols of Presburger's language $\mathcal{L}_{\text{Pres}}$ for $\mathbb{Z}$ (see 1.3), symbols to denote $+, -, \times, 0, 1$ in $\mathbb{C}((t))$ and in $\mathbb{C}$, and symbols for the valuation $\text{ord} : \mathbb{C}((t)) \setminus \{0\} \to \mathbb{Z}$ and for the function $\overline{\text{ac}} : \mathbb{C}((t)) \to \mathbb{C} : \gamma \mapsto$ the leading coefficient of the series γ. (We use the convention that $\overline{\text{ac}}(0) = 0$, $\text{ord}\, 0 = +\infty, (+\infty) + l = +\infty$ and $+\infty \equiv l \equiv d$, for all l in $\mathbb{Z} \cup \{+\infty\}$.)

A theorem of Pas [1989] states that $\mathbb{C}((t))$ has quantifier elimination in $\mathcal{L}_{\text{Pas}}$.

A subset of $\mathbb{C}((t))^n$ which is definable by a formula without quantifiers in $\mathcal{L}_{\text{Pas}}$ is called *semi-algebraic*.

PROPOSITION 4.6.1. *Let X be as in 4.1, and let $S \subset \mathcal{A}(X) \subset (\mathbb{C}[[t]])^n$ be semi-algebraic. Then $\pi_m(S)$ is a constructible subset of the algebraic variety $\mathcal{A}_m(X)$.*

PROOF. An easy application of the Theorem of Pas.   □

More generally one defines (in the obvious way) semi-algebraic subsets of $\mathcal{A}(X)$, for any algebraic variety X over $\mathbb{C}$. Obviously Proposition 4.6.1 remains valid.

4.7. Motivic integration on the arc space $\mathcal{A}(X)$. Motivic integration was recently introduced by Kontsevich [1995] and further developed and refined by Denef and Loeser [1999a].

THEOREM 4.7.1 [Denef and Loeser 1999a]. *Let X be as in Section 4.1 (or more generally any algebraic variety over $\mathbb{C}$). Let $S \subset \mathcal{A}(X)$ be semi-algebraic and $d = \dim X$. Then*

$$\mu(S) := \lim_{m \to +\infty} [\pi_m(S)]\mathbb{L}^{-md} \in \hat{\mathcal{M}}$$

exists in $\hat{\mathcal{M}}$. *Moreover* $S \mapsto \mu(S)$ *is an* $\hat{\mathcal{M}}$-*valued* σ-*additive measure on the boolean algebra of semi-algebraic subsets of* $\mathcal{A}(X)$.

We call μ the *motivic measure* on the arc space $\mathcal{A}(X)$ of X. This allows us to define

$$\int_S \mathbb{L}^{-\theta} d\mu := \sum_{m \in \mathbb{N}} \mathbb{L}^{-m} \mu(\theta^{-1}(m)) \in \hat{\mathcal{M}},$$

for any $\theta : \mathcal{A}(X) \to \mathbb{N}$ which is definable in $\mathcal{L}_{\text{Pas}}$. These motivic integrals have nice properties, such as an analogue of the classical change of variables formula; see [Denef and Loeser 1999a].

4.8. Some Ideas in the Proof of the Rationality Theorem. We only consider the weaker assertion that the image of $\tilde{P}(T)$ in $\hat{\mathcal{M}}[[T]]$ is rational. (The proof of the original statement is more difficult.) To prove this weaker assertion, we consider the motivic measure μ on the arc space $\mathcal{A}(\mathbb{A}^m) = (\mathbb{C}[[t]])^n$. For $m \in \mathbb{N} \setminus \{0\}$, put

$$S_m := \{\gamma \in (\mathbb{C}[[t]])^n \mid \exists \gamma' \in (\mathbb{C}[[t]])^n : f(\gamma') = 0, \gamma \equiv \gamma' \equiv t^m\}.$$

Then

$$[\tilde{A}_m(X)] = \mu(S_m)\mathbb{L}^{mn} \quad \text{in } \hat{\mathcal{M}},$$

where X is the locus of $f = 0$. The proof of the weaker assertion above proceeds now in close analogy with the proof of the rationality of $\tilde{P}(T)$ in the p-adic case (using resolution of singularities, but no cell decomposition).   □

4.9. Construction of New Invariants of Algebraic Varieties. p-adic integration was used by Denef and Loeser [1992] (see also [Denef 1991]) to obtain new geometric invariants, such as the topological zeta functions. These are calculated from a resolution of singularities using Euler characteristics and multiplicities. (Independence from the chosen resolution is proved by p-adic integration and use of the Grotendieck–Lefschetz trace formula.) See [Veys 1999] for related work.

Kontsevich [1995] obtained many more geometric invariants using motivic integration instead of p-adic integration. (This makes it possible to work with Hodge polynomials instead of Euler characteristics.) In the same paper he also used motivic integration to prove the conjecture that birationally isomorphic Calabi–Yau manifolds have the same Hodge numbers. (That they have the same Betti numbers was proved before by Batyrev [1997a] using p-adic integration.)

Denef and Loeser [1998; 1999a] have obtained some more geometric invariants by motivic integration. For example, if X is an algebraic variety over $\mathbb{C}$, one can consider

$$\chi(\mu(\mathcal{A}(X))) \in \mathbb{Q},$$

where χ denotes the Euler characteristic, since

$$\mu(\mathcal{A}(X)) \in \overline{M}_{\text{loc}}[((1 + \mathbb{L} + \mathbb{L}^2 + \cdots + \mathbb{L}^i)^{-1})_{i \in \mathbb{N}}].$$

(Recall that $\overline{\mathcal{M}}_{\mathrm{loc}}$ is the image of $\mathcal{M}_{\mathrm{loc}}$ in $\hat{\mathcal{M}}$.) When X is nonsingular we have $\chi(\mu(\mathcal{A}(X))) = \chi(X)$, but for singular X one gets something new which can be expressed in terms of Euler characteristics of the exceptional divisors (and their intersections) of a suitable resolution of singularities of X. Similar invariants can be defined using Hodge polynomials.

Very recently Batyrev [1998; 1997b] constructed some related new invariants (for algebraic varieties with "mild" singularities) which he calls the "string Euler characteristic" and the "string Hodge numbers". They play a role in quantum cohomology and mirror symmetry.

For connections with the theory of motives, see [Denef and Loeser 1998; 1999b].

References

[Ax and Kochen 1965a] J. Ax and S. Kochen, "Diophantine problems over local fields, I,", *Amer. J. Math.* **87** (1965), 605–630.

[Ax and Kochen 1965b] J. Ax and S. Kochen, "Diophantine problems over local fields, II", *Amer. J. Math.* **87** (1965), 631–648.

[Ax and Kochen 1966] J. Ax and S. Kochen, "Diophantine problems over local fields, III", *Ann. of Math.* (2) **83** (1966), 437–456.

[Batyrev 1997a] V. V. Batyrev, "On the Betti numbers of birationally isomorphic projective varieties with trivial canonical bundle", preprint, 1997. Available at http://xxx.lanl.gov/abs/alg-geom/9710020.

[Batyrev 1997b] V. V. Batyrev, "Stringy Hodge numbers and Virasoro algebra", preprint, 1997. Available at http://xxx.lanl.gov/abs/alg-geom/9711019.

[Batyrev 1998] V. V. Batyrev, "Stringy Hodge numbers of varieties with Gorenstein canonical singularities", pp. 1–32 in *Integrable systems and algebraic geometry* (Kobe/Kyoto, 1997), edited by M.-H. Saito et al., World Sci., River Edge, NJ, 1998.

[Batyrev 1999] V. V. Batyrev, "Non-Archimedean integrals and stringy Euler numbers of log-terminal pairs", *J. Eur. Math. Soc.* **1**:1 (1999), 5–33.

[Berkovich 1990] V. G. Berkovich, *Spectral theory and analytic geometry over non-Archimedean fields*, Math. surveys and monographs **33**, Amer. Math. Soc., Providence, RI, 1990.

[Bosch et al. 1984] S. Bosch, U. Güntzer, and R. Remmert, *Non-Archimedean analysis: a systematic approach to rigid analytic geometry*, Grundlehren der mat. Wiss. **261**, Springer, Berlin, 1984.

[Bourbaki 1967] N. Bourbaki, *Éléments de mathématique, fasc. XXXIII: Variétés différentielles et analytiques, fascicule de résultats, paragraphes 1 à 7*, Act. sci. et ind. **1333**, Hermann, Paris, 1967.

[Cohen 1969] P. J. Cohen, "Decision procedures for real and p-adic fields", *Comm. Pure Appl. Math.* **22** (1969), 131–151.

[Delon 1981] F. Delon, *Quelques propriétés des corps valués*, thèse d'état, Université Paris VII, 1981.

[Denef 1984] J. Denef, "The rationality of the Poincaré series associated to the p-adic points on a variety", *Invent. Math.* **77**:1 (1984), 1–23.

[Denef 1985] J. Denef, "On the evaluation of certain p-adic integrals", pp. 25–47 in *Séminaire de théorie des nombres* (Paris 1983–84), edited by C. Goldstein, Prog. in math. **59**, Birkhäuser, Boston, 1985.

[Denef 1986] J. Denef, "p-adic semi-algebraic sets and cell decomposition", *J. Reine Angew. Math.* **369** (1986), 154–166.

[Denef 1988] J. Denef, "Multiplicity of the poles of the Poincaré series of a p-adic subanalytic set", in *Séminaire de Théorie des Nombres de Bordeaux*, 1987–88, Univ. Bordeaux I, Talence, 1988. Exposé 43.

[Denef 1991] J. Denef, "Report on Igusa's local zeta function", pp. 359–386 in *Séminaire Bourbaki*, 1990/91, Astérisque **201-203**, Soc. math. France, Montrouge, 1991. Exposé 741.

[Denef and Loeser 1992] J. Denef and F. Loeser, "Caractéristiques d'Euler–Poincaré, fonctions zêta locales et modifications analytiques", *J. Amer. Math. Soc.* **5**:4 (1992), 705–720.

[Denef and Loeser 1998] J. Denef and F. Loeser, "Motivic Igusa zeta functions", *J. Algebraic Geom.* **7**:3 (1998), 505–537. Available at http://www.wis.kuleuven.ac.be/wis/algebra/denef.html.

[Denef and Loeser 1999a] J. Denef and F. Loeser, "Germs of arcs on singular algebraic varieties and motivic integration", *Invent. Math.* **135**:1 (1999), 201–232. Available at http://www.wis.kuleuven.ac.be/wis/algebra/denef.html.

[Denef and Loeser 1999b] J. Denef and F. Loeser, "Motivic exponential integrals and a motivic Thom–Sebastiani theorem", *Duke Math. J.* **99**:2 (1999), 285–309. Available at http://www.wis.kuleuven.ac.be/wis/algebra/denef.html.

[Denef and van den Dries 1988] J. Denef and L. van den Dries, "p-adic and real subanalytic sets", *Ann. of Math.* (2) **128**:1 (1988), 79–138.

[van den Dries 1978] L. van den Dries, *Model theory of fields*, Ph.D. thesis, Univ. of Utrecht, 1978.

[van den Dries 1992] L. van den Dries, "Analytic Ax–Kochen–Ersov theorems", pp. 379–398 in *Proceedings of the International Conference on Algebra, Part 3* (Novosibirsk, 1989), edited by L. A. Bokut et al., Contemporary mathematics **130**, Amer. Math. Soc., Providence, RI, 1992.

[Ershov 1965] Y. L. Ershov, "On the elementary theory of maximal normed fields, I", *Algebra i Logika (Seminar)* **4**:6 (1965), 31–69.

[Ershov 1966] Y. L. Ershov, "On the elementary theory of maximal normed fields, II", *Algebra i Logika (Seminar)* **5**:1 (1966), 5–40.

[Ershov 1967] Y. L. Ershov, "On the elementary theory of maximal normed fields, III", *Algebra i Logika (Seminar)* **6**:3 (1967), 31–73.

[Fresnel and van der Put 1981] J. Fresnel and M. van der Put, *Géométrie analytique rigide et applications*, Prog. in math. **18**, Birkhäuser, Boston, 1981.

[Gardener 2000] T. S. Gardener, "Local flattening in rigid analytic geometry", *Proc. London Math. Soc.* (3) **80**:1 (2000), 179–197.

[Gardener and Schoutens ≥ 2000] T. S. Gardener and H. Schoutens, "Flattening and subanalytic sets in rigid analytic geometry", *Proc. London Math. Soc.* (3). To appear.

[Gonzalez-Sprinberg and Lejeune Jalabert 1996] G. Gonzalez-Sprinberg and M. Lejeune Jalabert, "Sur l'espace des courbes tracées sur une singularité", pp. 9–32 in *Algebraic geometry and singularities* (La Rábida, 1991), edited by A. Campillo López and L. Narvèz Macarro, Prog. in math. **134**, Birkhäuser, Basel, 1996.

[Greenberg 1966] M. J. Greenberg, "Rational points in Henselian discrete valuation rings", *Publ. Math. Inst. Hautes Études Sci.* **31** (1966), 59–64.

[Grunewald, Segal, and Smith 1988] F. J. Grunewald, D. Segal, and G. C. Smith, "Subgroups of finite index in nilpotent groups", *Invent. Math.* **93**:1 (1988), 185–223.

[Hironaka 1964] H. Hironaka, "Resolution of singularities of an algebraic variety over a field of characteristic zero", *Ann. of Math.* (2) **79** (1964), 109–203, 205–326.

[Igusa 1974] J. Igusa, "Complex powers and asymptotic expansions, I: Functions of certain types", *J. Reine Angew. Math.* **268/269** (1974), 110–130.

[Igusa 1975] J. Igusa, "Complex powers and asymptotic expansions, II: Asymptotic expansions", *J. Reine Angew. Math.* **278/279** (1975), 307–321.

[Igusa 1978] J. Igusa, *Lectures on forms of higher degree*, Lectures on math. and phys. **59**, Tata Institute of Fundamental Research, Bombay, 1978. Notes by S. Raghavan.

[Igusa 1984] J. Igusa, "Some results on p-adic complex powers", *Amer. J. Math.* **106**:5 (1984), 1013–1032.

[Igusa 1987] J. Igusa, "Some aspects of the arithmetic theory of polynomials", pp. 20–47 in *Discrete groups in geometry and analysis* (New Haven, 1984), edited by R. Howe, Prog. in math. **67**, Birkhäuser, Boston, 1987.

[Igusa 1996] J. Igusa, "On local zeta functions", pp. 1–20 in *Selected papers on number theory and algebraic geometry*, Amer. Math. Soc. Transl. (2) **172**, Amer. Math. Soc., Providence, 1996. Translated from Sūgaku **46**:1 (1994), 23–38.

[Koblitz 1977] N. Koblitz, *p-adic numbers, p-adic analysis, and zeta-functions*, GTM **58**, Springer, New York, 1977.

[Kochen 1975] S. Kochen, "The model theory of local fields", pp. 384–425 in *ISILC Logic Conference* (Kiel, 1974), edited by G. H. Müller et al., Lecture Notes in Math. **499**, Springer, Berlin, 1975.

[Kontsevich 1995] M. Kontsevich, 1995. Lecture at Orsay, December 7, 1995.

[Lejeune-Jalabert 1990] M. Lejeune-Jalabert, "Courbes tracées sur un germe d'hypersurface", *Amer. J. Math.* **112**:4 (1990), 525–568.

[Lichtin ≥ 2000a] B. Lichtin, "On a question of Igusa : towards a theory of several variable asymptotic expansions, I", preprint.

[Lichtin ≥ 2000b] B. Lichtin, "On a question of Igusa : towards a theory of several variable asymptotic expansions, II", preprint.

[Lipshitz 1993] L. Lipshitz, "Rigid subanalytic sets", *Amer. J. Math.* **115**:1 (1993), 77–108.

[Lipshitz and Robinson 1996a] L. Lipshitz and Z. Robinson, "Rigid subanalytic sets II", Technical report, 1996. Available at http://www.math.purdue.edu/~lipshitz/.

[Lipshitz and Robinson 1996b] L. Lipshitz and Z. Robinson, "Rigid subanalytic subsets of the line and the plane", *Amer. J. Math.* **118**:3 (1996), 493–527.

[Lipshitz and Robinson 1999] L. Lipshitz and Z. Robinson, "Rigid subanalytic subsets of curves and surfaces", *J. London Math. Soc.* (2) **59**:3 (1999), 895–921.

[Lipshitz and Robinson ≥ 2000] L. Lipshitz and Z. Robinson, "Rings of separated power series", Technical report. Available at http://www.math.purdue.edu/~lipshitz/. To appear in *Astérisque*.

[Liu 1997] N. Liu, "Analytic cell decomposition and the closure of p-adic semianalytic sets", *J. Symbolic Logic* **62**:1 (1997), 285–303.

[Loeser 1989] F. Loeser, "Fonctions zêta locales d'Igusa à plusieurs variables, intégration dans les fibres, et discriminants", *Ann. Sci. École Norm. Sup.* (4) **22**:3 (1989), 435–471.

[Macintyre 1976] A. Macintyre, "On definable subsets of p-adic fields", *J. Symbolic Logic* **41**:3 (1976), 605–610.

[Macintyre 1990] A. Macintyre, "Rationality of p-adic Poincaré series: uniformity in p", *Ann. Pure Appl. Logic* **49**:1 (1990), 31–74.

[Meuser 1981] D. Meuser, "On the rationality of certain generating functions", *Math. Ann.* **256**:3 (1981), 303–310.

[Meuser 1986] D. Meuser, "The meromorphic continuation of a zeta function of Weil and Igusa type", *Invent. Math.* **85**:3 (1986), 493–514.

[Nash 1995] J. F. Nash, Jr., "Arc structure of singularities", *Duke Math. J.* **81**:1 (1995), 31–38. A celebration of John F. Nash, Jr.

[Oesterlé 1982] J. Oesterlé, "Réduction modulo p^n des sous-ensembles analytiques fermés de $\mathbb{Z}_p^N$", *Invent. Math.* **66**:2 (1982), 325–341.

[Pas 1989] J. Pas, "Uniform p-adic cell decomposition and local zeta functions", *J. Reine Angew. Math.* **399** (1989), 137–172.

[Pas 1990] J. Pas, "Cell decomposition and local zeta functions in a tower of unramified extensions of a p-adic field", *Proc. London Math. Soc.* (3) **60**:1 (1990), 37–67.

[Pas 1991] J. Pas, "Local zeta functions and Meuser's invariant functions", *J. Number Theory* **38**:3 (1991), 287–299.

[Presburger 1930] M. Presburger, "Über die Vollständigkeit eines gewissen Systems des Arithmetik", in *Sprawozdanie z 1. Kongresu matematyków krajów słowiańskich = Comptes-rendus du 1. Congrès des mathématiciens des pays slaves* (Warsaw, 1929), edited by F. Leja, Ksiaznica atlas, Warsaw, 1930.

[Robinson 1956] A. Robinson, *Complete theories*, North-Holland, Amsterdam, 1956.

[Robinson 1993] Z. Robinson, "Smooth points of p-adic subanalytic sets", *Manuscripta Math.* **80**:1 (1993), 45–71.

[Robinson 1997] Z. Robinson, "Flatness and smooth points of p-adic subanalytic sets", *Ann. Pure Appl. Logic* **88**:2-3 (1997), 217–225.

[Sato 1989] F. Sato, "On functional equations of zeta distributions", pp. 465–508 in *Automorphic forms and geometry of arithmetic varieties*, Academic Press, Boston, MA, 1989.

[du Sautoy 1993] M. P. F. du Sautoy, "Finitely generated groups, p-adic analytic groups and Poincaré series", *Ann. of Math.* (2) **137**:3 (1993), 639–670.

[Schoutens 1994a] H. Schoutens, "Rigid subanalytic sets", *Compositio Math.* **94**:3 (1994), 269–295.

[Schoutens 1994b] H. Schoutens, "Rigid subanalytic sets in the plane", *J. Algebra* **170**:1 (1994), 266–276.

[Schoutens 1994c] H. Schoutens, "Uniformization of rigid subanalytic sets", *Compositio Math.* **94**:3 (1994), 227–245.

[Schoutens 1997] H. Schoutens, "Closure of rigid semianalytic sets", *J. Algebra* **198**:1 (1997), 120–134.

[Schoutens 1999] H. Schoutens, "Rigid analytic flatificators", *Quart. J. Math. Oxford Ser.* (2) **50**:199 (1999), 321–353.

[Schoutens ≥ 2000] H. Schoutens, "Rigid analytic quantifier elimination", lecture notes.

[Serre 1981] J.-P. Serre, "Quelques applications du théorème de densité de Chebotarev", *Publ. Math. Inst. Hautes Études Sci.* **54** (1981), 323–401.

[Veys 1992] W. Veys, "Reduction modulo p^n of p-adic subanalytic sets", *Math. Proc. Cambridge Philos. Soc.* **112**:3 (1992), 483–486.

[Veys 1993] W. Veys, "Poles of Igusa's local zeta function and monodromy", *Bull. Soc. Math. France* **121**:4 (1993), 545–598.

[Veys 1996] W. Veys, "Embedded resolution of singularities and Igusa's local zeta function", 1996. Available at http://www.wis.kuleuven.ac.be/wis/algebra/veys.htm. To appear in *Academiae Analecta*.

[Veys 1997] W. Veys, "Zeta functions for curves and log canonical models", *Proc. London Math. Soc.* (3) **74**:2 (1997), 360–378.

[Veys 1999] W. Veys, "The topological zeta function associated to a function on a normal surface germ", *Topology* **38**:2 (1999), 439–456.

JAN DENEF
UNIVERSITY OF LEUVEN,
DEPARTMENT OF MATHEMATICS
CELESTIJNENLAAN 200B
3001 HEVERLEE
BELGIUM
  Jan.Denef@wis.kuleuven.ac.be
  http://www.wis.kuleuven.ac.be/wis/algebra/denef.html

Model Theory, Algebra, and Geometry
MSRI Publications
Volume **39**, 2000

Abelian Varieties and the Mordell–Lang Conjecture

BARRY MAZUR

ABSTRACT. This is an introductory exposition to background material useful to appreciate various formulations of the Mordell–Lang conjecture (now established by recent spectacular work due to Vojta, Faltings, Hrushovski, Buium, Voloch, and others). It gives an exposition of some of the elementary and standard constructions of algebro-geometric models (rather than model-theoretic ones) with applications (for example, via the method of Chabauty) relevant to Mordell–Lang. The article turns technical at one point (the step in the proof of the Mordell–Lang Conjecture in characteristic zero which passes from number fields to general fields). Two different procedures are sketched for doing this, with more details given than are readily found in the literature. There is also some discussion of issues of effectivity.

CONTENTS

The organizers of this MSRI workshop in Arithmetic and Model Theory gave me the agreeable task of lecturing on introductory background in the theory of abelian varieties, and especially those parts of the theory relevant to the Mordell–Lang Conjecture, which is the theme of some of the recent spectacular work (by Vojta, Faltings, Hrushovski, Buium, Voloch, and others). To keep this talk as "introductory" and as focussed as possible I will concentrate only on the version of the Mordell–Lang Conjecture that deals with abelian, rather than semi-abelian, varieties, and I will not discuss its elaborations that include "Manin–Mumford type" questions regarding torsion points. For this, see [Raynaud 1983b; 1983c; Coleman 1985b; Hindry 1988; McQuillan 1995], and the extensive bibliographies in these articles. For a general introduction to the model theory approach to Mordell–Lang, see [Bouscaren 1998], and particularly [Hindry 1998] therein. I have tried to make my expository article overlap as little as possible with Hindry's, given their nearly identical titles, and their similar missions.

We can look forward to further deep connections between model theory and arithmetic problems. For example, there is the model theory of difference fields which already has given rise to explicit bounds in certain arithmetic questions; see [Chatzidakis and Hrushovski 1999] and the bibliography there. Also there are important applications of model theory to aspects of the ABC Conjecture, in the work of Buium and that of Scanlon [1997a], which opens up a very promising avenue of research.

The "fundamentals" of the theory of abelian varieties are collected in Section 1. The main theorems are stated and discussed in Sections 2, 3, and 4. Motivated by the title of this conference, I also recorded a few of the much more modest construction of models, sometimes done explicitly, sometimes implicitly, in standard algebraic geometric arguments. These well known constructions are reviewed in Section 5. For fun, in Section 6 we put the "p-adic model" to work: we make use of the beautiful classical idea of Chabauty [1941] to give a proof of a (small) piece of Mordell–Lang. One can think of the strategy of Chabauty's method as being, first, to extend the groundfield to a rather large field (i.e., the p-adics) over which differential equations work well (i.e., giving us the Lie theory for p-adic analytic groups), and then reaping the benefits of being able to solve differential equations. Of course, I am saying it this way in order to force a kinship (albeit a distant one) between the Chabauty approach and the formidable methods (as in [Hrushovski 1996; Hrushovski and Zilber 1993; 1996]) that I am learning about in this conference, and therefore to justify including it in this article. Section 7 discusses the step in the proof of the Mordell–Lang Conjecture in characteristic 0 which passes from number fields to general fields; we give more details about this step than is found in the literature. Finally, Section 8 is devoted to some comments about effectivity.

Acknowledgements. I received signficant help from A. Buium, M. McQuillan, M. Nakamaye, A. Pillay, P. Vojta, and C. Wood in preparing this article; I'm grateful to them for the discussions we had, and I'm grateful that this article gave me the opportunity to have such enjoyable discussions with them.

Introduction

The central result in the constellation of theorems to be discussed in this article is the classical 1922 conjecture of Mordell, proved some sixty years later by Faltings [1983] (which asserts that a curve of genus greater than 1 defined over a number field has only a finite number of points rational over that number field). As an example of an application of this theorem, choose your favorite polynomial $g(x)$ with rational coefficients, no multiple roots, and of degree ≥ 5, for example

$$g(x) = x(x-1)(x-2)(x-3)(x-4),$$

and let K be any number field (that is, any field of finite degree over $\mathbb{Q}$). Then Faltings' Theorem implies:

COROLLARY. *Let $g(x)$ be a polynomial in $K[x]$ with no multiple roots, and of degree ≥ 5. Then there are only finitely many elements $\alpha \in K$ for which $g(\alpha)$ is a square in K.*

We can use this corollary to illustrate some of the "degrees of effectivity" that are of interest in this problem, and in similar problems. We distinguish three grades:

1. **Number-effectivity.** Is there a "directly computable" function of the coefficients of $g(x)$ and of the number field K which provides an upper bound for the *number* of such elements $\alpha \in K$ (i.e., for the number of α's such that $g(\alpha)$ is a square in K)?
2. **Size-effectivity.** Is there a "directly computable" function of the coefficients of $g(x)$ and of the number field K which provides an upper bound for the *heights* of elements $\alpha \in K$ for which $g(\alpha)$ is a square in K?
3. **Uniform number-effectivity.** Is there a "directly computable" function of the *degree* of $g(x)$ and of the number field K which provides an upper bound for the *number* of elements $\alpha \in K$ for which $g(\alpha)$ is a square in K?

There is at least one somewhat ambiguous term in all of the questions raised above; namely, what does one accept as "directly computable"? But by any reasonable standard of "direct computability", the answer to Question 1 regarding number-effectivity is yes: in fact the proof of Faltings' Theorem readily provides a quite large, but effective, upper bound for the number of K-valued points on a curve of genus greater than 1 defined over K, in terms of K and the curve. Question 2 is open, and is the focus of a good deal of activity. Question 3 is also open (see [Caporaso et al. 1997] for the connection between Question 3 and certain conjectures of Lang). For further discussion of these issues, see Section 8 below.

Our main interest in the present article will be higher-dimensional analogues of Faltings' theorem, these analogues being expressed in terms of abelian varieties.

1. Complex Tori and Abelian Varieties

An excellent reference for the basics of this theory is [Mumford 1974]. Let V be a finite dimensional complex vector space, and call its dimension d. Let $\Lambda \subset V$ be a discrete additive subgroup of rank $2d$. It follows that the natural homomorphism of real vector spaces $\Lambda \otimes_{\mathbb{Z}} \mathbb{R} \to V$ is an isomorphism, and that $2d$ is the maximal rank that a discrete additive subgroup of V can have. We refer to Λ as a *lattice* in V, and to the quotient (commutative, compact) complex Lie group $T := V/\Lambda$ as a *complex torus*. Given T we can reconstruct V up to unique isomorphism as the tangent space at the origin (i.e., the "Lie algebra") of the complex Lie group T. The elementary construction of passing to the quotient

$$(V, \Lambda) \mapsto T = V/\Lambda$$

is functorial from the category of lattices in finite-dimensional complex vector spaces to the category of complex tori. If $T = V/\Lambda$ we have the following diagram of endomorphism rings:

$$\mathrm{End}_{\mathrm{cx\,tori}}(T) = \mathrm{End}_{\mathrm{ab\,gp}}(\Lambda) \cap \mathrm{End}_{\mathrm{cx\,v.\,sp.}}(V)$$

$$\subset \mathrm{End}_{\mathrm{real\,v.\,sp.}}(\Lambda \otimes_{\mathbb{Z}} \mathbb{R}) = \mathrm{End}_{\mathrm{real\,v.\,sp.}}(V).$$

Since $\mathrm{End}_{\mathrm{cx\,tori}}(T) \subset \mathrm{End}_{\mathrm{ab\,gp}}(\Lambda)$ it follows that the underlying additive group of $\mathrm{End}_{\mathrm{cx\,tori}}(T)$ is a finitely generated and free abelian group.

If the complex manifold $T \hookrightarrow \mathbb{P}^N$ admits a complex analytic imbedding into projective N-space, then by Chow's Theorem, T is the locus of zeroes of homogeneous polynomial equations in $\mathbb{P}^N$ and therefore carries the structure of projective algebraic variety; its group law is algebraic. By a *projective variety* in this article, we do not impose the condition that it be *irreducible*; we mean what is sometimes referred to as *algebraic set*, i.e., cut out by a system of homogenous forms as in [Hartshorne 1977, Chapter 1, Section 2]. If our varieties are irreducible, we will signal this explicitly by the phrase *irreducible variety*. By an *abelian variety* over a field F we mean a group object in the category of geometrically irreducible, proper (i.e., "complete") , algebraic varieties (over F). Thus, a complex analytic imbedding of a complex torus T in projective space endows T with the structure of abelian variety over the field $\mathbb{C}$. Any abelian variety over F is a projective variety, i.e., is isomorphic over F to a subvariety of projective space. Any abelian variety over $\mathbb{C}$, when viewed with its underlying complex Lie group structure is a complex torus. The theory of Weierstrass guarantees that any complex 1-*dimensional* torus admits a complex analytic imbedding (as a plane cubic) in $\mathbb{P}^2$, and therefore admits the structure of an *elliptic curve*, i.e., the structure of an abelian variety (of dimension 1) over $\mathbb{C}$. When $d > 1$ the "Riemann period relations" impose definite conditions on lattices $\Lambda \subset V$ in order that $T = V/\Lambda$ be imbeddable in projective space, or equivalently, that T have the structure of abelian variety over $\mathbb{C}$.

When T is an abelian variety, its complex analytic endomorphisms are all "algebraic", in the sense that they are endomorphisms of the abelian variety T, defined over $\mathbb{C}$. If A is an abelian variety over any field K, we have the Poincaré complete reducibility theorem [Mumford 1974, Section 19, Theorem 1], which says that given any abelian subvariety $Y \subset A$ there is a "complementary" abelian subvariety $Z \subset A$ (in the sense that $Y \cap Z$ is a finite group and Y and Z taken together span A). In standard terminology, we have that the natural homomorphism $Y \oplus Z \rightarrow A$ is an *isogeny* in the sense that it is a surjective homomorphism with finite kernel. It follows directly that, "up to isogeny" , any abelian variety over a field F is a direct sum of (a finite number of) *simple* abelian varieties over F. (An abelian variety X over F is called *simple* (over F) if any abelian subvariety of X defined over F is either $\{0\}$ or X.) It follows easily [Mumford 1974, Corollaries 1 and 2 of Section 19] that the algebra of endomorphisms of the abelian variety A, after being tensored with $\mathbb{Q}$,

$$\mathrm{End}_{\mathrm{ab\ var}\ /F}(A) \otimes_{\mathbb{Z}} \mathbb{Q},$$

is a (finite-dimensional) semi-simple algebra over $\mathbb{Q}$.

2. Configurations and Configuration-Closure

Let F be a field and let V be a finite dimensional vector space over F. By a *configuration* in V we will mean a subset $W \subset V$ that is the (possibly empty) finite union of translates of vector subspaces (over the same field F) of V. That is, W is a configuration in V if

$$W = \bigcup_{j=1}^{n} (v_j + V_j)$$

for elements $v_j \in V$ and vector subspaces $V_j \subset V$. The collection of configurations in a finite dimensional vector space V is closed under finite union and arbitrary intersection — that is, it forms the collection of closed sets for a topology of V; moreover, this collection satisfies the "noetherian" (descending chain) condition. That is, given any decreasing sequence $\cdots \subset W_{j+1} \subset W_j \subset \cdots$ of configurations in V, indexed by natural numbers j, the sequence eventually stabilizes. This allows us to define the notion of *configuration-closure*:

DEFINITION. Given an arbitrary subset $S \subset V$, the *configuration-closure*, $\mathrm{Conf}_V(S)$, of S in V is the smallest configuration in V containing S. Equivalently, $\mathrm{Conf}_V(S)$ is the intersection of all configurations containing S.

A linear transformation of vector spaces, $\phi : V \to W$, brings configurations in the domain V to configurations in W, and the full inverse image of a configuration in W (under ϕ) is a configuration in V. If $S \subset V$ is a subset,

$$\phi\big(\mathrm{Conf}_V(S)\big) = \mathrm{Conf}_W(\phi S).$$

We can carry the notion of "configuration" and "configuration-closure" to abelian varieties, or to complex tori. We do this explicitly for abelian varieties over an algebraically closed field K. Recall that if A is an abelian variety over K, and $S \subset A(K)$ is a set of K-rational points of A, we can consider, the *Zariski closure* of S in A (which we denote $\mathrm{Zar}_A(S) \subset A$) which is defined to be the smallest closed subvariety of A whose set of K-valued points contain S. Equivalently, if you imbed A in $\mathbb{P}^N$, you may think of $\mathrm{Zar}_A(S)$ as the projective subvariety of $\mathbb{P}^N$ cut out by the ideal of all homogenous polynomial forms which contain the set S in their zero-locus.

Now, following the pattern set in our discussion of vector spaces above, define a *configuration* in the abelian variety A to mean a subset $B \subset A$ which is (possibly empty) finite union of translates of abelian subvarieties of A. That is, B is a *configuration* in A if

$$B = \bigcup_{j=1}^{n} (a_j + A_j)$$

for elements $a_j \in A$ and abelian subvarieties $A_j \subset A$.

As with vector spaces, the collection of configurations in an abelian variety A is closed under arbitrary intersection and finite union, and satisfies the descending chain condition. Therefore, we can define the "configuration-closure" of a subset $S \subset A(K)$ in A:

DEFINITION. Given an arbitrary subset $S \subset A(K)$, the *configuration-closure*, denoted $\mathrm{Conf}_A(S)$, of S in A is the smallest configuration C in A such that $S \subset C(K)$.

Clearly, $\mathrm{Conf}_A(S)$ is a closed subvariety of A containing S, and therefore

$$\mathrm{Zar}_A(S) \subset \mathrm{Conf}_A(S).$$

3. "Absolute" Mordell–Lang in Characteristic Zero: Theorems of Faltings and Vojta

Now we put these two notions of configuration (for abelian varieties and for vector spaces) together to give two equivalent formulations of the theorem of Faltings and Vojta. Let A be an abelian variety and $S \subset A(K)$ a subset of the group $A(K)$ of K-rational points of A.

THEOREM 3.1 ("ABSOLUTE" MORDELL–LANG IN CHARACTERISTIC 0, FIRST VERSION). *Let K be algebraically closed of characteristic 0. If $S \subset A(K)$ generates (or equivalently, is contained in) a finitely generated subgroup of $A(K)$, then the Zariski-closure of S in A is equal to its configuration-closure in A. In notation,*

$$\mathrm{Zar}_A(S) = \mathrm{Conf}_A(S).$$

This was first proved by Faltings [1994] (the proof in that reference is made explicit only for the case of $K = \bar{\mathbb{Q}}$; but see, for example, [McQuillan 1995] and Section 7 below). Earlier [1991] he had established the special case where $K = \bar{\mathbb{Q}}$ and $\mathrm{Zar}_A(S)$ contains no nontrivial translated abelian subvariety. The techniques of [Faltings 1991] have, as their starting point, Vojta's proof [1991] of the classical Mordell Conjecture. The original 1960 article of Lang, which formulated the special case of the Mordell–Lang conjecture proved in [Faltings 1991], is [Lang 1960], and this was followed by stronger versions of the conjecture [Lang 1965; 1974; 1986]. For further developments extending the proofs of the Mordell–Lang conjectures to include semi-abelian varieties see [Vojta 1996], and for the strongest statement of Mordell–Lang (on semi-abelian varieties combined with the Manin–Mumford conjecture) in characteristic 0, see [McQuillan 1995].

The "classical" application of Theorem 3.1. Let the abelian variety A be defined over a number field L and let K be an algebraically closed field containing L. Denote by $A(L)$ the group of L-rational points of A. Let $Z \subset A$ be a closed subvariety defined over L, and suppose that $S := Z(L)$, the set of L-rational points of Z, is *Zariski-dense* in Z. By the theorem of Mordell–Weil (see [Serre 1989, Section 4.3] or [Lang 1991, Chapter I, Theorem 4.1]), $A(L)$ is finitely generated, and so we have $Z = \mathrm{Zar}_A(S)$ and all the hypotheses of Theorem 3.1 above are satisfied ($S = Z(L) \subset A(L) \subset A(K)$, so S is contained in a finitely generated subgroup of $A(K)$). Therefore, Z is a configuration in A. To summarize:

COROLLARY. *Let A be an abelian variety defined over a number field L. Any closed subvariety of A defined over L which is the Zariski-closure of its set of L-rational points is a configuration in A; i.e., is a finite union of translates of abelian subvarieties of A.*

Bogomolov and Tschinkel [1999] introduced the term *potentially dense* to refer to algebraic varieties V defined over number fields which have the property that, over some (possibly larger) number field L, the variety V is the Zariski-closure of its set $V(L)$ of L-rational points. Another way of expressing the preceding corollary is to say that the closed subvarieties of an abelian variety that are *potentially dense* are precisely the configurations. One of the important projects in number theory these days is to understand in a more general context which algebraic varieties are potentially dense. There are many open issues here. It is even unknown, at present, whether there exist K3 surfaces which are not potentially dense! (But see [Bogomolov and Tschinkel 1999] for elliptic K3's!)

The "classical" Mordell Conjecture can be expressed as simply saying that a smooth projective irreducible algebraic curve defined over a number field is potentially dense (if and) only if its genus is 0 or 1. The "if" part of this statement is easy (and isn't usually packaged as part of the Mordell Conjecture). It is the "only if" assertion that is deep; it follows from the preceding corollary by noting that if C is a curve of positive genus defined over a number field, then by possible extension of the number field L we can suppose that C has an L-rational point $c \in C$. We then can view C as a subvariety of its jacobian, call it A, defined over L, by sending a point $x \in C$ to the linear equivalence class of the divisor $[x] - [c]$. The preceding corollary would then tell us that C is a *configuration* in A which can only be the case if its genus is 1.

The format of Theorem 3.1. Theorem 3.1 designates a class of algebraic varieties (i.e, abelian varieties) and says that if, within a variety in this class, S is a subset of points satisfying a certain "finiteness property" (i.e., is contained in a finitely generated subgroup of the Mordell–Weil group) then the Zariski-closure of S is a very restricted type of subvariety (i.e., is a configuration). Are there analogous results in other settings? For example, take, as class of algebraic varieties, smooth cubic hypersurfaces in $\mathbb{P}^N$. Define the "finiteness property" on a set S of points of a cubic hypersurface V to be: S is contained in a subset of V which is *finitely generated* in the sense of the chord-and-tangent process on V. I ask this question not because I have a sense that it is worthwhile to pursue (nor do I have any specific guess regarding the types of subvarieties of V that one can get as Zariski-closures of S's which are finitely generated in the above sense) but just in order to help us think, for moment, about the formal shape of Theorem 3.1. In the same vein, one can ask the analogous (Mordell–Lang) question about complex tori (which are not necessarily abelian varieties). I did so in an early draft of this article, and I am thankful to Anand Pillay for his affirmative answer to that question; see his proof [Pillay 1999] of the fact that if T is a complex torus and $S \subset T$ a subset which is contained in a finitely generated subgroup of T, then $\mathrm{Zar}_T(S) = \mathrm{Conf}_T(S)$ if we interpret $\mathrm{Zar}_T(S)$ as the smallest compact complex analytic subvariety of T containing S, and $\mathrm{Conf}_T(S)$ the smallest finite union of translates of complex subtori of T having the property that $S \subset \mathrm{Conf}_T(S)$.

Theorem 3.1 is equivalent to:

THEOREM 3.2 ("ABSOLUTE" MORDELL–LANG IN CHARACTERISTIC 0, SECOND VERSION). *Let K be algebraically closed of characteristic 0, A an abelian variety defined over K, and suppose that we are given a subset $S \subset \Gamma \subset A(K)$, where Γ is a finitely generated subgroup of $A(K)$. If Γ is Zariski-dense in A, and if the* *configuration-closure of the image of S in the real vector space $V = \Gamma \otimes \mathbb{R}$ is all of V, then S is Zariski-dense in A.*

PROOF OF EQUIVALENCE. Suppose the truth of Theorem 3.1 and the hypotheses of Theorem 3.2. We must prove that S is Zariski-dense in A. So

$$\mathrm{Zar}_A(S) = \mathrm{Conf}_A(S) = \bigcup_{j=1}^{n} (a_j + A_j) \subset A,$$

where the A_j's are abelian subvarieties of A and the $a_j \in A(K)$ are points. Put $\tilde{\Gamma}_j := (\Gamma \cap (a_j + A_j))$ so we have that $S \subset \bigcup_{j=1}^{n} \tilde{\Gamma}_j$. We have put the tilde on the $\tilde{\Gamma}_j$ to remind ourselves that $\tilde{\Gamma}_j$ isn't a subgroup of Γ but is rather a coset. Fix any element $\gamma_j \in \tilde{\Gamma}_j$ and we may write $\tilde{\Gamma}_j = \gamma_j + \Gamma_j$ for some subgroup $\Gamma_j \subset \Gamma$.

Therefore the image of S in $V = \Gamma \otimes \mathbb{R}$ is contained in the union of the translates by the image of γ_j of $\Gamma_j \otimes \mathbb{R}$ for $j = 1, \ldots, n$. Since S is configuration-dense in V we must have $V = \Gamma_j \otimes \mathbb{R}$ for some j, and for this j, $\Gamma_j \subset \Gamma$ must be a subgroup of Γ of finite index. But if Γ is Zariski-dense in A so is every (coset of every) subgroup of finite index in Γ. We deduce that $A_j = A$, and since a translate of A_j is contained in the Zariski-closure of S, it follows that S is Zariski-dense in A.

Now suppose the hypotheses of Theorem 3.1 and the truth of Theorem 3.2. We must prove that the conclusion of Theorem 3.1 holds. Consider the real vector space $V = \Gamma \otimes \mathbb{R}$. Let $\mathrm{Conf}_V(S) = \bigcup_{j=1}^{n}(v_j + V_j) \subset V$ be the configuration-closure of the image of S in $V = \Gamma \otimes \mathbb{R}$, where the V_j's are real vector subspaces of V and the $v_j \in V$ are points. Let $\tilde{S}_j \subset S$ be the inverse image of the affine subspace $v_j + V_j$, so that $S = \bigcup_{j=1}^{n} \tilde{S}_j$. Since Theorem 3.1 holds for S if it holds for each of the $\tilde{S}_j$'s, we need only prove Theorem 3.1 for each $\tilde{S}_j$. To do this we may, of course, take $\tilde{S}_j$ to be nonempty. Make a translation of $\tilde{S}_j$ by any element $\tilde{s}_j \in \tilde{S}_j$ to get a set which we denote $S_j := \tilde{S}_j - \tilde{s}_j$ and we have that $\mathrm{Conf}_V(S_j) = V_j$, i.e., is a vector subspace of V. If $\Gamma_j \subset \Gamma$ is the subgroup of all elements in Γ whose image in V lies in V_j, we have the inclusion $S_j \subset \Gamma_j$. By construction, $\mathrm{Conf}_{V_j}(S_j) = V_j$. Let $\tilde{A}_j \subset A$ be the Zariski-closure of the subgroup $\Gamma_j \subset A$.

If $\tilde{A}_j$ were connected, then it would be an abelian subvariety of A, and the triple $S_j, \Gamma_j, \tilde{A}_j$ would satisfy all the hypotheses required to apply Theorem 3.2, giving that

$$\mathrm{Zar}_{\tilde{A}_j}(S_j) = \tilde{A}_j,$$

thereby proving Theorem 3.1 for $\tilde{S}_j$. In general, let $A_j \subset \tilde{A}_j$ be the connected component containing the identity element, so that $\tilde{A}_j$ is a finite union of cosets of A_j. Breaking things up coset by coset, and applying the same argument as above, allows us to conclude the proof. $\qquad\square$

About positive characteristic. In the setting of characteristic $p > 0$, the statement analogous to "absolute" Mordell–Lang (Theorem 3.1 or 3.2 above) is no longer true, as the following well-known counter-example will make clear. If C is any curve (of positive genus) defined over a finite field k of characteristic p, and if $K = k(C)$ is the function field of C, then let $C_{/K}$ be the curve over K obtained from C by extending the field of scalars from k to K. We can think of $C_{/K}$ as the generic fiber of the constant family $\pi : C \times C \to C$, where π is projection to the second factor. There is a natural K-valued point, call it c, on $C_{/K}$, which can be described geometrically as the restriction to the generic fiber of the diagonal section $C \to C \times C$. If $F = F_{/k} : C_{/k} \to C_{/k}$ is the Frobenius endomorphism of $C_{/k}$ (defined on local rings by the rule: $f \mapsto f^q$ where $q = \mathrm{card}(k)$), denote by the same letter $F = F_{/K} : C_{/K} \to C_{/K}$ the base change of the Frobenius endomorphism to K. For any natural number n, $F^n(c)$ is the restriction to the generic fiber of $C \to C \times C$ of the graph of the n-th iterate of $F_{/k}$. The set, $S = \{F^n(c) \mid n = 1, 2, \dots\} \subset C(K)$, of images of c under these iterations of $F_{/K}$ is an infinite subset whose Zariski closure is therefore the entire curve $C_{/K}$. The Frobenius endomorphism F acting on $C_{/K}$ induces an endomorphism Φ of the jacobian $A_{/K} = \mathrm{Pic}^{\circ}(C_{/K})$. Let $\mathcal{E} := \mathrm{End}_K(A)$ denote the endomorphism ring of the abelian variety A over K, so that $\Phi \in \mathcal{E}$. Since C is of positive genus, we have an imbedding $\iota : C_{/K} \hookrightarrow A_{/K}$ defined by sending $x \in C_{/K}$ to the linear equivalence class of $[x] - [c]$ in $A_{/K}$. For any $m \geq 0$ we have

$$\Phi^m(\iota(Fc)) = \Phi^m([Fc] - [c]) = [F^{m+1}c] - [F^mc] = \iota(F^{m+1}c) - \iota(F^mc),$$

and therefore

$$\iota(F^nc) = \sum_{j=0}^{n-1} \Phi^j(\iota(Fc))$$

for any $n \geq 1$, so that the image of S under ι is contained in $\mathcal{E} \cdot \iota(Fc) \subset A(K)$ which is a finitely generated subgroup of $A(K)$ (the endomorphism ring $\mathcal{E}$ of the abelian variety A being a finitely generated abelian group).

So $\mathrm{Zar}_A(\iota S) = \iota C \subset A$, and therefore if C is of genus > 1, $\mathrm{Zar}_A(\iota S)$ is *not* a configuration in A despite the fact that ιS is contained in a finitely generated subgroup of $A(K)$. Therefore one needs to appropriately modify the statement of "absolute Mordell–Lang" if one wishes to obtain a result which is valid in characteristic p. This is the subject of the next section.

4. "Relative" Mordell–Lang in All Characteristics: Theorems of Manin, Grauert, Buium, Voloch, Hrushovski, etc.

THEOREM 4.1 ("RELATIVE" MORDELL–LANG IN ALL CHARACTERISTICS). *Let $k \subset K$ be an inclusion of algebraically closed fields. Let A be an abelian variety over K having the property that no positive dimensional factor abelian variety* *of A comes by base extension from an abelian variety over k. Let $S \subset A(K)$ be a subset generating a finitely generated subgroup of $A(K)$. Then*

$$\text{Zar}_A(S) = \text{Conf}_A(S).$$

There is, to be sure, a large literature dealing with the "classical version" of this theorem by which I mean the special case where $X := \text{Zar}_A(S)$ is a curve. This classical case was originally proved by Manin [1963] in characteristic zero. Another proof of it was given by Grauert [1965]. This latter proof was adapted by Samuel [1966] to make it work in characteristic p. In 1991, Voloch produced an extremely short, insightful, proof of this classical theorem under the auxiliary hypothesis that X is non-isotrivial, and its jacobian is ordinary [Voloch 1991].

To be sure, the statement of Theorem 4.1 in characteristic 0 is weaker than (and therefore follows immediately from) Theorem 3.1. So, it is only the characteristic p aspect of this theorem that is specifically new to our discussion. Progress towards the above theorem in characteristic p was made in [Abramovich and Voloch 1992], following on Voloch's approach. The full theorem is due to Hrushovski [1996], who writes that Buium's approach (to the characteristic 0 part of Theorem 4.1) inspired his own. Buium's ten-page paper [1992] is quite illuminating, and I would urge anyone who has not yet read it to do so! A certain universal jet space construction plays critical role in it. Briefly, given any affine smooth $\mathbb{C}$-scheme S with a derivation $\delta \in \text{Der}(\mathcal{O}_S)$, and any S-scheme X, Buium constructs a "pro-X-scheme" (i.e., a projective system of X-schemes) jet$(X/S, \delta)$ (call it J, for short) and he constructs a derivation δ_J on J lifting δ on S such that the pair (J, δ_J) satisfies the following universal property. For any pair $\eta = (Y, \delta_Y)$ where Y is an X-scheme and δ_Y is a derivation on Y lifting δ on S, there is a unique X-morphism jet$(\eta) : Y \to J = \text{jet}(X/S, \delta)$ which is *horizontal* in the sense that jet(η) intertwines the derivations δ_Y and δ_J. The method of Buium turns on the properties of jet$(X/S, \delta)$ where specifically X is a group scheme over S; this method depends especially on the manner in which finitely generated groups of S-sections in group schemes X lift to jet$(X/S, \delta)$. The fact that the theory of ordinary differential equations works so well in the complex analytic category is essential. In characteristic p, things aren't so smooth-going. People with some algebraic geometric background who wish to see a bridge between Buium's techniques and the model-theoretic techniques of Hrushovski might find it useful to read Chapter 2 of Scanlon's thesis [1997b], where he introduced the notion of $\mathcal{D}$-rings and $\mathcal{D}$-functors which serve as "jet-theoretic" technology suitable for use in characteristic p.

As with the Absolute Mordell–Lang Theorem, the "Relative" Mordell–Lang result we have just formulated has an alternate version:

THEOREM 4.2 ("RELATIVE" MORDELL–LANG IN ALL CHARACTERISTICS, SECOND VERSION). *Let $k \subset K$ be an inclusion of algebraically closed fields. Let A*

be an abelian variety over K having the property that no positive dimensional factor abelian variety of A comes by base extension from an abelian variety over k and suppose that we are given a subset $S \subset \Gamma \subset A(K)$ where Γ is a finitely generated subgroup of $A(K)$. If Γ is Zariski-dense in A and the configuration-closure of the image of S in the real vector space $V = \Gamma \otimes \mathbb{R}$ is all of V, then S is Zariski-dense in A.

The equivalence between Theorems 4.1 and 4.2 is proved in exactly the same way as the equivalence between Theorems 3.1 and 3.2 were proved.

5. Models in the Sense of Algebraic Geometry

If you start with the utterly general fields that appear in the statement of Theorems 3.1, 3.2, 4.1, and 4.2, here is the standard way of cutting down to the study of reasonably small fields. Consider, for example, the data of any of these theorems, say Theorem 4.2: that is, we are given (k, K, A, S) with $k \subset K$, an inclusion of algebraically closed fields, A an abelian variety over K having the property that no positive dimensional factor abelian variety of A comes by base extension from an abelian variety over k, and $S \subset \Gamma \subset A(K)$, a subset S generating Γ, a finitely generated subgroup of $A(K)$. Fix elements $\{\gamma_1, \ldots, \gamma_\nu\}$ that generate Γ.

If, contrary to the conclusion of Theorem 4.2, we were dealing with a counter-example to the assertion of that theorem, there would be some hypersurface $D \subset A$ which contains S but which does not contain some element $\gamma \in \Gamma - S$. Call such a pair (D, γ) a *witness to the fact that A, S, Γ is a counter-example* to Theorem 4.2, or just a *witness* for short. Similarly we can talk about the notion of "counter-example witness" to Theorem 3.2.

Let $k_0 \subset k$ denote the prime field (i.e., it is $\mathbb{Q}$ if we are in characteristic 0 and $\mathbb{F}_p$ if we are in characteristic p). Consider a specific set of equations defining the abelian variety A over K in projective space; for concreteness we can take A as given as an intersection of a finite collection of quadrics in some high-dimensional projective space, following Mumford [1966; 1967]). These equations have, all in all, only a finite number of coefficients and so do all the coordinates (in projective space) of the finitely many points $\gamma_1, \ldots, \gamma_\nu$. If we are presented with a *witness* (D, γ) to the fact that A, S, Γ is a counter-example, we add to this set of coefficients of these equations the coefficients of the equations defining the hypersurface D. Letting $\mathcal{C}$ denote the set of all the coefficents enumerated above, we see that $\mathcal{C} \subset K$ is finite. It follows that the subfield $K_0 := k_0(\mathcal{C}) \subset K$ is finitely generated over the prime field k_0, and moreover we have $S \subset \Gamma \subset A(K_0)$ (and in the case where we have prescribed witness (D, γ) we can get that D is defined over K_0 as well). We have then, a *finitely generated model* for our putative witnesses. For example, if we had a witnessed counter-example to Theorem 3.2 over any algebraically closed field K of characteristic 0 (that is, if we had A, $\Gamma \subset A(K)$, $S \subset \Gamma$, $S \subset D \subset A$, and $\gamma \in \Gamma$ with $\gamma \notin D$, satisfying the hypotheses required; i.e., that A is an abelian variety, Γ is a finitely generated subgroup of its group of K-rational points, S is configuration-dense in $\Gamma \otimes \mathbb{R}$, and D a hypersurface in A) the above discussion shows that we would also have such a witnessed counter-example all of whose ingredients are given over a field K_0 which is finitely generated over $\mathbb{Q}$.

COROLLARY 5.1. *To prove Theorems 3.1, 3.2, 4.1 and 4.2 it suffices to treat the case of fields K which are the algebraic closure of fields of finite transcendence degree over the prime fields; i.e., where K ranges through the algebraic closures of $k_0(x_1, \ldots, x_d)$ (for $d = 1, 2, \ldots$) where the k_0's are the prime fields (so that in the case of Theorems 3.1 and 3.2, $k_0 = \mathbb{Q}$), and where in the case of Theorem 4.1 and 4.2, k is an algebraic closure of k_0.*

Producing a "complex" or a "p-adic" model. In the case where K is of characteristic 0 it is sometimes useful to make use of complex analytic, or p-adic analytic methods. Imagine that we have given ourselves a witnessed counter-example to Theorem 3.2, which by Corollary 5.1 can be taken to be over a field K_0 which is of finite transcendence degree over $\mathbb{Q}$. Since any field of finite transcendence degree over $\mathbb{Q}$ is isomorphic to a subfield of $\mathbb{C}$, we may choose an imbedding $K_0 \hookrightarrow \mathbb{C}$ and make the base change (for our model) from K_0 to $\mathbb{C}$ gives us a model over the complex numbers. Similarly, let p be any prime number, and noting the fact that as an "abstract" field, $\bar{\mathbb{Q}}_p$ is the extension of $\mathbb{Q}$ given by adjoining an uncountable number of independent variables and then passing to the algebraic closure of the field so obtained, we see that there exists an imbedding $K_0 \subset \bar{\mathbb{Q}}_p$. Identify, then, K_0 with a subfield of $\bar{\mathbb{Q}}_p$ and form the compositum $E = \mathbb{Q}_p \cdot K_0 \subset \bar{\mathbb{Q}}_p$. Since K_0 is a finitely generated field extension of $\mathbb{Q}$, E is a finitely generated, algebraic, field extension of $\bar{\mathbb{Q}}_p$. It follows that $E/\bar{\mathbb{Q}}_p$ is of finite degree. As before, if we make the base change (for our model) from K_0 to E we get our sought-for model over E.

COROLLARY 5.2. *Let p be any fixed prime number. In proving Theorems 3.1 and 3.2 by reductio ad absurdum it suffices to assume that there is a counter-example defined over a finite field extension E of $\mathbb{Q}_p$, and then prove its nonexistence.*

Call such a putative counter-example a "p-adic model" of a counter-example. If we wish, we may make a further finite extension of our base field E so that the abelian variety A of our p-adic model over E has semi-stable reduction over the ring of integers of E (by a theorem of Grothendieck [1972]). As we shall see below, if we are willing to exclude a finite set of ("bad") primes p we may find finite field extensions $E/\mathbb{Q}_p$ and models over E for which the abelian variety A has good reduction over the ring of integers of E.

If our original putative counter-example is over a field of characteristic p, a similar argument as we have just given will allow us to produce a counter-example over the field $\mathbb{F}_q((t))$ of (finite-tailed) Laurent series with coefficients in a finite field $\mathbb{F}_q$ of cardinality $q = $ a power of p.

In our search for models, we needn't work only over fields: we can find subrings $r_0 \subset k_0$ and $r_0 \subset R_0 \subset K_0$ such that r_0 and R_0 are finitely generated rings over $\mathbb{Z}$, where $r_0 = \mathbb{F}_p$ if K_0 is of characteristic p, and $r_0 = \mathbb{Z}[1/m]$ for some positive integer m if K_0 is of characteristic zero, and such that

- the equations for A give us an abelian scheme, call it $\mathcal{A}_0$, such that
- the elements $\gamma_1, \ldots, \gamma_\nu$ are R_0-valued points of the R_0-scheme $\mathcal{A}_0$, and
- given a specific witness (D, γ), if it exists, the hypersurface D can be taken to be a relative Cartier divisor over R_0, and denoting $W_0 := \operatorname{Spec} R_0$ which is a scheme over $w_0 := \operatorname{Spec} r_0$ we may also arrange it so that (for fun)
- $W_0 \to w_0$ is a smooth surjective morphism of schemes.

We therefore have an abelian scheme

$$\mathcal{A}_0 \to W_0,$$

over the smooth w_0-scheme W_0 with the structures described above. The "picture", when $r_0 = \mathbb{Z}[1/m]$, looks as follows:

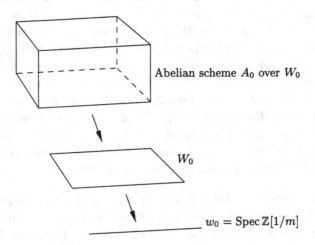

Abelian scheme $\mathcal{A}_0$ over W_0

W_0

$w_0 = \operatorname{Spec} \mathbb{Z}[1/m]$

At this point, here are some things we can do:

Producing a "p-adic model" with good reduction. If we are given a putative counter-example model over a field of characteristic 0, then $r_0 = \mathbb{Z}[1/m]$ for some nonzero integer m, and we have our geometric model $\mathcal{A}_0 \to W_0$, as above with witness (D, γ), where the hypersurface D is a relative Cartier divisor over R_0, and R_0 is as described above over $\mathbb{Z}[1/m]$, the mapping of schemes $\pi : \operatorname{Spec}(R_0) \to \mathbb{Z}[1/m]$ being smooth and surjective. After possibly "augmenting m", i.e., making the base change $\operatorname{Spec}(R_0) \mapsto \operatorname{Spec}(R_0 \otimes_{\mathbb{Z}[1/m]} \mathbb{Z}[1/m'])$ for m' a suitable nonzero multiple of m (we assume this done without changing our notation) we can prepare the ring R_0 via "Noether normalization" [Bourbaki 1972, Chapter V, Section 4, Corollary 1] so that R_0 contains a polynomial ring in a finite number of variables $\mathbb{Z}[1/m][x_1, \ldots, x_\nu] \subset R_0$, and such that R_0 is an integral extension of $\mathbb{Z}[1/m][x_1, \ldots, x_\nu]$. Let p be any prime number not dividing m. We choose any ring homomorphism $\xi : \mathbb{Z}[1/m][x_1, \ldots, x_\nu] \to \mathbb{F}_p$ (and, of course, there are some). Now find any system of ν elements $\alpha_j \in \mathbb{Z}_p$ (for $j = 1, \ldots, \nu$) which are transcendentally independent over $\mathbb{Z}$ taking care to choose them so that

$$\xi(x_j) = \alpha_j \mod p,$$

for $j = 1, \ldots, \nu$ (and we can do this). We use this system of α_j's to imbed $\mathbb{Z}[1/m][x_1, \ldots, x_\nu]$ in $\mathbb{Z}_p$. Call the imbedding $\alpha : \mathbb{Z}[1/m][x_1, \ldots, x_\nu] \hookrightarrow \mathbb{Z}_p$. Since R_0 is an integral domain, and finitely generated and integral (hence finite) over $\mathbb{Z}[1/m][x_1, \ldots, x_\nu]$, we may extend α to an imbedding of R_0 into a finite discrete valuation ring extension $\mathcal{O}$, of $\mathbb{Z}_p$. Making the base change from R_0 to $\mathcal{O}$, we transfer our model over R_0 to $\mathcal{O}$, and letting $E/\mathbb{Q}_p$ be the field of fractions of $\mathcal{O}$, we have constructed a "p-adic model" (over the discrete valued field E) for which the abelian variety $\mathcal{A}_0$ has good reduction over the residue field of E.

6. Proof of Absolute Mordell–Lang in Characteristic 0 for Curves of Rank 1

The beautiful classical argument of Chabauty may be adapted to give an elementary proof of Theorem 3.2 when the rank of Γ is equal to 1. Here is the briefest sketch of this argument; for a more general statement, see [Hindry 1988, Section 7]. If rank$(\Gamma) = 1$ the condition that the configuration-closure of the image of S in the real vector space $V = \Gamma \otimes \mathbb{R}$ is *all* of V boils down to simply saying that S is infinite. Since Theorem 3.2 would be immediate if A were of dimension ≤ 1 we may assume that the dimension of A is 2 or more. Suppose that Theorem 3.2 were false. Then there would be a hypersurface D in A containing S but not containing Γ. Now there is a "p-adic model" for this. That is, there is a prime number p and a finite extension K of $\mathbb{Q}_p$ admitting the same situation, i.e., an abelian variety A over K and a hypersurface $D \subset A$ over K, and $S \subset \Gamma \subset A(K)$ all with the same properties as before (e.g., $S \subset D(K)$). The existence of such "p-adic models" is explained in Corollary 5.2 above. Choose such a p-adic model, and note that $A(K)$ has the structure of (compact) p-adic Lie group. Let $\bar{\Gamma}$ denote the topological closure of $\Gamma \subset A(K)$, and note that (by the basic theory of the p-adic logarithm, which identifies an open neighborhood of $A(K)$ with an open "additive" subgroup in $K^{\dim A}$) we may identify $\bar{\Gamma}$ with a one-dimensional p-adic Lie subgroup of the p-adic Lie group $A(K)$. Since Γ is Zariski-dense in A, it follows that any open(-closed) subgroup of finite index in $\bar{\Gamma}$ is also Zariski-dense in A. Since S is infinite, it has at least one limit point in the p-adic Lie group $A(K)$ and all of the limit points of S lie in $D(K) \cap \bar{\Gamma} \subset A(K)$.

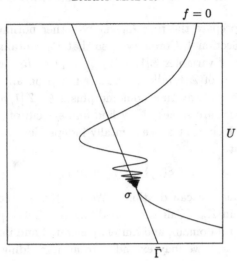

$$f = 0$$

U

σ

$\bar{\Gamma}$

Fix a limit point σ of S. Consider a local equation f for D in a neighborhood $\mathcal{U} \subset A(K)$ of σ. We view f as a p-adic analytic function on $\mathcal{U} \subset A(K)$, and restricting it to the intersection of $\mathcal{U}$ with the p-adic Lie subgroup $\bar{\Gamma}$, we see that f must *vanish* on some open neighborhood of σ in $\mathcal{U} \cap \bar{\Gamma}$ because it vanishes on the infinite set of points of $\mathcal{U} \cap S$ which has σ as accumulation point. It follows that $\bar{\Gamma}$ is contained in a finite union of translates of D, contradicting the Zariski-density of Γ.

About effectivity. Before we can discuss the status of *effectivity* of the above proof, we have to reformulate the theorem that it proves in a manner so that the result is *providing* something for us, and only then can we unambiguously ask whether or not our proof is providing that thing "effectively". One natural reformulation is to say that given the hypotheses of Theorem 3.2, in the case where V is of dimension one, and given any hypersurface $D \subset A$, there are only a finite number of points in the intersection $D \cap \Gamma$. Now even in this restricted setting– the above Chabauty-type argument seems to be giving us the same level of effectivity (and no more) that the general results of Vojta and Faltings provide. Namely, we can refine the above argument to give a *number-effective* statement, but not, it would seem, a *size-effective* one. But see [Coleman 1985a]!

7. The Reduction of the Proof of Mordell–Lang to Global Fields

It seems that there are (at least) two possible strategies for performing this reduction step. One approach is suggested in a few words in [Faltings 1994, Section 5]. We shall, up to a certain point, give a detailed account of how to follow this out. I say "up to a certain point" because, as the reader will see, we will be asserting that a demonstration given explicitly in the literature for number fields, also works for global fields. I am thankful to Paul Vojta for substantial help here. The alternative strategy requires, among other things, an excursus about the specialization properties of maximal abelian subvarieties in subvarieties in abelian varieties (sic). All the essential materials needed for this second strategy can be found in the literature. (See [Lang 1983, Chapter 9], especially Theorem 6.2 and Corollary 6.3, together with either [Hindry 1988, Appendice I] and, in particular, Lemma A there, or alternatively [Abramovich 1994; McQuillan 1994]. For further background I found [Raynaud 1983a, Section III] helpful.) We end this section with a sketch of how this second approach works, spelling out some (minor) aspects of the proof not found in the sources. I am thankful to Michael McQuillan and Dan Abramovich for substantial help with this.

First Approach. Recall that, as in Corollary 5.1 above, we know that to prove Theorems 3.1 and 3.2 it suffices to work over base fields K which are finitely generated over $\mathbb{Q}$. "Filtering" such a K by a sequence of subfields, each of transcendence degree 1 over the next, we shall see that it suffices to prove appropriately stated versions of Theorems 3.1 and 3.2:

(a) for number fields F, and

(b) for fields F of rational functions on curves C over base fields F_0, given that we have already proved (appropriately stated versions of) Theorems 3.1 and 3.2 for F_0.

In either case, our field F will therefore have the structure of a *global field* and therefore it may pay to recall this fundamental notion of *global field structure* (which was given center stage in Artin–Whaples' treatment of algebraic number theory and has kept that position in every subsequent treatment of the subject). We will be referring specifically to the discussion in [Lang 1983, Chapter 2, Section 1]. Recall that a valuation v on a field F of characteristic 0 is called *proper* if it is nontrivial, and its restriction to $\mathbb{Q}$ is either trivial, the negative logarithm of ordinary absolute value or a p-adic valuation for some prime number p. (If F is of positive characteristic, for the valuation v to be *proper*, it is also required to be "well-behaved", a condition automatically satisfied in characteristic 0; see [Lang 1983].) By a *global field structure* on a field F we mean a collection of "proper" valuations M_F of F, and of multiplicities $v \mapsto \lambda_v > 0$ for $v \in M_F$ such that for every $x \in F^*$, we have the *summation formula with the multiplicities* $v \mapsto \lambda_v$,

$$\sum_{v \in M_F} \lambda_v \cdot v(x) = 0.$$

Here we have written things additively, rather than multiplicatively as was done in [Lang 1983]. Our valuations $v : F^* \to \mathbb{R}$ are related to the absolute values $|\ |_v$ there by the standard formula $v(x) = -\log |x|_v$.

Recall that if F is endowed with a global field structure, and E/F is a finite field extension, there is a unique global field structure on E extending that on F (for details, again consult loc. cit.).

The examples of global field structures are precisely the fields that enter into the cases (a) and (b) above. More precisely, if F is a number field, we take M_F to be the standard collection of normalized absolute values corresponding to the archimedean and non-archimedean places of F and λ_v the collection of multiplicities as set out in [Lang 1983, Chapter 2, Section 1]. If F is expressed as a field of rational functions of a proper smooth curve C defined over a subfield $F_0 \subset F$, i.e., if we can write $F = F_0(C)$, we may impose a global field structure on K, again in the standard manner (see loc. cit.). Note that if F is of transcendence degree at least 2 over the prime field, there are *many* possible global field structures we can impose on F. Nevertheless (when no confusion can arise) a field F endowed with a specific global field structure will be called a "global field" and we will denote it (along with its global field structure) simply F.

Given a global field F, with $\bar{F}/F$ an algebraic closure, it is standard (see [Lang 1983, Chapters 3–5] for the details) to define the associated *height function* on $\bar{F}$-rational points of N-dimensional projective space (any N)

$$h_F : \mathbb{P}^N(\bar{F}) \to \mathbb{R}.$$

If $\iota : X \hookrightarrow \mathbb{P}^N$ is a projective variety defined over F, we can restrict the above height function to $\bar{F}$-rational points of X and divide by the degree of X to obtain a height function "on X" which we denote

$$h_{F,X,\iota} : X(\bar{F}) \to \mathbb{R};$$

that is, $h_{F,X,\iota}(P) = \frac{1}{d} \cdot h_F(\iota(P))$ where d = the degree of the projective variety $\iota(F) \subset \mathbb{P}^N$. [In the special case where $\mathrm{Pic}(F) = \mathbb{Z}$ (e.g., if $F = C$ is a smooth projective curve), dividing by d has the effect of making the real-valued function $h_{F,X,\iota}$ on the set $F(\bar{K})$ independent (modulo bounded functions) of the projective imbedding ι; i.e., it depends only on the global field F and X (mod $O(1)$).]

Given a global field F, an abelian variety A defined over K, and a line bundle L over A defined over F, we also have the normalized Néron–Tate height function on $\bar{F}$-rational points of A [Lang 1983, Chapter 5, Section 3]

$$\hat{h}_{F,A,L} : A(\bar{K}) \to \mathbb{R}.$$

All of these height functions depend upon the global field structure of F.

The Ueno–Kawamata structures on subvarieties of abelian varieties.

For a compendium of the results we are about to cite, and for complete references to the literature containing their proofs, see [Lang 1991, Chapter I, Section 6]. Recall *Ueno's Theorem* (loc. cit.), which says if $X \subset A$ is an irreducible subvariety of an abelian variety (over a field) and if $B \subset A$ is the connected component of the subgroup of translations of A that preserve X, then if $Y := X/B \subset A/B$

is the quotient variety of X under the action of the group B we have that Y is a variety of general type (or, in the terminology of [Lang 1991], is *pseudo-canonical*). Refer to the morphism $X \to Y$ as the *Ueno fibration* of X. Recall *Kawamata's Structure Theorem* that says if $X \subset A$ is an irreducible subvariety of general type (alias: pseudo-canonical) and the base field is of characteristic 0, then there is a finite number of proper subvarieties $Z_i \subset X$ $(i = 1, \ldots, \nu)$ whose Ueno fibrations $Z_i \to Y_i$ have positive fiber dimension, and such that any finite union of translates of nontrivial abelian subvarieties in X is actually contained in

$$Z := \bigcup_{i=1}^{\nu} Z_i \subset X.$$

Refer to $Z \subset X$ as the *Kawamata locus* of X.

THEOREM 7.1 (FALTINGS, EXTENDING THE METHOD OF VOJTA). *Let F be a global field of characteristic 0, A an abelian variety over F, and L any line bundle over A (defined over F). Let $X \subset A$ be an irreducible subvariety of general type (defined over F) and $Z \subset X$ its Kawamata locus. Then the set of F-rational points on X not lying in Z has bounded (Néron–Tate) height. That is, there is a bound B such that*

$$\hat{h}_{F,A,L}(x) \leq B$$

for all $x \in X(F) - Z(F)$.

For a proof of this the reader might consult [Faltings 1994]; in that reference Faltings only states the theorem for F a number field, but his proof goes through, word-for-word, for global fields of characteristic 0. Vojta [1993] has given another account of the proof of this same result (also stated only for number fields, but he assures me that his account of the proof works as well, with no change, in the context of global fields).

We shall now sketch the proof of why the preceding theorem implies Theorem 3.1 (and therefore also Theorem 3.2). Firstly, let A be an abelian variety over F and $S \subset A(F)$ as hypothesized in Theorem 3.1, where F is a field which is finitely generated over $\mathbb{Q}$. We shall prove Theorem 3.1, in effect, by induction on the transcendence degree of F, and the dimension of A. We view F with a given global field structure, coming either because F is a number field, or else by writing $F = F_0(C)$ where C is a curve. In the latter case, we may assume by induction that Theorem 3.1 holds for abelian varieties A_0 over the field F_0, and for subsets $S_0 \subset A_0(F_0)$ generating finitely generated subgroups. We may also assume that Theorem 3.1 holds for abelian varieties over F of strictly lower dimension than the dimension of A. Consider $\text{Zar}_A(S)$, the Zariski-closure of the set S.

Step 1. Reduction to the case where $\mathrm{Zar}_A(S)$ is an irreducible variety of general type. Writing $\mathrm{Zar}_A(S) = \bigcup_j X_j$ as a finite union of irreducible varieties, we may cover the set S by a finite union of subsets $S_j = S \cap X_j(F)$ and note that it suffices to prove Theorem 3.1 for each of these S_j's separately. We may assume, therefore, that $X := \mathrm{Zar}_A(S)$ is irreducible, and then, using Ueno's Theorem, we pass to an appropriate quotient abelian variety of A in which X is of general type. Here, note that in this reduction step we have possibly reduced (but we have not increased) the dimension of A.

Step 2. Applying the theorem. We apply the above theorem for a choice of ample line bundle L over A to get that there is a bound B such that S breaks up into the union of two sets, the part of S contained in Z, $S_1 := S \cap Z(F)$ and the part of small height,

$$S_2 := \{x \in S \mid \hat{h}_{F,A,L}(x) \leq B\}.$$

Now we can apply Step 1 again to S_1, and note that here the application of Step 1 is guaranteed to *reduce* the dimension of the ambient abelian variety A; our inductive hypothesis therefore proves Theorem 3.1 for S_1. We may assume, then, that

$$S = S_2 = \{x \in S \mid \hat{h}_{F,A,L}(x) \leq B\}.$$

If F is a number field, it follows that S is finite, and so again Theorem 3.1 follows. We have reduced ourselves, therefore, to the case where $F = F_0(C)$ and S is of bounded height. After applying Step 1 again, we may assume that $X := \mathrm{Zar}_A(S)$ is irreducible (of general type). Let $B \subset A$ be the F/F_0-trace (that is, the "largest abelian subvariety of A defined over F_0"; [Lang 1991, Chapter I, Section 4]) and apply [Lang 1983, Chapter 6, Theorem 5.3] to guarantee that S is contained in a finite union of cosets of $B(F_0)$ (in $A(F)$). In particular, S is contained in a configuration in A consisting of a finite union of translates of B. We may assume that one of these translates covers X, and translating back to the origin, we may simply assume now that $S \subset B(F_0)$. Since, by our inductive hypothesis, Theorem 3.1 holds for the abelian variety B over the field F_0 and for $S \subset B(F_0)$, we are done.

Remarks regarding the second method. Here, as mentioned above, the relevant literature is [Lang 1983, Chapter 9] and either [Hindry 1988, Appendice I] or [Abramovich 1994; McQuillan 1994]. To be brief, we start by performing Step 1 of the first method, and therefore we have reduced ourselves to proving, simply, that if A is an abelian variety over $\mathcal{K}$ a field finitely generated over $\mathbb{Q}$, $\Gamma \subset A(\mathcal{K})$ is a finitely generated group, and $X \subset A$ is an irreducible subvariety of general type, then $X \cap \Gamma$ is not Zariski dense in X. For short, refer to the triple (A, Γ, X) as ξ. Also, by induction we assume that this has already been proved for all triples $\xi' = (A', \Gamma', X')$ satisfying the same properties (meaning that A' an abelian variety over $\mathcal{K}$, Γ' a finitely generated subgroup of the Mordell–Weil group of

A' ,and $X' \subset A'$ a subvariety of general type) where X' is of dimension strictly smaller than X and $\dim(A') \leq \dim(A)$, or A' is of dimension strictly smaller than A. Moreover, we assume Mordell–Lang for number fields!

Also, we will be availing ourselves of a *good* model of ξ over R, an integral domain and regular, finitely generated over $\mathbb{Q}$, whose field of fractions is $\mathcal{K}$. We will be explaining what we mean by *good* presently. But we begin with a model of the following form: an abelian scheme A over R, and $\Gamma \subset A(R)$ a finitely generated subgroup of the Mordell–Weil group over R, and $X \subset A$ a closed irreducible subvariety flat over R, such that the fiber of the triple $\xi_R = (A, \Gamma, X)$ over $\mathcal{K}$ is ξ.

Ueno–Kawamata collections. For any homomorphism $R \to K$, with K a field, let $\xi_{/K} = (A_{/K}, \Gamma_{/K}, X_{/K})$ denote the "fiber" of our triple over K. We may apply the Ueno–Kawamata theory to the subvariety $X_{/K} \subset A_{/K}$ obtaining the Kawamata locus $Z(X_{/K}, A_{/K}) \subset X_{/K}$ and after possible finite field extension of K we may write $Z(X_{/K}, A_{/K}) = \bigcup_{j=1}^{\nu} Z_j(X_{/K}, A_{/K})$ with Z_j geometrically irreducible, and having Ueno fibrations denoted $\eta_j : Z_j \to X_j$ where the mapping η_j is induced from the natural projection $A \to A/A_j$ for A_j some abelian variety of positive dimension, and such that $X_j \subset A/A_j$ is of general type. We define inductively, a (finite) set $\mathcal{C}(A_{/K}, X_{/K})$ of pairs (B, Y), such a pair consisting of an abelian subvariety of positive dimension $B \subset A$ and an irreducible subvariety of general type $Y \subset A/B$. For this purpose we note that single points are deemed to be of general type (and, of course, the Kawamata locus of a point is empty). An inductive definition of the *Ueno–Kawamata Collection* can be culled from the following axioms:

0. If $A_{/K} = 0$ then $\mathcal{C}(A_{/K}, X_{/K})$ is empty.

1. If the Kawamata locus of $X_{/K}$ is empty, and $A_{/K}$ is of positive dimension, then $\mathcal{C}(A_{/K}, X_{/K})$ consists of the single element $(A_{/K}, X_{/K})$.

2. Keep the notation as in the preceding paragraph, but drop the subscript $/K$ in some of the terms to make it easier to read. For each index $1 \leq j \leq \nu$ and each pair $(B', Y') \in \mathcal{C}(A/A_j, X_j)$ form the pair (B, Y) where $B \subset A$ is the abelian subvariety which is the inverse image in A of $B' \subset A/A_j$ under the homomorphism $A \to A/A_j$, and where $Y \subset A/B$ is the subvariety $Y' \subset (A/A_j)/B' = A/B$. The elements of the set $\mathcal{C}(A_{/K}, X_{/K})$ are precise all these pairs (B, Y) together with the pair $(A_{/K}, X_{/K})$.

We have a partial ordering on Ueno–Kawamata collections where, by definition, $(B', Y') \leq (B, Y)$ if $B \subset B'$ and the inverse image of Y' in A/B is contained in Y.

"Good" Models. Consider $\xi_R = (A, \Gamma, X)$, as described above, and the Ueno–Kawamata collections $\mathcal{C}(A_{/K}, X_{/K})$ for each homomorphism $R \to K$ where K is a field. We say that our model is *good* if there is a collection which we can denote $\mathcal{C}(A_{/R}, X_{/R})$ of pairs $(B_{/R}, Y_{/R})$ where $B_{/R} \subset A_{/R}$ is an abelian subscheme over

R, and $Y_{/R} \subset (A/B)_{/R}$ is a closed subscheme, flat over R, with the following property:

"THE UENO–KAWAMATA COLLECTION SPECIALIZES WELL": *For each $R \to K$, with K a field, the "fiber" of the collection of pairs $\mathcal{C}(A_{/R}, X_{/R})$ over K (i.e., the set of pairs $(B_{/K}, Y_{/K})$ obtained by restriction to the fiber over K of the pairs $(B_{/R}, Y_{/R})$ in $\mathcal{C}(A_{/R}, X_{/R})$) is the Ueno–Kawamata collection $\mathcal{C}(A_{/K}, X_{/K})$.*

Here one has a choice. One may use the results in [Hindry 1988, Appendix 1] (in particular, Lemme A), which guarantee that the generic Kawamata locus specializes to the special Kawamata locus outside a proper closed subscheme in $\mathrm{Spec}(R)$. An alternative argument for this using a method of Abramovich [1994] is found in [McQuillan 1994], where (loc. cit., Theorem 1.2) it is shown that there is a closed subscheme of $X_{/R}$ whose fibers (over R) are the Kawamata loci of the fibers of $X_{/R}$.

An inductive application of these results to the successive tiers in the Ueno–Kawamata collection allows us to conclude that we have such a "good model".

Specialization of Γ. Fix a "good model" as described above.

LEMMA 1. *There exists a closed point u of $\mathrm{Spec}(R)$ such that for every pair $(B_{/R}, Y_{/R}) \in \mathcal{C}(A_{/R}, X_{/R})$ the specialization mappings of $\Gamma / \Gamma \cap B \subset (A/B)(R)$ to $(A/B)(k(u))$ is injective; i.e., the specialization mapping allows us to identify $\Gamma / \Gamma \cap B$ with a subgroup $\Gamma / \Gamma \cap B \hookrightarrow (A/B)(k(u))$. Here $k(u)$ is the residue field of the point $u \in \mathrm{Spec}(R)$.*

PROOF. Apply [Lang 1974, Chapter 9, Theorem 6.2] to the finitely generated subgroup $\prod \Gamma / \Gamma \cap B$ of the Mordell–Weil group of the abelian scheme $\prod A/B$, where the products range over all abelian subschemes B occurring in pairs in the Ueno–Kawamata collection $\mathcal{C}(A_{/R}, X_{/R})$. $\square$

Specializing points in the complement of the Kawamata locus. Fix such a closed point $u \in \mathrm{Spec}(R)$ with the properties guaranteed by Lemma 1. Let S denote the set of points in $\Gamma_{/\mathfrak{K}} \cap X_{/\mathfrak{K}}$ which lie outside the Kawamata locus $Z(X_{/\mathfrak{K}}, A_{/\mathfrak{K}})$. (We are eventually aiming to prove that S is finite.) For a pair $(B, Y) \in \mathcal{C}(A_{/R}, X_{/R})$ consider the subsets $S(B, Y) \subset S$ consisting of those elements of S, which when projected to A/B and specialized to $k(u)$ land in $Y(k(u))$. Clearly,

$$S = \bigcup S(B, Y),$$

where the union is taken over all pairs in the Ueno–Kawamata collection. Given a point $y \in Y(k(u))$ let $S(B, Y, y) \subset S$ consist of the elements of $S(B, Y)$, which when projected to A/B and specialized to $k(u)$ map to y.

LEMMA 2. *The sets $S(B, Y, y)$ are finite.*

PROOF. By the definition of $S(B, Y, y)$ and the injectivity guaranteed by Lemma 1, we have that the projection of the subset $S(B, Y, y) \subset \Gamma$ to $\Gamma / \Gamma \cap B$ consists of a single point, and therefore $S(B, Y, y)$ lies in a coset of the abelian subvariety B. Since B is of strictly smaller dimension than A, our inductive hypothesis guarantees that the Zariski-closure of $S(B, Y, y)$ is a union of translates of abelian subvarieties of B, and since $S(B, Y, y)$ is external to the Kawamata locus of X this union must only contain abelian varieties of dimension zero, i.e., $S(B, Y, y)$ is finite. $\square$

Conclusion of the proof by downwards induction on the dimension of the Y's.

PROPOSITION. *The set S is finite.*

PROOF. Consider the following statement:

P(N): The union of the subsets $S(B, Y)$ where (B, Y) range through all pairs in the Ueno–Kawamata locus such that $\dim(Y) \leq N$ has *finite* complement in S.

By Lemma 2, we see that **P**(0) implies that S is finite. Clearly **P**(d) is true for $d = \dim(X)$. For $N > 0$, suppose **P**(N) and we shall show **P**$(N-1)$. There are only finitely many pairs (B, Y) in the Ueno–Kawamata locus with $\dim(Y) = N$. Fix one such pair (B, Y). Now invoke Mordell–Lang for $Y_{/k(u)}$: there are only a finite number of $k(u)$-rational points of $Y_{/k(u)}$ which lie outside its Kawamata locus. By Lemma 2, only a finite subset of S which specializes to this finite set of points. Excluding those points, the remainder specialize to the Kawamata locus of Y, and therefore the complement of a finite subset of S lies in the union of the subsets $S(B', Y')$ where (B', Y') range through all pairs in the Ueno–Kawamata locus such that $\dim(Y') \leq N - 1$, giving our proposition. $\square$

8. Number-Effectivity Revisited

In the introduction we asked briefly about possible levels of effectivity we might hope for in connection with the finiteness of numbers of solutions of a certain class of Diophantine equations (we chose *hyper-elliptic curves* as our illustrative class). In this section I would like to discuss number-effectivity questions in the general context of curves of genus at least 2 but with some attention paid to uniformity over families of curves and the base fields. As Teresa de Diego [1997] has explained, the methods of Vojta, Faltings, and Bombieri produce number-effective upper bounds of a very precise nature. Let X be a smooth projective curve of genus > 0 defined over $\bar{\mathbb{Q}}$, and consider the "canonical" mapping of X to its jacobian (abelian) variety A,

$$j : X \hookrightarrow A$$

by the rule $x \mapsto$ the linear equivalence class of the divisor $(2g-2)[x] - \kappa$, where κ is (a choice of) canonical divisor for the curve X. Now consider the Néron–Tate height function $\hat{h} : A(\bar{\mathbb{Q}}) \to \mathbb{R}$, where $\hat{h} := \hat{h}_{\bar{\mathbb{Q}}, A, \mathcal{L}}$ in the notation that we introduced earlier, where $\mathcal{L}$ is the "Poincaré" line bundle on the jacobian

$A = \mathrm{jac}(X)$. Composing the $\hat{h} : A(\bar{\mathbb{Q}}) \to \mathbb{R}$ with $j : X(\bar{\mathbb{Q}}) \to A(\bar{\mathbb{Q}})$, gives us the height function on algebraic points of X, which we will continue to refer to as "Néron–Tate height":

$$\hat{h}(X, P) := \hat{h}(j(P)) \in \mathbb{R}.$$

Our height function $\hat{h}(X, P)$ is "canonical", in the sense that it depends upon nothing more than X and P. For our purposes below we need to make a further *normalization* of this height function, scaling it relative to the "height" of the curve X itself. At present, we must do this in a slightly ad hoc way. Eventually we might want a thoroughly *canonical* normalization (e.g., perhaps the correct thing to do is to replace $\hat{h}(X, P)$ by the ratio

$$\frac{\hat{h}(X, P)}{\sqrt{\omega_X^2}},$$

where ω_X is the canonical divisor attached to a semi-stable model of the curve X over the ring of integers in an appropriate number field, as given in Arakelov's Theory; see [Lang 1988]). But a perfectly serviceable, and more down-to-earth if less canonical, normalization, which works uniformly for families of curves parametrized by specific quasi-projective varieties, is as follows. Let $\iota : T \hookrightarrow \mathbb{P}^N$ be an irreducible quasi-projective variety which is a parameter space for a family of curves of genus > 1. That is, we give ourselves a smooth proper family $\mathfrak{X} \to T$ with fibers $\mathfrak{X}_t$ equal to (smooth proper) curves of genus > 1 as t ranges through the $\bar{\mathbb{Q}}$-rational points of T. Now, for an algebraic point $P \in \mathfrak{X}_t(\bar{\mathbb{Q}})$, define

$$h(\mathfrak{X}_t, P) := \frac{\hat{h}(\mathfrak{X}_t, P)}{1 + \sqrt{h_{\bar{\mathbb{Q}}, T, \iota}}}.$$

For example, we can take the hyper-elliptic family $y^2 = g(x)$ discussed in the introduction, where $g(x) = \sum_{j=0}^{d} g_j x^j$ is a polynomial of degree d (at least 5) with no multiple roots, and the variety $T = T_d$ is the open variety in affine $(d + 1)$-space over $\mathbb{Q}$ (parametrized by $g_0, \ldots, g_d$) which is the complement of $g_d = 0$ and the discriminant locus.

But fix any quasi-projective variety $\iota : T \hookrightarrow \mathbb{P}^N$ over a number field $k \subset \bar{\mathbb{Q}}$ which parametrizes a family, $\mathfrak{X} \to T$, of (smooth projective) curves of genus ≥ 2. For C a positive number, define a *C-small point of $\mathfrak{X}_t$* to be an algebraic point $P \in \mathfrak{X}_t(\bar{\mathbb{Q}})$ such that

$$h(\mathfrak{X}_t, P) \leq C,$$

and a *C-large point of X* to be one for which the reverse inequality holds, i.e.,

$$h(\mathfrak{X}_t, P) > C.$$

For $\mathfrak{X}_t$ be a curve in our family, and an intermediate field $k \subset L \subset \bar{\mathbb{Q}}$, partition the set, $\mathfrak{X}_t(L)$, of L-rational points on $\mathfrak{X}_t$ into the set of C-small and C-large points,

$$\mathfrak{X}_t(L) = \mathfrak{X}_t(L)_{\mathrm{small}} \coprod \mathfrak{X}_t(L)_{\mathrm{large}}.$$

We visibly have a *size-estimate* on $\mathfrak{X}_t(L)_{\text{small}}$ and, for appropriate choice of C, we seek a *number-estimate* on $\mathfrak{X}_t(L)_{\text{large}}$.

PROBLEM. Given our parametrized family $\mathfrak{X} \to T \hookrightarrow \mathbb{P}^N$ over the number field k, as above, find three constants C, D and E for which an estimate of the following form holds. For *any* number field L containing k, and $t \in T(L)$,

$$\left| \mathfrak{X}_t(L)_{C-\text{large}} \right| \le D \cdot E^{r(\mathfrak{X}_t, L)},$$

where $r(\mathfrak{X}_t, L)$ is the rank of the Mordell–Weil group of the jacobian of $\mathfrak{X}_t$ over L.

As mentioned, de Diego [1997], who establishes uniform estimates, developing upon methods as given, for example, in [Bombieri 1990], shows that there exist constants C, D, and E that solve this problem. In fact, de Diego shows that (for any family as above) there is a choice of C for which:

$$\left| \mathfrak{X}_t(L)_{C-\text{large}} \right| \le \tfrac{55}{2} \cdot 7^{r(\mathfrak{X}_t, L)},$$

for all number fields L containing k and points $t \in T(L)$. Of course, what is missing here is an explicit evaluation of a constant $C = C(\mathfrak{X}, T, \iota)$ that does the above job.

In the recent work of A. Pacheco, there are similar estimates for curves over finite fields replacing number fields (but with further complications due to inseparability phenomena).

In view of a solution to the preceding problem, we might hope for an upper bound with better uniformity in the sense that it would involve only the *genus* of the curve and the quantity $r(X, L)$. We might ask for the following stronger assertion than the above (this being, at the same time, weaker than the question about uniform number-effectivity posed on page 201 (Question 3).

QUESTION. If X is a (smooth projective) curve over a field L, let $g(L)$ denote its genus; let $X(L)$ denote the set of its L-rational points, and $r(X, L)$ the rank of the group of L-rational points of the jacobian of X. Is there a function $N(g, L, r)$ with the property that

$$|X(L)| \le N\big(g(X), L, r(X, L)\big),$$

where L ranges through all number fields, and X ranges through all (smooth projective) curves over L of genus ≥ 2?

In the direction of achieving effective results in characteristic p, Buium and Voloch [1996] establish an affirmative answer to the question analogous to the question above in the case where X is a curve defined over a field K (of characteristic p) and the jacobian of X is *ordinary* and such that no non-trivial factor of the jacobian is definable over K^p. Specifically, they show that

$$|X(L)| \le p^{r(X,K)} \cdot (3p)^g \cdot (8g - 2)g!,$$

where $g = g(X) \ge 2$.

Returning to characteristic 0, Buium [1993; 1994] has shown:

PROPOSITION (BUIUM). *Let X be a closed (possibly singular) curve in an abelian variety A (over the complex numbers) and let Γ be a subgroup of $A(\mathbb{C})$ of rank r. Assume that X is of geometric genus $g \geq 2$ and its normalisation is not defined over the field of algebraic numbers. Set $N = \max\{g, r, 4\}$. Then the cardinality of $X(\mathbb{C}) \cap \Gamma \subset A(\mathbb{C})$ is at most $N(!)^{6N+6}$, where $(!)^m$ means factorial iterated m times.*

In e-mail correspondence, Buium noted to me the curious fact that the bound just quoted for characteristic zero is much worse than the one proved in [Buium and Voloch 1993] for characteristic p. In the higher-dimensional situation (but over number fields) one has the very interesting preprint [Hrushovski and Pillay 1998], which gives a related uniform upper bound. This work makes use of the powerful methods explained in this conference; their context is as follows. One is given a subvariety $X \subset A$ in a (semi-) abelian variety both defined over a number field contained in $\mathbb{C}$ and one assumes that X contains no subvarieties of the form $X_1 + X_2 \subset A$, for X_1, X_2 positive dimensional varieties. One considers finitely generated subgroups $\Gamma \subset A(\mathbb{C})$ and seeks an upper bound for the number of *nonalgebraic* points in $X \cap \Gamma$, i.e., for $|X \cap \Gamma - X(\bar{\mathbb{Q}})|$. Letting $r := \text{rank}(\Gamma)$, the upper bound given in [Hrushovski and Pillay 1998] for $|X \cap \Gamma - X(\bar{\mathbb{Q}})|$ is doubly exponential in r. More specificially, the upper bound is of the form

$$a^{(br)^{cr}},$$

where a, b, c are explicit, and quite computable, functions of X and A.

The double exponential upper bound in this result of Hrushovski and Pillay raises the hope (as Buium mentioned to me in e-mail correspondence) that the iterated factorial upper bound obtained in the non-isotrival context (i.e., in the proposition of Buium quoted above) will eventually be significantly improved.

References

[Abramovich 1994] D. Abramovich, "Subvarieties of semiabelian varieties", *Compositio Math.* **90**:1 (1994), 37–52.

[Abramovich and Voloch 1992] D. Abramovich and J. F. Voloch, "Toward a proof of the Mordell–Lang conjecture in characteristic p", *Internat. Math. Res. Notices* **1992**:5 (1992), 103–115.

[Bogomolov and Tschinkel 1999] F. A. Bogomolov and Y. Tschinkel, "Density of rational points on elliptic K3 surfaces", preprint, 1999. Available at http://xxx.lanl.gov/abs/alg-geom/9902092. To appear in *Asian Journ. of Math.*

[Bombieri 1990] E. Bombieri, "The Mordell conjecture revisited", *Ann. Scuola Norm. Sup. Pisa Cl. Sci.* (4) **17**:4 (1990), 615–640. Errata in **18**:3 (1991), 473.

[Bourbaki 1972] N. Bourbaki, *Commutative algebra*, Hermann, Paris, 1972.

[Bouscaren 1998] E. Bouscaren (editor), *Model theory and algebraic geometry: An introduction to E. Hrushovski's proof of the geometric Mordell–Lang conjecture*, edited by E. Bouscaren, Lecture Notes in Math. **1696**, Springer, Berlin, 1998.

[Buium 1992] A. Buium, "Intersections in jet spaces and a conjecture of S. Lang", *Ann. of Math.* (2) **136**:3 (1992), 557–567.

[Buium 1993] A. Buium, "Effective bound for the geometric Lang conjecture", *Duke Math. J.* **71**:2 (1993), 475–499.

[Buium 1994] A. Buium, "On a question of B. Mazur", *Duke Math. J.* **75**:3 (1994), 639–644.

[Buium and Voloch 1993] A. Buium and J. F. Voloch, "Integral points of abelian varieties over function fields of characteristic zero", *Math. Ann.* **297**:2 (1993), 303–307.

[Buium and Voloch 1996] A. Buium and J. F. Voloch, "Lang's conjecture in characteristic p: an explicit bound", *Compositio Math.* **103**:1 (1996), 1–6.

[Caporaso et al. 1997] L. Caporaso, J. Harris, and B. Mazur, "Uniformity of rational points", *J. Amer. Math. Soc.* **10**:1 (1997), 1–35.

[Chabauty 1941] C. Chabauty, "Sur les points rationnels des courbes algébriques de genre supérieur à l'unité", *C. R. Acad. Sci. Paris* **212** (1941), 882–885.

[Chatzidakis and Hrushovski 1999] Z. Chatzidakis and E. Hrushovski, "Model theory of difference fields", *Trans. Amer. Math. Soc.* **351**:8 (1999), 2997–3071.

[Coleman 1985a] R. F. Coleman, "Effective Chabauty", *Duke Math. J.* **52**:3 (1985), 765–770.

[Coleman 1985b] R. F. Coleman, "Torsion points on curves and p-adic abelian integrals", *Ann. of Math.* (2) **121**:1 (1985), 111–168.

[de Diego 1997] T. de Diego, "Points rationnels sur les familles de courbes de genre au moins 2", *J. Number Theory* **67**:1 (1997), 85–114.

[Faltings 1983] G. Faltings, "Endlichkeitssätze für abelsche Varietäten über Zahlkörpern", *Invent. Math.* **73**:3 (1983), 349–366.

[Faltings 1991] G. Faltings, "Diophantine approximation on abelian varieties", *Ann. of Math.* (2) **133**:3 (1991), 549–576.

[Faltings 1994] G. Faltings, "The general case of S. Lang's conjecture", pp. 175–182 in *Barsotti Symposium in Algebraic Geometry* (Abano Terme, 1991), edited by V. Christante and W. Messing, Perspectives in Math. **15**, Academic Press, San Diego, CA, 1994.

[Grauert 1965] H. Grauert, "Mordells Vermutung über rationale Punkte auf algebraischen Kurven und Funktionenkörper", *Publ. Math. Inst. Hautes Études Sci.* **25** (1965), 131–149.

[Grothendieck 1972] A. Grothendieck, *Modèles de Néron et monodromie*, edited by A. Grothendieck, Lecture Notes in Math. **288**, Springer, Berlin, 1972.

[Hartshorne 1977] R. Hartshorne, *Algebraic geometry*, Graduate Texts in Mathematics **52**, Springer, New York, 1977.

[Hindry 1988] M. Hindry, "Autour d'une conjecture de Serge Lang", *Invent. Math.* **94**:3 (1988), 575–603.

[Hindry 1998] M. Hindry, "Introduction to Abelian varieties and the Mordell–Lang conjecture", pp. 85–100 in *Model theory and algebraic geometry*, edited by E. Bouscaren, Lecture Notes in Math. **1696**, Springer, Berlin, 1998.

[Hrushovski 1996] E. Hrushovski, "The Mordell–Lang conjecture for function fields", *J. Amer. Math. Soc.* **9**:3 (1996), 667–690.

[Hrushovski and Pillay 1998] E. Hrushovski and A. Pillay, "Effective bounds for generic points on curves in semi-abelian varieties", preprint, 1998.

[Hrushovski and Zilber 1993] E. Hrushovski and B. Zilber, "Zariski geometries", *Bull. Amer. Math. Soc.* (*N.S.*) **28**:2 (1993), 315–323.

[Hrushovski and Zilber 1996] E. Hrushovski and B. Zilber, "Zariski geometries", *J. Amer. Math. Soc.* **9**:1 (1996), 1–56.

[Lang 1960] S. Lang, "Some theorems and conjectures in diophantine equations", *Bull. Amer. Math. Soc.* **66** (1960), 240–249.

[Lang 1965] S. Lang, "Division points on curves", *Ann. Mat. Pura Appl.* (4) **70** (1965), 229–234.

[Lang 1974] S. Lang, "Higher dimensional diophantine problems", *Bull. Amer. Math. Soc.* **80** (1974), 779–787.

[Lang 1983] S. Lang, *Fundamentals of Diophantine geometry*, Springer, New York, 1983.

[Lang 1986] S. Lang, "Hyperbolic and Diophantine analysis", *Bull. Amer. Math. Soc.* (*N.S.*) **14**:2 (1986), 159–205.

[Lang 1988] S. Lang, *Introduction to Arakelov theory*, Springer, New York, 1988.

[Lang 1991] S. Lang, *Number theory III: Diophantine geometry*, Encyclopaedia of Math. Sciences **60**, Springer, Berlin, 1991.

[Manin 1963] J. I. Manin, "Rational points on algebraic curves over function fields", *Izv. Akad. Nauk SSSR Ser. Mat.* **27** (1963), 1395–1440. In Russian; translated in *Amer. Math. Soc. Translations* (2) **50** (1966) 189–234. See also letter to the editor, *Izv. Akad. Nauk. USSR* **34** (1990) 465–466.

[McQuillan 1994] M. McQuillan, "Quelques compléments à une démonstration de Faltings", *C. R. Acad. Sci. Paris Sér. I Math.* **319**:7 (1994), 649–652.

[McQuillan 1995] M. McQuillan, "Division points on semi-abelian varieties", *Invent. Math.* **120**:1 (1995), 143–159.

[Mumford 1966] D. Mumford, "On the equations defining abelian varieties, I", *Invent. Math.* **1** (1966), 287–354.

[Mumford 1967] D. Mumford, "On the equations defining abelian varieties, II and III", *Invent. Math.* **3** (1967), 75–135, 215–244.

[Mumford 1974] D. Mumford, *Abelian varieties*, 2nd ed., Tata Institute Studies in Mathematics **5**, Oxford U. Press, London, 1974.

[Pillay 1999] A. Pillay, "Mordell–Lang for complex tori", preprint, 1999.

[Raynaud 1983a] M. Raynaud, "Around the Mordell conjecture for function fields and a conjecture of Serge Lang", pp. 1–19 in *Algebraic geometry* (Tokyo/Kyoto, 1982), edited by M. Raynaud and T. Shioda, Lecture Notes in Math. **1016**, Springer, Berlin, 1983.

[Raynaud 1983b] M. Raynaud, "Courbes sur une variété abélienne et points de torsion", *Invent. Math.* **71**:1 (1983), 207–233.

[Raynaud 1983c] M. Raynaud, "Sous-variétés d'une variété abélienne et points de torsion", pp. 327–352 in *Arithmetic and geometry, I*, edited by M. Artin and J. Tate, Progress in math. **35**, Birkhäuser, Boston, 1983.

[Samuel 1966] P. Samuel, "Compléments à un article de Hans Grauert sur la conjecture de Mordell", *Publ. Math. Inst. Hautes Études Sci.* **29** (1966), 55–62.

[Scanlon 1997a] T. Scanlon, "The *abc* theorem for commutative algebraic groups in characteristic *p*", *Internat. Math. Res. Notices* **1997**:18 (1997), 881–898.

[Scanlon 1997b] T. Scanlon, *Model theory of valued D-Fields*, Ph.D. thesis, Harvard University, Cambridge, MA, 1997.

[Serre 1989] J.-P. Serre, *Lectures on the Mordell–Weil theorem*, Vieweg, Braunschweig, 1989.

[Vojta 1991] P. Vojta, "Siegel's theorem in the compact case", *Ann. of Math.* (2) **133**:3 (1991), 509–548.

[Vojta 1993] P. Vojta, "Applications of arithmetic algebraic geometry to Diophantine approximations", pp. 164–208 in *Arithmetic algebraic geometry* (Trento, 1991), edited by E. Ballico, Lecture Notes in Math. **1553**, Springer, Berlin, 1993.

[Vojta 1996] P. Vojta, "Integral points on subvarieties of semiabelian varieties, I", *Invent. Math.* **126**:1 (1996), 133–181.

[Voloch 1991] J. F. Voloch, "On the conjectures of Mordell and Lang in positive characteristics", *Invent. Math.* **104**:3 (1991), 643–646.

BARRY MAZUR
DEPARTMENT OF MATHEMATICS
HARVARD UNIVERSITY
CAMBRIDGE, MA 02138
UNITED STATES
mazur@math.harvard.edu